Cellulose: Diverse Applications

Cellulose: Diverse Applications

Edited by **Dwight Cowan**

NY RESEARCH PRESS

P R E S S

New York

Published by NY Research Press,
23 West, 55th Street, Suite 816,
New York, NY 10019, USA
www.nyresearchpress.com

Cellulose: Diverse Applications
Edited by Dwight Cowan

International Standard Book Number: 978-1-63238-072-2 (Hardback)

Printed in the United States of America.

Contents

Preface

Over the recent decade, advancements and applications have progressed exponentially. This has led to the increased interest in this field and projects are being conducted to enhance knowledge. The main objective of this book is to present some of the critical challenges and provide insights into possible solutions. This book will answer the varied questions that arise in the field and also provide an increased scope for furthering studies.

This book offers an in-depth look at cellulose, elucidating its diverse applications. It discusses both the existing and potential applications of cellulose. It displays the features of cellulose that makes it versatile component to be applied in varied field of science such as pharmaceuticals, bioelectronics etc. This book would help readers to explore this domain and they will be able to use their innovativeness to introduce some novel applications of cellulose.

I hope that this book, with its visionary approach, will be a valuable addition and will promote interest among readers. Each of the authors has provided their extraordinary competence in their specific fields by providing different perspectives as they come from diverse nations and regions. I thank them for their contributions.

Editor

Cellulose Expression in
Pseudomonas fluorescens SBW25
and Other Environmental Pseudomonads

Andrew J. Spiers, Yusuf Y. Deeni, Ayorinde O. Folorunso,
Anna Koza, Olena Moshynets and Kamil Zawadzki

Additional information is available at the end of the chapter

1. Introduction

Bacterial cellulose was first isolated from the air-liquid (A-L) interface biofilm produced by *Bacterium xylinum* in 1886 [1], an acetic acid bacterium strain which would probably now be recognised as *Gluconacetobacter xylinus* (formerly *Acetobacter xylinum*) or a related species. Over the following century, more acetic acid bacteria and additional *Proteobacter* were found to produce cellulose (reviewed in [2-3]). Cellulose-producing bacteria include a mixture of gut commensals, plant and animal pathogens (these are listed in Table 1), and all share soil as a common secondary habitat. It is likely that cellulose provides protection against physical disturbance, predation or other environmental stresses common to these diverse environments. The biochemistry of bacterial cellulose expression has been studied extensively for *Gluconacetobacter*, and this understanding has been used as a model for enteric bacteria and pseudomonads [4-5] (for a range of bacterial cellulose reviews, see [2-3, 6-9]). Experimental reports of bacteria expressing cellulose are increasing, as well as the annotation of putative cellulose synthase-like operons in bacterial whole-genome sequences, suggesting that an increasingly wider range of bacteria may be capable of producing cellulose.

Our interest in bacterial cellulose began with the experimental evolution of the soil and plant-associated pseudomonad, *Pseudomonas fluorescens* SBW25 [10-12]. This resulted in a novel biofilm–forming adaptive mutant known as the Wrinkly Spreader (WS) and shown in Figure 1. Subsequent investigation of the WS phenotype identified partially-acetylated cellulose as the main matrix component of the biofilm. The pseudomonads are a highly diverse genus (see recent reviews by [13-14]), and biofilm-formation and cellulose-expression are now known to be common amongst the water, soil, plant-associated and plant-pathogenic environmental pseudomonads [15]. However, the ecological role of cellulose and the fitness advantage it confers to these bacteria is poorly understood.

Class	Order	Family	Genus	Key habitat
Clostridia	Clostridales	Clostridiaceae	Sarcina	Mammalian intestine commensals
α-Proteobacter	Rhizobiales	Rhizobiaceae	Agrobacterium Rhizobium	Plant pathogens Plant symbionts
	Rhodospirillales	Acetobacteraceae*	Gluconacetobacter	Rotting fallen fruits
β-Proteobacter	Burkholderiales	Alcaligenaceae	Alcaligenese	Opportunistic human pathogens
γ-Proteobacter	Enterobacteriales	Enterobacteriaceae†	Enterobacter Escherichia Salmonella	}Mammalian } intestinal } commensals and pathogens
	Pseudomonadales	Pseudomonadaceae	Pseudomonas	Water, soil and plant-associated, including plant, fungal and animal pathogens

Adapted from [2-3]. *, Also known as the acetic acid bacteria; †, Referred to here as the enteric bacteria.

Table 1. Cellulose-expressing bacterial genera

Here we provide a review of our work focussing on biofilm-formation and cellulose expression by SBW25 and other environmental pseudomonads. We do not provide an extensive list of primary literature or current reviews, but hope that the citations we have made will allow others to access the growing wealth of publications relevant to the subjects raised in this review.

2. Bacterial assemblages and biofilms

The formation of biofilms by bacteria is a key strategy in the colonisation of many environments, though biofilms are only one of a range of bacterial assemblages involved in this process. Bacterial assemblages range from isolated surface-attached bacteria, monolayers of associated bacteria forming micro-colonies, larger and more complex structures including differentiated biofilms, as well as poorly-attached or free-floating flocs and slime. At times the differences between assemblage types may be minor and will depend on local environmental conditions. These differences are frequently ignored by many who prefer the simple dichotomy of individual, free-swimming planktonic bacteria verses the structurally complex and genetically-determined biofilms. Here we use the term 'biofilm' to include partially and fully-saturated aggregations growing on solid surfaces, as well as those that are poorly-attached or 'free-floating', after the early and broad definition of Costerton *et al.* [16].

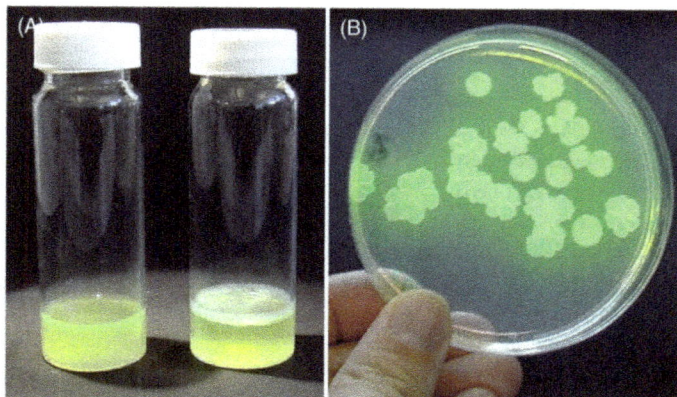

Figure 1. **The Wrinkly Spreader mutant of *Pseudomonas fluorescens* SBW25.** The Wrinkly Spreader (WS) mutant was isolated from evolving populations of wild-type SBW25 in static King's B microcosms. **(A)** Wild-type SBW25 (*left*) grows throughout the liquid column; in comparison, the WS (*right*) occupies the air-liquid (A-L) interface by producing a robust biofilm 1-2mm thick. **(B)** Wild-type SBW25 (smooth and rounded) and WS (wrinkled) colonies are readily differentiated on agar plates. Images from A. Spiers.

The importance of biofilms (aggregations) in nature is reflected by their prevalence in aquatic, soil, fungal, plant and animal ecosystems, and their role in many chronic human diseases and antibiotic resistance. Many natural biofilms are multi-species structures with complex interactions, and in earlier literature they were often referred to as zoogleal mats. Bacteria found within biofilms are profoundly different from those growing in suspension, differing in both gene expression and physiology and more resistant to desiccation, physical disturbance and predation. A range of biofilm reviews are provided by [16-31].

3. Archetypal 'flow-cell' biofilms

Biofilm research has largely focussed on submerged, solid-liquid (S-L) interface biofilms to provide archetypal models of biofilm structure, function and allow genetic investigation (e.g. *Pseudomonas aeruginosa* PA01 flow-cell biofilms). In these, a surface-attached exopolysaccharide (EPS) polymer matrix-based structure develops away from the solid surface, into the flow of a nutrient and O_2-rich growth medium, and where fluid flow and mass transfer affects biofilm development, structure and rheology (for reviews, see [19, 28-29]).

Biofilm formation begins when planktonic bacterial cells initiate attachment to a solid surface. Attached bacteria start to move across the surface, grow and form micro-colonies, which then develop slowly into the mature biofilm structure in which bacterial cells are embedded in an exopolysaccharide polymer matrix. When conditions become unfavourable within the biofilm, single bacteria or large lumps of biofilm material detach and move away to colonise new surfaces in more favourable environments (reviewed in [22]). Biofilms of mixed bacterial communities and of individual species that develop on solid surfaces exposed to a continuous flow of nutrients form a thick layer generally described as consisting of differentiated mushroom and pillar-like structures separated by water-filled spaces.

A defining feature of many biofilms is the exopolysaccharide polymer 'slime' that encapsulate the bacteria and provide the main structural component or matrix of the biofilm [20, 22, 24-25]. Although generally assumed to be primarily composed of polysaccharides, e.g. alginate, PEL (a glucose-rich polymer) and PSL (a repeating pentasaccharide containing d-mannose, d-glucose and l-rhamnose) produced by *P. aeruginosa* PA01, PIA (a 28 kDa soluble linear β(1-6)-N-acetylglucosamine) and related PNAG polymer produced by *Staphylococcus aureus* MN8m and *S. epidermidis* 13-1, and PIA-like polymers produced by *Escherichia coli* K-12 MG1655, biofilm matrices can also contain proteins and nucleic acids having significant structural roles (reviewed in [30]). Exopolysaccharides are typically viewed as a shared resource that provides a benefit to the biofilm community by maintaining structure, facilitating signalling, and protecting residents from predation, competition, and environmental stress [20, 22, 32-35].

A second characteristic common to many S-L interface biofilms has been the involvement of quorum sensing in micro-colony development, exopolysaccharide expression, and dispersal. For example, the quorum signalling molecule, acyl-homoserine lactone (AHL), functions as a signal for the development of *P. aeruginosa* PA01 and *Pseudomonas fluorescens* B52 biofilms [36-37]. However, mathematical models based on O_2 and nutrient transport (diffusion) limitation result in similar biofilm architecture (reviewed in [38]), suggesting biofilm development is equally sensitive to environmental conditions as it may be to genetically-determined regulation. Although quorum sensing is important in the development of some biofilms, the bacterial community will exploit all available mechanisms to adapt to local environmental conditions. In order to further understand the development and role of biofilms, the local environment should be considered in terms of ecological landscape theory in which the spatial configuration of the biofilm biomass is shaped by multiple physical and biological factors [39]. It is therefore likely that biofilm formation is the net result of many independent interactions, rather than the result of a unique pathway initiating attachment and terminating with dispersal of mature biofilm communities.

4. Air-liquid (A-L) interface biofilms

In contrast to the archetypal S-L interface biofilms, bacterial biofilms also form at the air-liquid (A-L) interface of static liquids and are sometimes referred to as 'pellicles' [30]. Perhaps the earliest experimental observations of these were made for *Bacterium aceti* and *B. xylinum* in 1886 [1, 40]. Both bacteria were isolated from beer undergoing acetic fermentation in which alcohol is converted into acetic acid. *B. aceti*, an acetic acid bacterium whose modern name is unclear, was found to produce a greasy-looking biofilm which varied in thickness from an 'almost invisible film' to a paper-thick structure depending on the growth medium [40]. In contrast, the *B. xylinum* isolate, which would probably now be recognised as a *Gluconacetobacter* spp. produced a 'vinegar plant' described as a jelly-like transparent mass at the bottom of the liquid, but under favourable conditions it could also produce a robust gelatinous A-L interface biofilm up to 25 mm thick [1].

Vinegar plants are generally a consortia of acetic acid bacteria and yeasts which produce a zoogleal mat or mixed-species biofilm, and were traditionally used to produce vinegar from beer, cider or wine. Acetic fermentation is initiated by a starter culture known as the 'mother' and obtained from a previous vinegar in a process known as back-slopping [41]. A similar starter often referred to as a 'tea fungus' is used today to produce Kombucha, a carbonated cider-like drink from a sugary solution containing black tea (see the description given in [42]). Acetic acid bacteria, including *Gluconacetobacter* spp., can be isolated from these and similar consortia where they are responsible for the cellulose matrix-based biofilm (see an early review of the acetic acid bacteria by [43]). These artificially-maintained *Gluconacetobacter* spp. are probably better adapted to growth in static liquid conditions than environmental isolates recovered from rotting fallen fruit [44] and under the right conditions, some can produce a gelatinous 'plug' up to 20 mm deep in 10-12 days [45]. In these, cellulose expression and probably growth, is restricted to a thin 50-100 μm deep zone at the top, where it is limited by O_2 diffusing from above and nutrients diffusing through the mature biofilm from below [45]. The growing biofilm is maintained in position by the accumulation of small CO_2 bubbles and by pressing against the walls of the container as it develops.

We expect that smaller-scale A-L interface biofilms might also occur in a wide range of natural environments, such as the partially-saturated fluid-filled pore networks of soils, in temporary puddles collecting on plants and other surfaces after rainfall, water-logged leaf tissues, or in small protected bodies of water such as ponds where the surface is not disturbed by wind or currents. In these environments, biofilm development would be restricted by a combination of nutrient availability, O_2 diffusion, physical disturbance, as well as microbial competition and predation by protists and nematodes.

A-L interface biofilms are readily produced in experimental static liquid-media microcosms [5, 11, 15], and an example of the *P. fluorescens* SBW25 Wrinkly Spreader A-L interface biofilm is shown in Figure 1. In a survey of environmental pseudomonads using nutrient-rich liquid King's B microcosms, we categorised A-L interface biofilms on the basis of phenotype and physical robustness into the physically cohesive (PC), floccular mass (FM), waxy aggregate (WA) and viscous mass (VM)-class biofilms [15, 46]. The characteristics of these biofilm-types are summarised in Table 2 (see also Figure 2). A-L interface biofilm formation appears to be an evolutionary deep-rooted ability amongst bacteria, presumably with significant ecological advantages. In experimental microcosms, increases in competitive fitness of biofilm-formers have been observed compared to non-biofilm–forming strains, whilst the cost to being a biofilm-forming mutant in an environment not suited to these structures is also measurable [5, 47-49].

	Waxy aggregate (WA)	Floccular mass (FM)	Physically cohesive (PC)	Viscous mass (VM)
Occurrence	Rare	Common	Common	Common
Structure	Single-piece rigid and brittle structure	Multiple flocs	Single-piece flexible and elastic structure	Large viscous mass
Strength	Strong	Medium	Strong	Weak
Resilience	Good, disruption produces smaller fragments	Good, disruption produces flocs that are hard to destroy	Very good, hard to break into smaller fragments	Very poor, disruption solubilises the structure
Attachment	High	Medium	High	Poor
Matrix	No evidence for EPS, possible cell-to-cell interactions	Observed	Observed	Observed

Biofilm attributes compiled from [15, 46, 73]. Strength, ability to withstand weight applied to the top of the biofilm; Resilience, response to applied physical disturbance such as gentle or vigorous mixing; Attachment, connection to the microcosm vial walls in the meniscus region; Matrix, evidence of EPS from behaviour of samples during microscopy; Cellulose, evidence from Calcofluor-staining and fluorescent microscopy.

Table 2. Different classes of air-liquid (A-L) interface biofilms produced by environmental pseudomonads

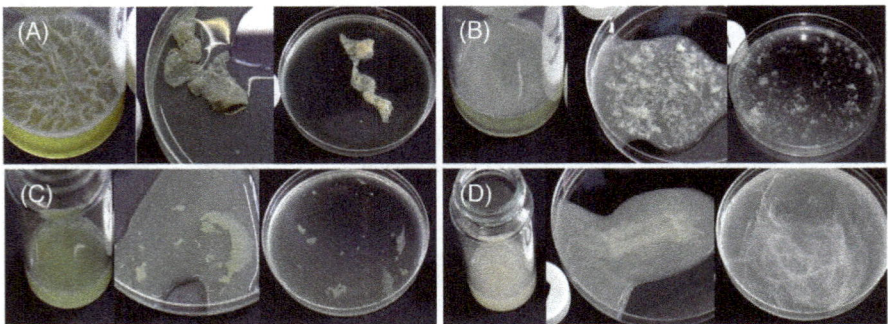

Figure 2. Air-liquid (A-L) interface biofilms. A-L interface biofilms produced by environmental pseudomonads can be categorised into four biofilm types according to visual phenotype, robustness and resistance to physical disturbance. These are the **(A)** Physically cohesive (PC), **(B)** Floccular mass (FM), **(C)** Waxy aggregate (WA), and **(D)** Viscous mass (VM) types. Shown are biofilms in static King's B microcosms (*left*), after pouring into petri dishes (*middle*), and after vigorous mixing (*right*). Figure adapted from [15].

5. Experimental evolution and the Wrinkly Spreader

Many aspects of the ecological and mechanistic bases of evolution have been investigated by the experimental evolution of bacteria (reviewed in [50-52]). The adaptive radiation of the soil and plant-associated pseudomonad, *P. fluorescens* SBW25 [10, 12], has been investigated in some detail using experimental King's B microcosms (see Figure 1) following the first report by Rainey and Travisano [11]. These can be incubated with shaking to provide a homogenous environment, or statically without physical disturbance to provide a heterogeneous environment. The initial wild-type SBW25 colonists of static microcosms rapidly establish a gradient in which O_2 drops to < 0.05% of normal levels below a depth of 200μm [53]. This gradient produces heterogeneity in the microcosm and defines three niches for colonisation and adaptation: the A-L interface, the liquid column, and the vial bottom. In contrast, microcosms subject to constant and vigorous mixing do not develop an O_2 gradient or different niches. Wild-type SBW25 rapidly radiates to produce a range of phenotypically distinguishable mutants (morphs or morphotypes) to occupy the different niches [11]. This diversification is reproducible and occurs rapidly, typically within ~100 generations and 1-3 days. The main morphotypes recovered from evolving populations of wild-type SBW25 in static microcosms include the Wrinkly Spreaders (WS) which produce a wrinkled colony morphology and colonise the A-L interface through the formation of a biofilm (Figure 1); the Smooth (SM) morphs, including wild-type SBW25, which produce round, smooth colonies and colonise the liquid column, and the Fuzzy Spreaders (FS) which are characterised by fuzzy-topped colonies and colonise the anoxic bottom of static microcosms [11].

In an effort to understand the mechanistic basis of the adaptive leap of wild-type SBW25 from the liquid column-colonising SM-morph to the WS A-L interface niche specialist, the underlying molecular biology of the WS phenotype was investigated. This work, described in the following section, ultimately showed that the evolutionary innovation was the use of cellulose to produce a physically robust and resilient biofilm which allowed the colonisation of the A-L interface. Competitive fitness experiments have demonstrated that the WS has a significant fitness advantage over non-biofilm–forming strains in static microcosms [5, 11, 49]. Simplistically, WS cells were able to intercept O_2 diffusing across the A-L interface from the atmosphere before non-biofilm–forming competitors could do so lower down the liquid column, and as a result, WS populations could grow more rapidly than non-WS populations [53]. In contrast however, the WS do not enjoy a fitness advantage in shaken microcosms where the O_2 concentrations are uniform or on agar plates where the WS phenotype is unstable [5, 48-49].

6. Cellulose expression in *P. fluorescens* SBW25

In order to understand the underlying mechanistic basis of the WS phenotype, a mini-transposon screening approach using mini-Tn5 was adopted to identify critical genes and pathways [5]. Mini-transposon insertions typically destroy the function of the targeted gene, and the disruption of critical genes in the WS would be expected to result in mutants that produced rounded, smooth (SM)-like colonies rather than the typical WS colony. Plates containing hundreds or thousands of WS colonies could be easily screened for a few SM-like colonies which could then be isolated for further examination.

This approach allowed the identification and sequencing of the SBW25 *wss* operon containing ten genes (*wssA-J*) required for the WS phenotype and is shown in Figure 3 (*wss* is an acronym for <u>WS</u> <u>s</u>tructural locus, responsible for the production of the main structural component required for the WS phenotype) [5]. Overall, the *wss* operon showed strong similarity to the cellulose biosynthetic clusters originally identified as the *acs* operon in *Gluconacetobacter hansenii* (formerly *Acetobacter xylinus*) ATCC 23769 [54] and subsequently annotated as the *yhj* operon in the whole-genome sequence of *Escherichia coli* K-12 [55]. Most *acs* (<u>A</u>cetobacter <u>c</u>ellulose-<u>s</u>ynthesizing) homologues are now referred to as *bcs* (<u>b</u>acterial <u>c</u>ellulose <u>s</u>ynthesizing) genes as we do here (*yhj* has no meaning). The degree of homology between the *wss*, *bcs* and *yhj* genes at the amino acid level strongly suggested that the SBW25 *wss* operon encoded a functional cellulose synthase, and the predicted functions of the Wss proteins are listed in Table 3.

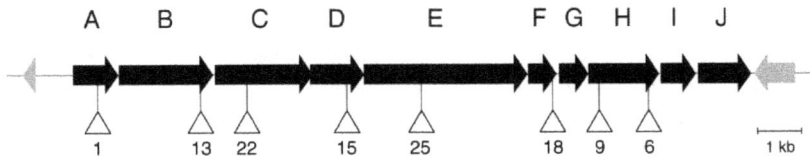

Figure 3. Structure of the cellulose biosynthesis operon. The *Pseudomonas fluorescens* SBW25 cellulose synthase is encoded by the *wss* operon (*wssA-J*, black arrows). The core synthase is composed of WssBCDE subunits and the associated acetylation activity produced by WssFGHI. WssA and WssJ may be involved in the correct cellular localisation of the Wss complex, though WssJ is functionally redundant. The locations of key mini-Tn5 transposon insertions are indicated (open triangles). WS-1, 13, 22, 15 & 25 mutants are unable to express cellulose. WS-18, 9 & 6 mutants express un-acetylated cellulose. Upstream of the *wss* operon is tRNA^Thr and downstream a hypothetical protein of unknown function (grey arrows). Scale bar: 1 kb. Figure adapted from [5].

However, the SBW25 *wss* operon showed two notable differences to the *G. xylinus bcs* and *E. coli yhj* operons. First, the *wss* operon contains two MinD-like homologues, WssA and WssJ, not previously recognised as having a role in cellulose synthesis. WssJ shows 51% identity at the amino acid level with WssA, but only short sections of similarity at the nucleotide level and does not appear to be a simple repeat of the *wssA* gene sequence. As MinD is involved in cell division and determining cell polarity [56], WssA and WssJ were proposed to ensure the correct spatial localization of the cellulose synthase complex at the cell poles [5]. Subsequently, the WssA-homologue, YhjQ (BcsQ), has been shown to be essential for this in *E. coli* K-12 [57]. Secondly, the *wss* operon includes three genes, *wssGHI*, that shares homology with the alginate acetylation proteins of *P. aeruginosa* FRD1, AlgFIJ [58].

In order to demonstrate that the SBW25 *wss* operon encoded a functional cellulose synthase, and to determine the role of the alginate acetylation-like *wssGHI* genes, cellulose expression in the WS and mini-Tn5 mutants was examined by a variety of techniques, including Congo red colony staining, fluorescent microscopy, enzymatic digestion and structural analysis of purified matrix material [5, 59-60].

Protein (synonyms)	Function (Accession No.)
WssA (BcsQ, YhjQ)	Cellulose synthase-associated positioning subunit (CAY46577.1): MinD–like ATPase involved in the appropriate spatial localisation of the cellulose synthase complex.
WssB (BcsA, YhjO)	Cellulose synthase subunit (CAY46578.1): catalytically-active subunit responsible for the polymerisation of UDP-Glucose into cellulose. Predicted integral transmembrane protein, contains conserved D residue, QXXRW, HAKAGN and QTP motifs, a PilZ domain. Binds c-*di*-GMP.
WssC (BcsB, YhjN)	Cellulose synthase subunit (CAY46579.1): unknown function (originally thought to bind c-*di*-GMP). Often fused with WssB.
WssD (Orf1, YhjM)	Cellulose synthase subunit (CAY46580.1): Endo-1,4-D-glucanase (D-family cellulase).
WssE (BcsC/S, YhjI)	Cellulose synthase subunit (CAY46581.1): unknown function. Includes a putative signal peptide.
WssF (BcsX)	Cellulose synthase-associated acetylation subunit (CAY46582.1): suggested function is to present acyl groups to WssGHI.
WssG	Cellulose synthase-associated acetylation subunit (CAY46583.1): AlgF-like protein involved in the acetylation of cellulose. Includes a putative signal peptide.
WssH	Cellulose synthase-associated acetylation subunit (CAY46584.1): AlgI-like protein involved in the acetylation of cellulose. Predicted integral transmembrane protein.
WssI	Cellulose synthase-associated acetylation subunit (CAY46585.1): AlgJ-like protein involved in the acetylation of cellulose. Localised to the periplasm.
WssJ	Cellulose synthase-associated positioning subunit (CAY46586.1): MinD-like ATPase like WssA but apparently functionally redundant.

G. xylinus Bcs and *E. coli* Yhj synonyms are provided in parentheses. Function suggested from Wss experiments and Wss homologue investigations. AlgFIJ homologues are from *P. aeruginosa* FRD1 [58].

Table 3. Predicted functions of the *Pseudomonas fluorescens* SBW25 Wss proteins

The diazo dye, Congo red (CR), had been used previously to stain bacterial colonies expressing cellulose [61], and we used this technique to show that WS, WS-18, WS-6 and WS-9, but not WS-1, WS-13, WS-22, WS-15 and WS-25 mutants, appeared to express cellulose on King's B plates [5, 59]. WS and WS-18 biofilm material were subsequently stained with the more specific fluorescent dye, Calcofluor, and examined by fluorescent microscopy (Figure 4). This showed that the biofilm was dominated by an extensive network of extracellular cellulose, with fibres ranging from 0.02 μm to over 100 μm thick. In places, the fibres appeared to form large clumps of material and in other places forming thin films, with bacterial cells associated with the fibres and found within the voids [59]. In comparison, in colonies the cellulose fibres appear to collect above the mass of the colony (Figure 5). Scanning electron microscopy images of WS biofilms suggest a lattice-work of pores (Figure 6) which might be the result of constant growth at the top surface of the biofilm which slowly displaces older strata deeper into the liquid column [53]. Rough calculations of the density of WS biofilms suggest that they were > 97% liquid, which is in agreement with the finding that microbial amorphous celluloses are very hydrophilic, with gels having a water holding capacity of 148 – 309 g water / g dry cellulose [62-63]. Recent rheological tests have shown that the WS biofilm structure is a classic viscoelastic solid (gel-like) material (AK & AJS, unpublished observations). The structural integrity of WS biofilms could be destroyed by incubation with cellulase, adding support to initial conclusions that the major matrix component of the WS biofilm, expressed by the *wss* operon, was cellulose or a cellulose-like polymer [5, 59].

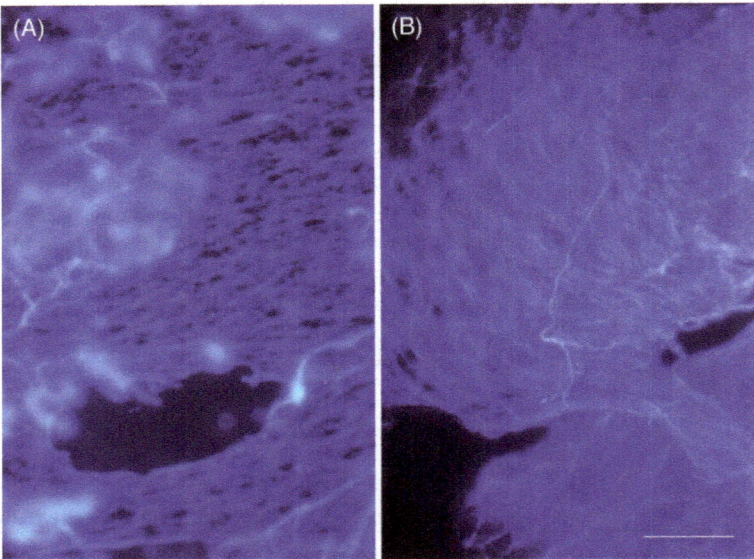

Figure 4. Fluorescent microscopy of WS biofilms. The cellulose fibre matrix of the Wrinkly Spreader (WS) biofilm can be visualised by staining with Calcofluor and fluorescent microscopy. Shown are two images showing the highly hydrated and fibrous nature of the WS biofilm. Scale bar: 100 μm. Images from A. Spiers.

WS and WS-18 biofilms were subsequently purified in order to determine the chemical identity of the matrix component. Carbohydrate analysis indicated that both samples contained ~75% glucose (Glu) and ~25% rhamnose (Rha) [59]. The latter could be explained as coming from contaminating Rha-containing A-band LPS which is highly conserved amongst the pseudomonads [64]. Linkage analyses of derivatized WS samples by GC-MS identified a major peak corresponding to 4-Glu, and minor peaks corresponding to 2,4-Glu, 3,4-Glu and 4,6-Glu, which is consistent with a β(1-4)-linked glucose polymer, i.e. cellulose. In contrast, the WS-18 material did not contain 2,4-Glu, 3,4-Glu or 4,6-Glu derivatives, suggesting that the *wss* alginate acetylation-like genes were responsible for the acetylation of glucose residues at the 2, 3, and 6 Carbon positions. This was further supported by [¹H]-NMR analysis which confirmed the presence of acetylated hexose residues in the WS extract, with 14% of the glucose residues estimated to be modified with one acetyl group [59]. Although cellulose is readily acetylated by chemical treatment, we are not aware of any other reports of biologically-produced acetylated cellulose.

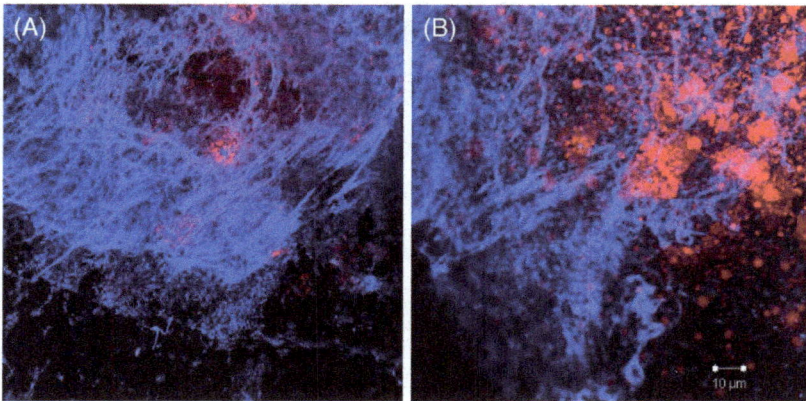

Figure 5. Inducing cellulose expression with c-*di*-GMP. An increase in c-*di*-GMP levels can induce cellulose expression in some pseudomonads. Shown are confocal laser scanning microscopy (CLSM) images of colony material from **(A)** *Pseudomonas fluorescens* SBW25 and **(B)** *P. syringae* DC3000 expressing the constitutively-active DGC response regulator WspR19 *in trans* which increases c-*di*-GMP levels, visualised with Calcofluor for cellulose (blue) and ethidium bromide for bacteria (red). Scale bar: 10 μm. Images from O. Moshynets.

The partial acetylation of the cellulose fibres expressed by the WS clearly had an impact on colony morphology and biofilm strength. Colonies produced by WS-18 were readily differentiated from WS and wild-type SM-like colonies, whilst the WS-18 biofilm was ~ 4x weaker than the partially-acetylated structure produced by the WS, suggesting that acetylation increased the connectivity of cellulose fibres within colonies and the biofilm matrix [59-60]. Incubation with EDTA reduced WS-18 biofilm strength, whilst incubation with some diazo dyes increased WS-18 biofilm strength compared to WS biofilms, suggesting that fibre interactions could be altered by sequestering Mg^{2+} and coating cellulose fibres with dyes [60]. A lipopolysaccharide-deficient mutant which produces a very weak biofilm compared to both WS-18 and WS, was also affected by EDTA and diazo dyes, indicating that the partially-acetylated cellulose fibres also interacted with the lipopolysaccharide on the surface of cells or associated with cell debris to further strengthen the biofilm structure [60].

Individual cellulose polymers can also interact directly to produce a number of different forms or allomorphs. *G. xylinus* produces two crystalline allomorphs, known as cellulose I and II, which requires the cellulose synthase-associated BcsD subunit that couples cellulose polymerisation and crystallization [54, 65]. However, SBW25 lacks a BcsD homologue and therefore can only produce non-crystalline amorphous cellulose.

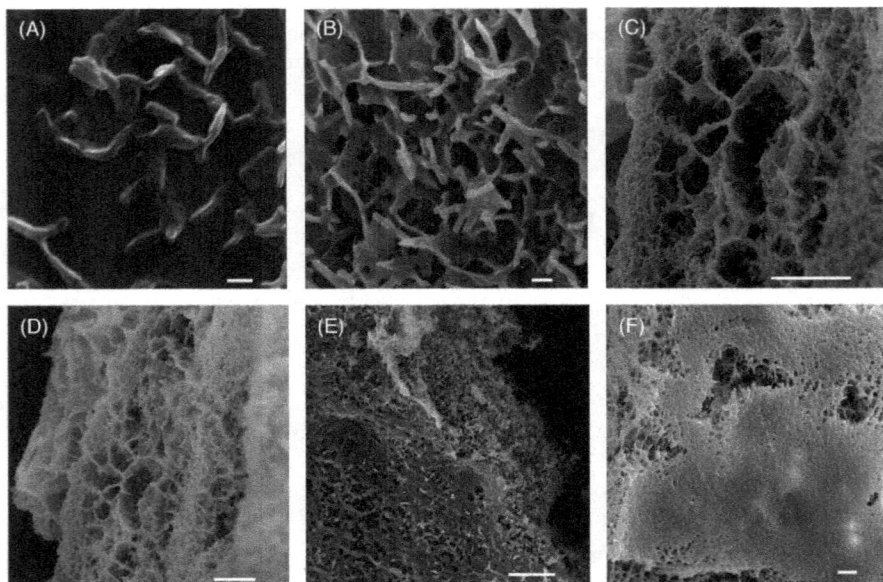

Figure 6. WS biofilm ultrastructure. Scanning electron microscopy (SEM) images of Wrinkly Spreader (WS) biofilms suggest a porous but robust structure. Shown here are a series of SEM images of decreasing magnification, from **(A)** single cells to **(F)** large pieces of biofilm. Images were obtained after freeze-drying and shadowing with gold. Scale bars: A & B, 1µm; C – F, 10 µm. Images from O. Moshynets.

Following the analysis of the WS mini-Tn5 mutants, a further round of mini-transposon mutagenesis was undertaken using IS*phoA*/hah [66]. This mini-transposon allowed both the impact of polar and non-polar insertions to be assessed; the former destroy the function of the disrupted gene as well as any down-stream expression, whilst the latter destroys gene function but leaves down-stream expression unaffected (this is possible after Cre-mediated deletion of the central portion of the IS*phoA*/hah cassette which leaves a 63-codon in-frame insertion). WS IS*phoA*/hah mutants were recovered for each of the *wss* genes, except *wssJ*. Each of the polar IS*phoA*/hah insertions and corresponding non-polar Cre-deletions in *wssA-E* resulted in the loss of cellulose expression, whilst polar and non-polar mutants in *wssF-I* resulted in a WS-18–like phenotype [47, 59]. Finally, a WS *wssJ* deletion mutant was constructed and shown to have no impact on cellulose expression, suggesting that the final gene of the *wss* operon might be functionally redundant [47].

Examination of other mini-Tn5 mutants also lead to the identification of the *wsp* regulatory operon of seven genes, *wspA-E & R* (*wsp* is an acronym for <u>WS</u> phenotype locus, responsible for the regulation of the WS phenotype) [67]. The function of these have been modelled on the *Escherichia coli* Che chemosensory system (reviewed in [68]) to provide a mechanistic explanation of the induction of the WS phenotype [67] (a schematic of this is shown in Figure 7). In this the methyl-accepting chemotaxis protein (MCP) WspA forms a membrane-bound complex with two scaffold proteins, WspB and WspD, plus the histidine kinase WspE. In the absence of an appropriate environmental signal, the complex is silent and does not activate the associate response regulator, WspR, by phosphorylation. The system is controlled by a negative feedback loop mediated by the WspC methyltransferase and WspF methylesterase. WspC constitutively antagonises WspF, and in wild-type SBW25 the activities of the two are balanced, preventing the activation of WspR and allowing the Wsp complex to oscillate between active and inactive states. WspR is a di-guanylate cyclase (DGC) response regulator, and the phosphorylated active form, WspR-P, synthesizes c-*di*-GMP (bis-(3'-5')-cyclic dimeric guanosine monophosphate) from GTP [69]. In this model, we hypothesised that mutations in WspR which stimulated DGC activity without requiring phosphorylation, or mutations inhibiting WspF function, would result in an increase in c-*di*-GMP production. This would then lead to the activation of the WS phenotype through the direct stimulation of the cellulose synthase complex, rather than up-regulated *wss* transcription [5] (the second component required for the WS phenotype, the pilli-like attachment factor, has not yet been identified and it is not known how it might be regulated by c-*di*-GMP). Several mutants of WspR had been engineered, and the effect of the constitutively-active mutant WspR19 [70] on cellulose expression by wild-type SBW25 is shown in Figure 5.

This model for the activation of the WS phenotype has been confirmed through the identification and testing of WspF mutations found in a number of independently-isolated Wrinkly Spreaders [67]. Interestingly, no naturally occurring WspR mutants have been identified yet, despite the fact that engineered WspR mutants like WspR19 were found to show the predicted phenotype [60, 69-71]. Mutations in other operons leading to the activation of the AwsR and MwsR DGCs can also induce the WS phenotype [47, 72]. These different routes activating the WS phenotype can be seen as an example of parallel evolution leading to new A-L interface biofilm-forming genotypes in static microcosms [72].

During the molecular investigation of the WS phenotype, the non-biofilm–forming wild-type SBW25 was modified by the insertion of a constitutive promoter to increase the levels of *wss* operon transcription. This mutant, JB01, was found to produce a very weak biofilm, poorly attached to the microcosm vial walls and to express similar amounts of cellulose as the WS [5, 60]. Subsequently, we found that wild-type SBW25 could be non-specifically induced by exogenous Fe to produce a phenotypically-similar biofilm referred to as the VM biofilm [73] (see below for a description of Viscous mass (VM)-class biofilms). However, VM biofilm-formation was not the result of an increase in *wss* transcription, and as yet, no link has been identified between Fe regulation and cellulose expression. We speculate that the induction of the VM biofilm is due to a minor perturbation of the Wsp system or the cellulose synthase complex itself that allows activation despite sub-critical levels of c-*di*-GMP.

(A) Wild-type SBW25

(B) Wrinkly Spreader

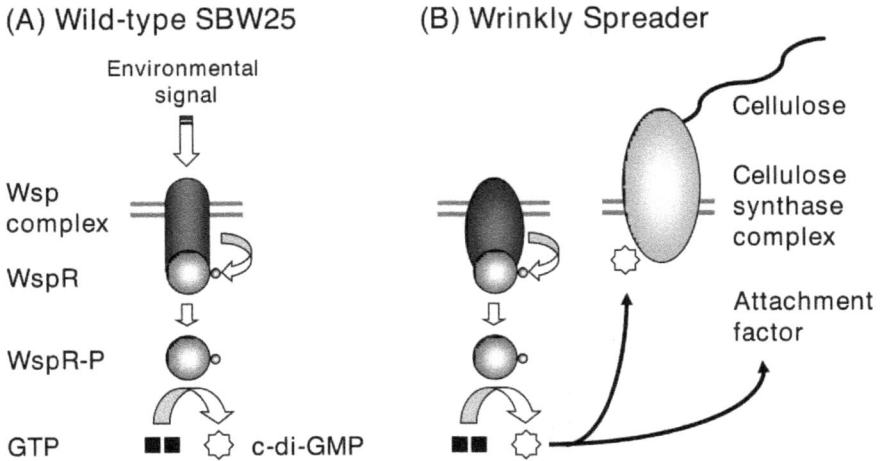

Figure 7. Activation of the WS phenotype. The Wrinkly Spreader (WS) phenotype is controlled by the membrane-associated Wsp complex and associated DGC response regulator WspR. **(A)** In wild-type *Pseudomonas fluorescens* SBW25, when an appropriate environmental signal is received the Wsp complex phosphorylates WspR which then results in the production of c-*di*-GMP. However, in the absence of signal, the Wsp complex is silent and c-*di*-GMP levels remain low. **(B)** In the Wrinkly Spreader a mutation in a Wsp subunit (WspF) alters the sensitivity of the Wsp complex such that it activates WspR in the absence of the environmental signal. The resulting increase in c-*di*-GMP activates the membrane-associated cellulose synthase complex to produce cellulose, and also activates the unidentified attachment factor that is also required for the WS phenotype.

7. Biofilm formation and cellulose expression amongst other pseudomonads

Having discovered that *P. fluorescens* SBW25 could express cellulose in the WS biofilm (and subsequently in the VM biofilm), we were interested to see if related environmental pseudomonads could produce similar cellulose-matrix–based A-L interface biofilms. We therefore undertook a survey of environmental pseudomonads, including water, soil, plant-associated and plant pathogenic isolates (we did not include human or other animal pathogens in this survey) [15]. The ability of each to produce an A-L interface biofilm was assessed in static King's B liquid media microcosms. Importantly, this assay did not differentiate between isolates that constitutively produced biofilms, with those that might utilise quorum sensing-like signalling to initiate biofilm-formation, or those that had mutated into a biofilm-forming genotype. Of the 185 environmental pseudomonads tested, 76% were found to produce observable biofilms within 15 days incubation. The phenotypes of these were variable, with biofilm strengths ranging 1500x, but could be categorised into the physically-cohesive (PC), floccular mass (FM), waxy aggregate (WA) and viscous mass (VM)-class biofilms described in Table 2 (see also Figure 2) [15, 46].

Calcofluor-fluorescent microscopy identified cellulose as the matrix component of 20% of the biofilm-forming isolates, indicating that at least seven *Pseudomonas* species were capable of expressing cellulose under the conditions tested. These included *P. corrugata* (tomato pathogens), *P. fluorescens* (plant-associated isolates), *P. marginalis* (alfalfa and parsnip pathogens), *P. putida* (rhizosphere and soil isolates), *P. savastanoi* (olive pathogens), *P. stutzeri* (represented by a single clinical isolate), and *P. syringae* (celery, cucumber, tobacco, and tomato isolates or pathogens) (isolates from another eleven *Pseudomonas* spp. were tested, including *P. aeruginosa* PA01 and PA14, and were not found to produce cellulose). For two of the cellulose-expressing isolates, *P. putida* KT2440 and *P. syringae* DC3000, the whole genome sequences were available and SBW25 *wss*-like cellulose synthase operons had been annotated [74-75], though no experimental reports of either expressing cellulose had been made.

Many environmental pseudomonads can also be induced to form A-L interface biofilms and to express cellulose using WspR19. When expressed *in trans* in wild-type SBW25 it produces the WS phenotype [60, 70], though in other pseudomonads the impact was found to be more variable. In a test of 16 pseudomonads known to form biofilms and express cellulose, WspR19 was found to significantly increase biofilm attachment, strength, and cellulose expression in *P. fluorescens* 54/96, *P. syringae* DC3000, *P. syringae* T1615 and *P. syringae* 6034 [15] (WspR19 induction of cellulose production by SBW25 and DC3000 is shown in Figure 5). WspR19 also induced a WS-like phenotype in *P. putida* KT2440, despite the fact that biofilm-formation or cellulose expression in this pseudomonad had not been observed in the initial survey (cellulose expression was subsequently reported for both wild-type and WspR19-carrying strains under different experimental conditions by [76]). Similarly, nine of ten non-biofilm–forming and non-cellulose expressing *P. syringae* isolates were found to produce biofilms when induced with WspR19, and two of these also expressed detectable levels of cellulose [15]. These findings suggest that biofilm-formation and cellulose expression in pseudomonads closely related to *P. fluorescens* SBW25 are probably controlled by the same c-*di*-GMP–mediated regulatory system or are sensitive to non-specific increases in c-*di*-GMP levels.

Habitat (sample size)	A-L interface biofilms	Evidence of cellulose
Plant pathogens (n = 57)	6%	26%
Plant & soil associated (n = 28)	82%	39%
Scottish soil (n = 73)[a]	95%	76%*
River (n = 57)	82%	5%
Indoor & outdoor ponds (n = 50)[b]	94%	56%
Pitcher plants (*Sarracenia* spp.) (n = 50)[c]	74%	68%
Spoilt cold-stored meat (n = 60)[d]	77%	28%
Mushroom pathogens (n = 26)[e]	77%	69%

* Estimated from a sub-sample of 25 isolates. Data compiled from [15] and unpublished research from **a,** R. Ahmed, AK & AJS; **b,** B. Varun, AK & AJS; **c,** D.S. Kumar, AK & AJS; **d,** M. Robertson & AJS; and **e,** AK & AJS.

Table 4. Prevalence of A-L interface biofilm formation and cellulose expression amongst environmental pseudomonads

We have conducted additional surveys of pseudomonads isolated from other habitats, including pond water, pitcher plant (*Sarracenia* spp.) deadfall trap-water, spoilt cold-stored meat and mushrooms (Table 4). These confirm the wide-spread ability of environmental pseudomonads to form A-L interface biofilms and to express cellulose under the experimental conditions used previously [15]. It is also evident that pseudomonads are capable of producing a wide range of EPS in addition to cellulose, including alginate, levan, marginalan, PEL, PSL, and a number of other polymers, which may be utilised as biofilm matrix components [77-80].

8. Distribution of *wss*-like cellulose synthase operons amongst the proteobacteria

We are undertaking a bioinformatics analysis of all publicly-available fully-sequenced bacterial genomes in order to determine the phylogenetic distribution and structural variation of *P. fluorescens* SBW25 *wss*-like cellulose biosynthetic operons amongst the proteobacteria. Protein (TBLASTN) homology searches were run against the GenBank complete genome database [81] using the SBW25 Wss proteins as the query sequences in October of 2011. From this, we have identified over 50 bacteria with gene clusters that showed significant protein sequence homology (≥ 25% ID) to three or more Wss proteins, including a minimum of two key cellulose synthase subunits, WssB, WssC, or WssE. Putative SBW25 *wss*-like operons were then manually curated for accuracy to provide Wss homologue protein sequences and operon structures. Phylogentic trees were constructed using the multiple sequence alignment program ClustalW 2.0 [82], and neighbour-joining and minimal evolution methods implemented by Geneious 5.5.5 (Biomatters Ltd, NZ).

Although this bioinformatics analysis is on-going and the final results expected to be published elsewhere, we make the following preliminary observations. First, whilst *wssB* tends to be followed by *wssC* in the *wss*-like cellulose synthase operons as found previously [5, 8, 83], there are examples within the *Burkholderia* and in *Cupriavidus metallidurans* CH34 where *wssB* is separated from the rest of the operon. Second, we note that only the *P. fluorescens* SBW25 and *P. syringae* DC3000 *wss* operons include the *wssG-I* alginate acetylation-like genes, but not the closely-related pseudomonad *P. putida* KT2440. This suggests that DC3000 may also be able to express partially-acetylated cellulose, and that *wssF-I* genes may be narrowly restricted to the *P. fluorescens–syringae* complex. Third, there is considerable variation in cellulose synthesis operon structure amongst the *Enterobacteriacea*, with many having two clusters of genes (e.g. *Erwinia billingiae* Eb66, *Klebsiella pneumoniae* MGH78578, *Pantoea* sp. At-9b). Many enteric bacteria also include the additional genes *bcsEFG* which have no *wss* homologues (e.g. *Escherichia coli* K-12 MG1655, *Salmonella enterica* Typhimurium LT2, *Vibrio fisheri* MJ11). These have been reported to be associated with cellulose production in *Salmonella enteritidis* 3934 [4]. However, *Escherichia coli* K-12 DH10B contains only *bcsFG*, raising the possibility that *bcsE-G* are not essential for cellulose production in these bacteria. Although the *Gluconacetobacter* are not closely related to the enteric bacteria, we note that *G. xylinus* NBRC 3288 has a small *wssBCE* operon plus a larger *wssDBCE* operon. It is possible that such duplications might enable higher levels of cellulose expression under some environmental conditions, or that such gene duplications may persist for some time before deletion.

Finally, the clustering of WssB homologue sequences (Figure 8) generally follows the 16S phylogenetic relationships between bacteria. However, we are surprised to find that *P. fluorescens* SBW25 and *P. syringae* DC300 cluster with many of the *Burkholderia* and *Xanthomonas*, whilst *P. putida* and *P. stutzeri* strains cluster with the enteric bacteria. We have yet to compare the clustering patterns of the WssC, WssD and WssE homologues, where conserved patterns may reflect different functional roles for cellulose and host lifestyles, whilst aberrant placements of single proteins might reflect the random mutation of a phenotype no longer of functional value or under positive selection.

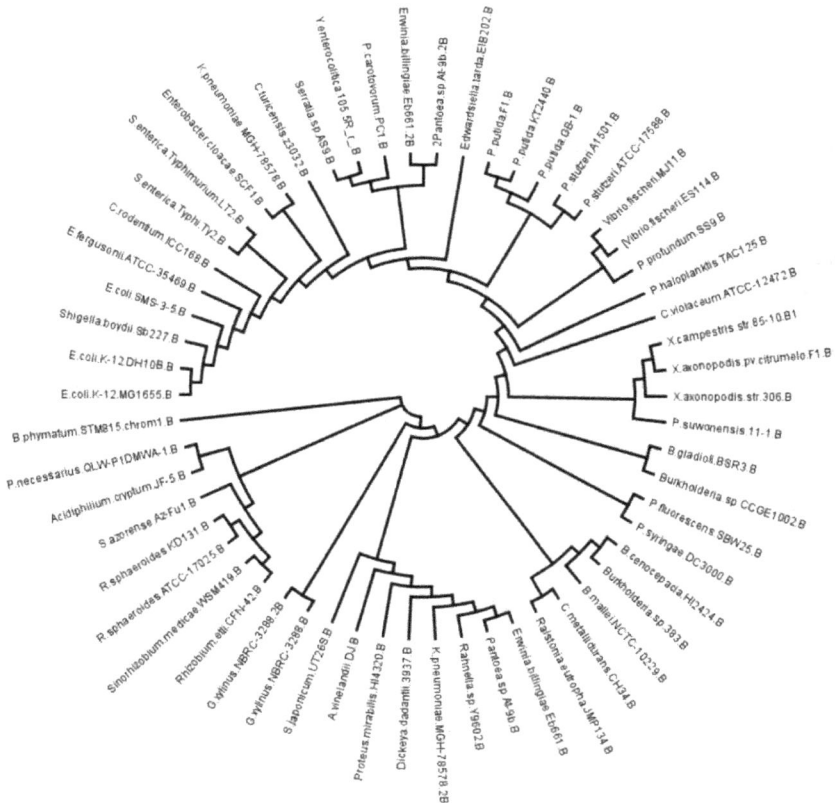

Figure 8. Cladogram of WssB homologues. The structure of the WssB cladogram is similar to that constructed using 16s rRNA sequences, with the enteric bacteria and pseudomonads forming two distinct clusters. Within the pseudomonads, the *P. fluorescens-syringae* complex has diverged earlier than the *P. putida-stutzeri* group. The cladogram was constructed using Geneious 5.5.5 (Biomatters Ltd, NZ) default parameters after multiple sequence alignment of 58 WssB proteins by ClustalW 2.0 [82].

9. Ecological role and fitness advantage of cellulose

Attempts to understand the importance of cellulose expression by *P. fluorescens* SBW25 in soil and the phytosphere have involved investigation of the regulators controlling *wss* operon transcription [5, 84], and measurement of fitness advantage using plant microcosms [85]. Cellulose expression is clearly regulated at two levels in SBW25. First, the activity of the cellulose synthase complex is regulated by c-*di*-GMP levels, but it is not known what environmental signals control WspR, other DGCs or their antagonists, though c-*di*-GMP is known to be involved in a range of surface colonisation and pathogenicity systems in a variety of bacteria (reviewed in [86-87]). Second, cellulose expression is regulated at the level of *wss* operon transcription, with mini-transposon analysis identifying AlgR, AwsR and WspR as positive regulators, and AmrZ and FleQ as negative regulators [5, 84].

AwsXR was a previously unrecognised regulatory system, first identified in SBW25 where the mutational activation of the DGC AwsR results in the WS phenotype [47, 72], though the normal means of regulating AwsR activity remains unknown. FleQ is a c-*di*-GMP–responsive transcriptional regulator, and in *P. aeruginosa* PA01 it controls the hierarchical regulatory cascade for flagella biosynthesis and the repression of the PEL biosynthesis genes [88-89]. AlgR and AmrZ (also referred to as AlgZ) are transcription factors involved in the regulation of a number of systems including alginate biosynthesis and twitching motility [79, 90]. It is possible that AlgR, AmrZ, and AwsR directly repress SBW25 *wss* transcription, whilst AwsX, FleQ, and WspR act indirectly to regulate transcription and therefore cellulose expression [84]. However, no environmental signals have been identified that induce cellulose expression via these repressors.

Sugar beet (*Beta vulgaris*) seedlings have been used to determine the competitive fitness advantage cellulose-expression may provide wild-type SBW25 compared to a cellulose-deficient mutant [85]. In these experiments, seeds were first inoculated with a mixture of wild-type SBW25 and SM-13, a mutant containing a mini-Tn5 insertion cassette derived from WS-13 [5]. These were then germinated and grown for four weeks in an artificial soil substrate. Bacteria were then recovered from the stems and leaves (the phyllosphere), from roots and adherent vermiculite (the rhizosphere), and from un-planted containers ('bulk soil') to allow the calculation of competitive fitness (W) [91] (we report W for SM-13 *cf* wild-type SBW25 here for clarity). In the phyllosphere and rhizosphere, SBW25 was found to have a significant fitness advantage over SM-13, with W ≈ 1.8 and W ≈ 1.11, respectively, but not in bulk soil where W ≈ 1.05. These findings suggest that the appropriately-controlled expression of cellulose by wild-type SBW25 provides some benefit on plant surfaces. It is possible that the mechanistic nature of this benefit may be an improved tolerance to water-limiting conditions rather than resistance to physical disturbance and predation, as the seedlings were watered using a tray rather than from a sprinkler, and vermiculite was used instead of natural soil.

Comparisons of the survival of wild-type SBW25 and a cellulose-deficient mutant similar to SM-13 under water-limiting conditions have shown that the loss of cell viability is faster for the mutant than for wild-type SBW25 (A Koza & A Spiers, unpublished observations). A similar observation has been made for *P. putida* mt-2 (the progenitor of KT2440) where water stress was also found to increase cellulose expression [76]. Many bacteria respond to desiccation by producing exopolysaccharides, many of which are hygroscopic and retain water entropically [92], and amorphous cellulose is more hygroscopic and retains more liquid than crystalline cellulose [62]. Support for an anti-predation role for cellulose comes from the finding that the competitive fitness of WS genotypes in static microcosms increases in the presence of the grazing protist *Tetrahymena thermophile* [93].

10. Concluding statement

Bacterial cellulose production and air-liquid (A-L) interface biofilm-formation was first described 1886 for *Bacterium xylinum*, and subsequently observed by ourselves and colleagues in the evolution of the *Pseudomonas fluorescens* SBW25 Wrinkly Spreader (WS) some 116 years later. It is clear that this type of biofilm-formation is common-place amongst the environmental pseudomonads, many of which also utilise cellulose as the main matrix component of the biofilm. The fitness advantage of cellulose matrix-based biofilm-formation by SBW25 in static microcosms is well-proven, but the fitness in natural environments, and the true function of cellulose, is poorly researched and not yet understood. It may be that bacterial cellulose is used to form small biofilms in water bodies, acting to retain bacteria at the A-L interface or to maintain them on solid surfaces against water flow. Appositely, cellulose fibres may resist desiccation stress in water-limited environments, or even provide protection from protist and nematode predation. It is of course possible that cellulose performs a number of functions, which might explain the wide distribution of cellulose synthase operons amongst the *Proteobacter* inhabiting a diverse array of environments.

Author details

Andrew J. Spiers*, Ayorinde O. Folorunso and Kamil Zawadzki
The SIMBIOS Centre, University of Abertay Dundee, Dundee, UK

Yusuf Y. Deeni
School of Contemporary Sciences, University of Abertay Dundee, Dundee, UK

Anna Koza
Novo Nordisk Foundation Center for Biosustainability, Hørsholm, Denmark

Olena Moshynets
Laboratory of Microbial Ecology, Institute of Molecular Biology and Genetics of the National Academy of Sciences of Ukraine, Kiev, Ukraine

Acknowledgement

AJS is a member of the Scottish Alliance for Geoscience Environment and Society (SAGES) and AK was a SAGES-associated PhD student with AJS. KZ undertook his MSc research

---

* Corresponding Author

project with AJS. AF is currently undertaking his MSc with AJS and YYD; his preliminary bioinformatics results are presented here and will be published at a later date. Funding from the Royal Society of Edinburgh through the International Exchange Programme supported the collaboration between AJS and OM for this project. The University of Abertay Dundee is a charity registered in Scotland, No: SC016040.

11. References

[1]   Brown AJ. XLIII. On an acetic ferment which forms cellulose. J Chem Soc, Trans 1886;49 432-439.

[2]   Brown RM. Bacterial cellulose. In: Phillips GO, Williams PA, Kennedy JF. (eds) Cellulose: structural and functional aspects. Ellis Horwood Series in Polymer Science and Technology. UK: Ellis Horwood Ltd.; 1989. p145-151.

[3]   Ross P, Mayer R, Benziman M. Cellulose biosynthesis and function in bacteria. Microbiol Rev 1991;55(1) 35-58.

[4]   Solano C, García B, Valle J, Berasain C, Ghigo J.-M, Gamazo C, Lasa I. Genetic analysis of *Salmonella enteritidis* biofilm formation: critical role of cellulose. Mol Microbiol 2002;43(3) 793-808.

[5]   Spiers AJ, Kahn SG, Bohannon J, Travisano M, Rainey PB. Adaptive divergence in experimental populations of *Pseudomonas fluorescens*. I. Genetic and phenotypic bases of Wrinkly Spreader fitness. Genetics 2002;161 33-46.

[6]   Cannon RE, Anderson SM. Biogenesis of Bacterial Cellulose. Crit Rev Microbiol 1991;17(6) 435-447.

[7]   Iguchi M, Yamanaka S, Budhiono A. Bacterial cellulose - a masterpiece of nature's arts. J Materials Sci 2000;35 261-270.

[8]   Römling U. Molecular biology of cellulose production in bacteria. Res Microbiol 2002;153 205-212.

[9]   Chawla PR, Bajaj IB, Survase SA, Singhal RS. Fermentative Production of Microbial Cellulose. Food Technol Biotechnol 2009;47(2) 107-124.

[10] Rainey PB, Bailey MJ. Physical map of the *Pseudomonas fluorescens* SBW25 chromosome. Mol Microbiol 1996;19(3) 521-533.

[11] Rainey PB, Travisano M. Adaptive radiation in a heterogeneous environment. Nature 1998;394 69-72.

[12] Silby MW, Cerdeño-Tárraga AM, Vernikos G, Giddens SR, Jackson R, Preston G, Zhang X-X, Godfrey S, Spiers AJ, Harris S, Challis GL, Morningstar A, Harris D, Seeger K, Murphy L, Rutter S, Squares R, Quail MA, Saunders E, Anderson I, Mavromat K, Brettin TS, Bentley S. Thomas CM, Parkhill J, Levy SB, Rainey PB, Thomson NR. Genomic and functional analyses of diversity and plant interactions of *Pseudomonas fluorescens*. Genome Biology 2009;10 R51.

[13] Peix A, Ramírez-Bahena M-H, Velázquez E. Historical evolution and current status of the taxonomy of genus *Pseudomonas*. Infect Genet Evol 2009;9 1132-1147.

[14] Silby MW, Winstanley C, Godfrey SAC, Levy SB, Jackson RW. *Pseudomonas* genomes: diverse and adaptable. FEMS Microbiol Rev 2011;35(4) 652-680.

[15] Ude S, Arnold DL, Moon CD, Timms-Wilson T, Spiers AJ. Biofilm formation and cellulose expression among diverse environmental *Pseudomonas* isolates. Environ Microbiol 2006;8(11) 1997-2011.

[16] Costerton JW, Lewandowski Z, Caldwell D, Korber D, Lappin-Scott HM. Microbial biofilms. Annu Rev Microbiol 1995;49 711-745.

[17] Costerton JW, Lewandowski Z, De Beer D, Caldwell D, Korber D, James G. Biofilms, the customized microniche. J Bacteriol 1994;176(8) 2137-2142.

[18] Stickler D. Biofilms. Curr Opin Microbiol 1999;2 270-275.

[19] Dalton HM, March PE. Molecular genetics of bacterial attachment and biofouling. Curr Opin Biotechnol 1998;9(3) 252-255.

[20] Davey ME, O'Toole, GA. Microbial biofilms: from ecology to molecular genetics. Microbiol Mol Biol Rev 2000;64(4) 847-867.

[21] Kuchma SL, O'Toole GA. Surface-induced and biofilm-induced changes in gene expression. Curr Opin Biotechnol 2000;11 429-433.

[22] Watnick P, Kolter, R. Biofilm, city of microbes. J Bacteriol 2000;182(10) 2675-2679.

[23] Stewart PS, Costerton JW. Antibiotic resistance of bacteria in biofilms. Lancent 2001;358 135-138.

[24] Sutherland IW. Biofilm exopolysaccharides: a strong and sticky framework. Microbiology 2001;147 3-9.

[25] Sutherland IW. The biofilm matrix- an immobilized but dynamic microbial environment. TRENDS Microbiol 2001;9(5) 222-227.

[26] Dunne WM. Bacterial adhesion: seen and good biofilms lately? Clinical Microbiol Rev 2002;15(2) 155-166.

[27] Morris CE, Monier J-M. The ecological significance of biofilm formation by plant-associated bacteria. Annu Rev Phytopathol 2003;41 429-453.

[28] Webb JS, Givskov M, Kjelleberg S. Bacterial biofilms: prokaryotic adventures in multicellularity. Curr Opin Microbiol 2003;6(6) 578-585.

[29] O'Toole GA. To build a biofilm. J Bacteriol 2003;185(9) 2687-2689.

[30] Branda SS, Vik A, Friedman L, Kolter R. Biofilms: the matrix revisited. TRENDS Microbiol 2005;13(1) 20-26.

[31] Hall-Stoodley L, Stoodley P. Evolving concepts in biofilm infections. Cellular Microbiol 2009;11(7) 1034-1043.

[32] West SA, Diggle SP, Buckling A, Gardner A, Griffin AS. The Social Lives of Microbes. Ann Rev Ecol Evol S 2007;38 53-77.

[33] Xavier JB, Foster KR. Cooperation and conflict in microbial biofilms. PNAS (USA) 2007;104(3) 876-881.

[34] Kolter R, Greenberg EP. Microbial sciences: the superficial life of microbes. Nature 2006;441 300-302.

[35] Chang WS, Van De Mortel M, Nielsen L, Nino De Guzman G, Li X, Halverson LJ. Alginate production by *Pseudomonas putida* creates a hydrated microenvironment and contributes to biofilm architecture and stress tolerance under water-limiting conditions. J Bacteriol 2007;189(22) 8290-8299.

[36] Allison DG, Ruiz B, Carmen S, Jaspe A, Gilbert P. Extracellular products as mediators of the formation and detachment of *Pseudomonas fluorescens* biofilms. FEMS Microbiol Lett 1998;167(2) 179-184.

[37] Davies DG, Parsek MR, Pearson JP, Iglewski BH, Costerton JW, Greenberg EP. The Involvement of cell-to-cell signals in the development of a bacterial biofilm. Science 1998;280(5361) 295-298.

[38] Van Loosdrecht MCM, Heijnen JJ, Eberl H, Kreft J, Picioreanu C. Mathematical modelling of biofilm structures. Antonie van Leeuwenhoek 2002;81 245–256.

[39] Battin T, Sloan WT, Kjelleberg S, Daims H, Head IM, Curtis TP, Eberl L. Microbial landscapes: new paths to biofilm research. Nature Rev Microbiol 2007;5 76-81.

[40] Brown AJ. XIX. The chemical action of pure cultivations of *Bacterium aceti*. J Chem Soc Trans 1886;49 172-187.

[41] Holzapfel WH. (1997). Use of starter cultures in fermentation on a household scale. Food Control 1997;8(5) 241–258.

[42] Greenwalt CJ, Steinkraus KH, Ledford, RA. Kombucha, the fermented tea: microbiology, composition, and claimed health effects. J Food Protection 2000;63(7) 976-981.

[43] Raghavendra MRR. Acetic acid bacteria. Annu Rev Microbiol 1957;11 317-338.

[44] Nguyen VT, Flanagan B, Gidley MJ, Dykes GA. Characterization of cellulose production by a Gluconacetobacter xylinus strain from Kombucha. Curr Microbiol 2008;57 449-453.

[45] Verschuren, PG, Cardona TD, Nout MHR, De Gooijer KD, Van Den Heuvel JC. Location and limitation of cellulose production by *Acetobacter xylinum* established from oxygen profiles. J Biosc Bioeng 2000;89(5) 414-419.

[46] Spiers AJ, Arnold DL, Moon CD, Timms-Wilson TM. A survey of A-L biofilm formation and cellulose expression amongst soil and plant-associated *Pseudomonas* isolates. In: Bailey MJ, Lilley AK, Timms-Wilson TM. (eds.) Microbial Ecology of Aerial Plant Surfaces. UK, CABI; 2006. p121-132.

[47] Gehrig SM. Adaptation of *Pseudomonas fluorescens* SBW25 to the air-liquid interface: a study in evolutionary genetics. DPhil thesis. University of Oxford; 2005.

[48] Spiers AJ. Wrinkly-Spreader fitness in the two-dimensional agar plate microcosm: maladaptation, compensation and ecological success. PLoS ONE 2007;2(8) e740.

[49] Green JH, Koza A, Moshynets O, Pajor R, Ritchie MR, Spiers AJ. Evolution in a test tube: rise of the Wrinkly Spreaders. J Biol Educ 2011;45(1) 54–59.

[50] Rainey PB, Buckling A, Kassen R, Travisano M. The emergence and maintenance of diversity: insights from experimental bacterial populations. Trends Ecol Evol 2000;15(6) 243-247.

[51] Buckling A, Wills MA, Colegrave N. Adaptation limits diversification of experimental bacterial populations. Science 2003;302(5653) 2107-2109.

[52] Buckling A, Maclean CR, Brockhurst MA, Colegrave N. The Beagle in a bottle. Nature 2009;457 824-829.

[53] Koza A, Moshynets O, Otten W, Spiers AJ. Environmental modification and niche construction: Developing O₂ gradients drive the evolution of the Wrinkly Spreader. ISME J 2011;5(4) 665-673.

[54] Saxena IM, Kudlicka K, Okuda K, Brown RM. Characterization of genes in the cellulose-synthesizing operon (*acs* operon) of *Acetobacter xylinum*: implications for cellulose crystallization. J Bacteriol 1994;176(18) 5735-5752.

[55] Blattner FR, Plunkett G. Bloch CA, Perna NT, Burland V, Riley M, Collado-Vides J, Glasner JD, Rode CK, Mayhew GF, Gregor J, Davis NW, Kirkpatrick HA, Goeden MA, Rose DJ, Mau B, Shao Y. The complete genome sequence of *Escherichia coli* K-12. Science 1997;277(5331) 1453-1462.

[56] Sullivan SM, Maddock JR. Bacterial division: finding the dividing line. Curr Biol 2000;10(6) R249-252.

[57] Le Quéré B, Ghigo JM. BcsQ is an essential component of the *Escherichia coli* cellulose biosynthesis apparatus that localizes at the bacterial cell pole. Mol Microbiol 2009;72(3) 724-740.

[58] Franklin MJ, Ohman DE. Identification of *algI* and *algJ* in the *Pseudomonas aeruginosa* alginate biosynthetic gene cluster which are required for alginate O acetylation. J Bacteriology 1996;178(8) 2186-2195.

[59] Spiers AJ, Bohannon J, Gehrig SM, Rainey PB. Biofilm formation at the air–liquid interface by the *Pseudomonas fluorescens* SBW25 wrinkly spreader requires an acetylated form of cellulose. Mol Microbiol 2003;50(1) 15-27.

[60] Spiers AJ, Rainey PB. The *Pseudomonas fluorescens* SBW25 wrinkly spreader biofilm requires attachment factor, cellulose fibre and LPS interactions to maintain strength and integrity. Microbiology 2005;151(9) 2829-2839.

[61] Zevenhuizen LP, Bertocchi C, Van Neerven AR. Congo red absorption and cellulose synthesis by *Rhizobiaceae*. Antonie Van Leeuwenhoek 1986;52(5) 381-386.

[62] Hoshino E, Wada Y, Nishizawa K. Improvements in the hygroscopic properties of cotton cellulose by treatment with an endo-type cellulase from *Streptomyces* sp. KSM-26. J Biosci Bioeng 1999;88(5) 519-525.

[63] Schrecker ST, Gostomski PA. Determining the water holding capacity of microbial cellulose. Biotechnol Lett 2005;27(19) 1435-1438.

[64] Rocchetta HL, Burrows LL, Lam JS. Genetics of O-antigen biosynthesis in *Pseudomonas aeruginosa*. Microbiol Molec Biol Rev 1999;63(3) 523-553.

[65] Benziman M, Haigler CH, Brown RM, White AR, Cooper KM. Cellulose biogenesis: polymerization and crystallization are coupled processes in *Acetobacter xylinum*. PNAS (USA) 1980;77(11) 6678-6682.

[66] Bailey J, Manoil C. Genome-wide internal tagging of bacterial exported proteins. Nat Biotechnol 2002;20(8) 839-842.

[67] Bantinaki B, Kassen R, Knight CG, Robinson Z, Spiers AJ, Rainey PB. Adaptive divergence in experimental populations of *Pseudomonas fluorescens*. III. Mutational origins of Wrinkly Spreader diversity. Genetics 2007;176(1) 441-453.

[68] Bren A, Eisenbach M. How signals are heard during bacterial chemotaxis: protein-protein interactions in sensory signal propagation. J Bacteriol 2000;182(24) 6865-6873.

[69] Malone JG, Williams R, Christen M, Jenal U, Spiers AJ, Rainey PB. The structure–function relationship of WspR, a *Pseudomonas fluorescens* response regulator with a GGDEF output domain. Microbiology 2007;153(4) 980-994.

[70] Goymer PJ. The role of the WspR response regulator in the adaptive evolution of experimental populations of *Pseudomonas fluorescens* SBW25. DPhil thesis. University of Oxford; 2002.

[71] Goymer P, Kahn SG, Malone JG, Gehrig SM, Spiers AJ, Rainey PB. Adaptive divergence in experimental populations of *Pseudomonas fluorescens*. II. Role of the GGDEF regulator WspR in evolution and development of the Wrinkly Spreader phenotype. Genetics 2006;173(2) 515-526.

[72] McDonald MJ, Gehrig SM, Meintjes PL, Zhang X, Rainey PB. Adaptive Divergence in experimental populations of *Pseudomonas fluorescens*. IV. Genetic constraints guide evolutionary trajectories in a parallel adaptive radiation. Genetics 2009;183(3) 104-1053.

[73] Koza A, Hallett PD, Moon CD, Spiers AJ. Characterization of a novel air–liquid interface biofilm of *Pseudomonas fluorescens* SBW25. Microbiology 2009;155(5) 1397-1406.

[74] Nelson KE, Weinel C, Paulsen IT, Dodson RJ, Hilbert H, Martins Dos Santos VA, Fouts DE, Gill SR, Pop M, Holmes M, Brinkac L, Beanan M, Deboy RT, Daugherty S, Kolonay J, Madupu R, Nelson W, White O, Peterson J, Khouri H, Hance I, Chris Lee P, Holtzapple E, Scanlan D, Tran K, Moazzez A, Utterback T, Rizzo M, Lee K, Kosack D, Moestl D, Wedler H, Lauber J, Stjepandic D, Hoheisel J, Straetz M, Heim S, Kiewitz C, Eisen JA, TImmis KN, Düsterhöft A, Tümmler B, Fraser CM. Complete genome sequence and comparative analysis of the metabolically versatile *Pseudomonas putida* KT2440. Environ Microbiol 2002;4(12) 799-808.

[75] Buell CR, Joardar V, Lindeberg M, Selengut J, Paulsen IT, Gwinn ML, Dodson RJ, Deboy RT, Durkin AS, Kolonay JF, Madupu R, Daugherty S, Brinkac L, Beanan MJ, Haft DH, Nelson WC, Davidsen T, Zafar N, Zhou L, Liu J, Yuan Q, Khouri H, Fedorova N, Tran B, Russell D, Berry K, Utterback T, Van Aken SE, Feldblyum TV, D'Ascenzo M, Deng WL, Ramos AR, Alfano JR, Cartinhour S, Chatterjee AK, Delaney TP, Lazarowitz SG, Martin GB, Schneider DJ, Tang X, Bender CL, White O, Fraser CM, Collmer A. The complete genome sequence of the Arabidopsis and tomato pathogen *Pseudomonas syringae* pv. *tomato* DC3000. PNAS (USA) 2003;100(18) 10181-10186.

[76] Nielsen L, Li X, Halverson LJ. Cell-cell and cell-surface interactions mediated by cellulose and a novel exopolysaccharide contribute to *Pseudomonas putida* biofilm

formation and fitness under water-limiting conditions. Environ Microbiol 2011;13(5) 1342-1356.

[77] Fett WF, Osman SF, Dunn MF. Characterization of exopolysaccharides produced by plant-associated fluorescent pseudomonads. Appl Envir Microbiol 1989;55(3) 579-583.

[78] Friedman L, Kolter R. Genes involved in matrix formation in *Pseudomonas aeruginosa* PA14 biofilms. Molec Microbiol 2004;51(3) 675-690.

[79] Ramsey DM, Wozniak DJ. Understanding the control of *Pseudomonas aeruginosa* alginate synthesis and the prospects for management of chronic infections in cystic fibrosis. Molec Microbiol 2005;56(2) 309-322.

[80] Byrd MS, Sadovskaya I, Vinogradov E, Lu H, Sprinkle AB, Richardson SH, Ma L, Ralston B, Parsek MR, Anderson EM, Lam JS, Wozniak DJ. Genetic and biochemical analyses of the *Pseudomonas aeruginosa* Psl exopolysaccharide reveal overlapping roles for polysaccharide synthesis enzymes in Psl and LPS production. Mol Microbiol 2009;73(4) 622-638.

[81] National Center for Biotechnology Information (USA). http://www.ncbi.nlm.nih.gov.

[82] Larkin MA, Blackshields G, Brown NP, Chenna R, McGettigan PA, McWilliam H, Valentin F, Wallace IM, Wilm A, Lopez R, Thompson JD, Gibson TJ, Higgins DG. Clustal W and Clustal X version 2.0. Bioinformatics 2007;23(21) 2947-2948.

[83] Kimura S, Chen HP, Saxena IM, Brown RM, Itoh T. Localization of c-*di*-GMP-binding protein with the linear terminal complexes of *Acetobacter xylinum*. J Bacteriol 2001;183(19) 5668-5674.

[84] Giddens SR, Jackson RW, Moon CD, Jacobs MA, Zhang X-X, Gehrig SM, Rainey PB. Mutational activation of niche-specific genes provides insight into regulatory networks and bacterial function in a complex environment. PNAS (USA) 2007;104(46) 18247-18252.

[85] Gal M, Preston GM, Massey RC, Spiers AJ, Rainey PB. Genes encoding a cellulosic polymer contribute toward the ecological success of *Pseudomonas fluorescens* SBW25 on plant surfaces. Molec Ecol 2003;12(11) 3109-3121.

[86] Jenal U, Malone J. Mechanisms of cyclic-di-GMP signaling in bacteria. Ann Rev Genet 2006;40 385-407.

[87] Hengge R. Principles of c-*di*-GMP signalling in bacteria. Nat Rev Microbiol 2007;7(4) 263-273.

[88] Dasgupta N, Ferrell EP, Kanack KJ, West SE, Ramphal R. *fleQ*, the gene encoding the major flagellar regulator of *Pseudomonas aeruginosa*, is sigma70 dependent and is downregulated by Vfr, a homolog of *Escherichia coli* cyclic AMP receptor protein. J Bacteriol 2002;184(19) 5240-5250.

[89] Hickman JW, Harwood CS. Identification of FleQ from *Pseudomonas aeruginosa* as a c-*di*-GMP-responsive transcription factor. Mol Microbiol 2008;69(2) 376-389.

[90] Baynham PJ, Ramsey DM, Gvozdyev BV, Cordonnier EM, Wozniak DJ. The *Pseudomonas aeruginosa* ribbon-helix-helix DNA-binding protein AlgZ (AmrZ) controls twitching motility and biogenesis of type IV pili. J Bacteriol 2006;188(1) 132-140.

[91] Lenski RE, Rose MR, Simpson SC, Tadler SC. Long-term experimental evolution in *Escherichia coli*. I. Adaptation and divergence during 2,000 generations. Am Nat 1991;138(6) 1315-1341.

[92] Flemming H-C, Wingender J. The biofilm matrix. Nat Rev Microbiol 2010;8(9) 623-633.

[93] Meyer R, Kassen R. The effects of competition and predation on diversification in a model adaptive radiation. Nature 2007;446(7134) 432-435.

Chapter 2

Cellulose-Based Bioelectronic Devices

Ana Baptista, Isabel Ferreira and João Borges

Additional information is available at the end of the chapter

1. Introduction

The integration of biomolecules with electronic elements to form multifunctional devices has been recently the subject of intense scientific research. The need of new sensors exhibiting a high selectivity and a total reliability in connection with smart systems and actuators for real time diagnostic and monitoring of diseases has driven wonderful developments in sensors and particularly in biosensors. Biosensors can be regarded as complementary tools to classical analytical methods due to their inherent simplicity, relative low cost, rapid response and proneness to miniaturization, thereby allowing continuous monitoring. They can integrate portable and implantable devices and be used in biological and biomedical systems. However, the development of biocompatible, nontoxic and lightweight power sources devices is still challenging. It would enable the production of various functional devices mechanically flexible and auto-sustained, allowing their integration into a wide range of innovative products such as in implantable medical devices.

2. Bioelectronics

Bioelectronics is a new multidisciplinary scientific area that results from the combination of biology, electronics and nanotechnology. Multifunctional devices can be made by integrating biological materials with electronic elements providing a novel and broad platform for biochemical and biotechnological processes. These functional devices can be used to develop sensing devices, such as enzyme-based biosensors [1], DNA-sensors [2], immunosensors [3], and to develop implantable biofuel cells [4] for biomedical applications, self-powered biosensors [1], autonomously operated devices, among others.

2.1. Biosensors

Functional devices can successfully convert (bio)chemical information into electronic one by means of an appropriate transducer which contains specific molecular recognition structures. In this way, biosensors can be described as integrated receptor-transducer devices which provide selective quantitative or semi-quantitative analytical information using biological recognition elements. The main advantages of biosensors, over traditional analytical detection techniques, are their cost-effectiveness, fast and portable detection, which makes *in situ* and real time monitoring possible. Implantable biosensors can made a continuous monitoring of metabolites providing an early signal of metabolic balances and assist in the prevention and cure of various disorders, for instance diabetes and obesity [5].

Enzymes are well-known biological sensing materials used in the development of biosensors due to their specificity. However, since they have poor stability in solution, enzymes need to be stabilized by immobilization. Enzyme immobilization can be made by covalent linkage, physical adsorption, cross-linking, encapsulation or entrapment [6, 7]. The choice of the immobilization method depends on the nature of the biological element, the type of transducer used, the physicochemical properties of the analyte and the conditions in which the biosensor should operate [8, 9]. Moreover, it is essential that the biological element exhibit maximum activity in its immobilized environment.

As a result, the development of a sensing device based on enzymes is in a good agreement with the present concerns of Green Chemistry due to inherently being a clean process. Notwithstanding some shortcomings such as high sensitivity to environmental factors (like pH, ionic strength and temperature), dependence on some cofactors and limited lifetime hinder the utilization of enzymes in some specific situations.

To overcome the drawbacks, enzyme-free biosensors have been actively developed owing to their simple fabrication, stability and reproducible characteristics. Novel nanoparticle (NP)-modified electrodes and other functionalized electrodes have been tested in the design of enzyme-free biosensors [10, 11]. Nanostructured materials have the advantage to be easily functionalized exhibiting high electrocatalytic activity and stability. For instance, carbon-based nanostructures have been widely studied as a platform which can hybridize with other functionalized materials, such as metal and metal oxides, forming nanocomposites with improved electrochemical properties [12]. Overall, these nanostructures can provide optimal composite electrode materials for high-performance enzyme-free biosensors.

2.2. Implantable energy harvesting devices

The rising interest in Micro Electrical Mechanical Systems (MEMS) due to expanding application areas and new products opportunities, gave rise to the need for reliable and cost effective MEMS, especially in areas such as biosensors, energy harvesting, and drug delivery [13, 14].

Biomedical technology usually requires various portable, wearable, easy-to-use, and implantable devices that can interface with biological systems. Currently, implantable medical microsystems are powered by small batteries with limited lifetime. Although, the scientific progress in this area has enabled a decrease in the electrical requirements of the miniaturized devices, the development of a suitable power source remains a major challenge for many devices in the bioengineering and medical fields. (MEMS)-based electrical power generation devices can allow the autonomous operation of implantable biosensors by direct power supply or supplement the existing battery-based power systems. Harvesting energy directly from the environment is one of the most effective and promising approaches for powering nanodevices. Mechanical energy surrounds us in our daily life, taking the form of sonic waves, mechanical vibrations and impacts. These vibrations can be converted into electricity via electrostatic, electromagnetic, and piezoelectric microgenerators [15-17]. For instance, harvesting energy from the human body can be possible by converting hydraulic energy from blood flow, heart beats and blood vessels contraction [18]. Another consideration is to use body heat to generate electricity using a thermoelectric generator [19].

More recently, biofuel cells have also been considered for energy harvesting. Implantable fuel cell systems, convert endogenous substances and oxygen into electricity by means of a spatially separated electrochemical reaction. Unlike conventional fuels cells, which rely on expensive rare metal catalyst and/or operate on reformed fossil fuels, biofuel cells rely on the chemical reactions driven by diverse biofuels and biological catalyst. Biofuel cells can be classified according to the biocatalyst. Almost all biochemical processes are catalyzed by enzymes. Systems using specific isolated enzymes at least for a part of their operation are known as enzymatic fuel cells [20], while those utilizing whole organisms containing complete pathways are known as microbial fuel cells [21].

After all, energy harvesting devices and their applications are expanding and becoming more attractive especially with advance in microelectronics and MEMS. MEMS-based generation techniques have many characteristics that make them appealing for biological applications, including the ability to control their physical and chemical characteristics on the micrometer and nanometer scale.

3. Cellulose

The demand for products made from renewable and sustainable resources, non-petroleum based, and with low environmental safety risk is persistently increasing. For that reason, renewable materials have been widely explored by consumers, industry, and government. Half of the biomass produced by photosynthetic organisms such as plants, algae, and some bacteria is made up of cellulose, which is the most abundant molecule on the planet. Natural cellulose-based materials, such as wood and cotton, have been used by our society as engineering materials for thousands of years. Cellulose exhibit excellent characteristics, which include hydrophilicity, chirality, biodegradability, capacity for broad chemical modification, and ability to form semicrystalline fiber morphologies, which drawn considerably increased interest and encouraged interdisciplinary research on cellulose-based materials.

3.1. Cellulose source materials

Cellulose plays a significant role in the structural support of wood, plants, and composites because of its high mechanical properties. Wood remains the most important raw material source of cellulose. The structure of wood is highly complex due to the presence of lignin, a three-dimensional polymer network that binds to carbohydrates (hemicellulose and cellulose) to form a tight and compact structure. The compact structure of wood biomass is particularly challenging because in its native state is impossible to dissolve it in conventional solvents. Traditionally, cellulose is extracted from wood through the Kraft pulping process [22] which involves toxic chemicals and the intensive processing conditions. Recently, research studies focused on a "greener" process which uses Ionic Liquids (ILs) for wood dissolution [23]. A wide variety of plant materials have been studied for the extraction of cellulose including cotton, potato tubers, sugar beet pulp, soybean stock, and banana rachis[24, 25]. Furthermore, cellulose microfibrils can be produced by several species of algae, such as green, gray, red, and yellow-green. Among the algae species, differences in cellulose microfibrils structures can be obtained due to the different biosynthesis process [26]. The cellulose obtained from algal species contains porous or spongy like structure, which is substantially different from the higher plant cellulose. Cellulose microfibrils can also be segregate by bacteria under special culturing conditions. Bacteria can produce a thick gel composed by cellulose microfibrils and water (97% of water content). The major advantage found in bacterial cellulose is the possibility to modify microfibrils structure by changing the culture conditions [27].

3.2. Cellulose functionalization

The solubility of cellulose depends on many factors especially on its structure, molecular weight and source. Polysaccharides are well-known to manifest a strong tendency to aggregate or to incomplete solubilization due to the formation of hydrogen bonds. The hydrogen bonding patterns in cellulose are considered as one of the most relevant factors on its physical and chemical properties. The solubility, crystalinity and hydroxyl reactivity can be directed affected by intra- and intermolecular bond formation (Figure 1) [28].

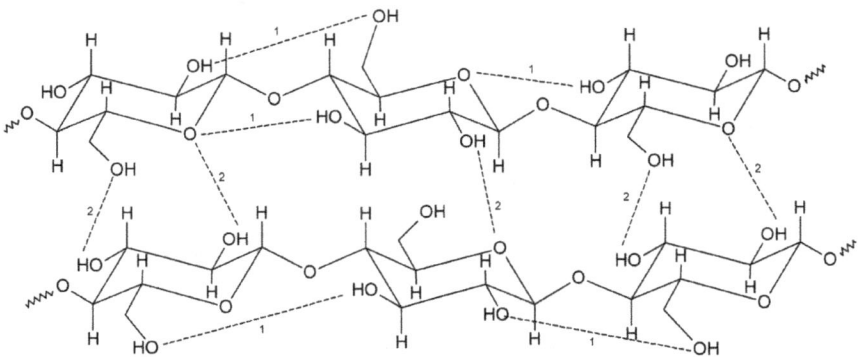

Figure 1. The structure and intra- (1) and interchain (2) hydrogen bonding pattern in cellulose.

Moreover, cellulose can be chemically modified to yield cellulose derivatives. The cellulose derivatives were designed and fine-tuned to obtain certain desired properties and the chemical functionalization of cellulose is done by changing the inherent hydrogen bond network and by introducing different substituents (Figure 2). Indeed, the properties of cellulose derivatives are mainly determined by the group of substituents and the degree of substitution. These substituents can prevent spontaneous formation of hydrogen bonding or even create new interactions between the cellulose chains. With this insight, recent progress has been made in cellulose chemical modification achieving new routes that are now

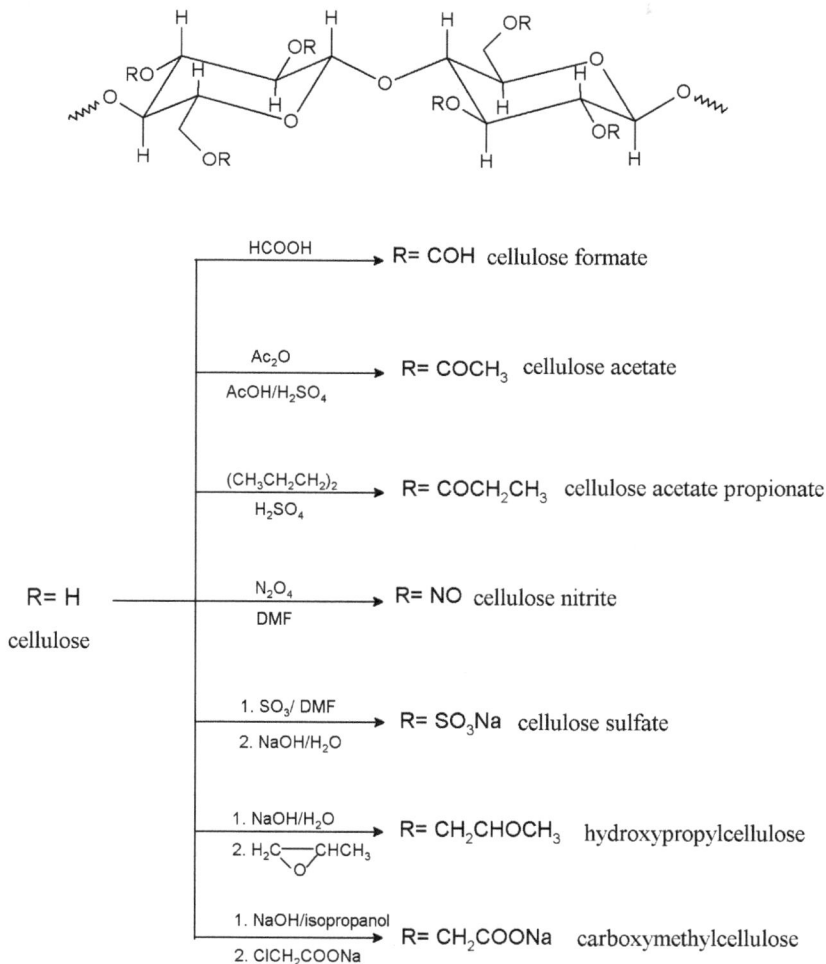

Figure 2. The most relevant cellulose derivatives and their synthesis pathways.

available for the production of functional and sustainable cellulose–based materials [29]. The chemical modification of cellulose surface is a classical approach to transform the polar hydroxyl groups sitting at the surface of cellulose into moieties able to enhance interactions with the matrix. Indeed, the high density of free hydroxyl groups in cellulose makes it a helpful solid substrate that can undergo functionalization to come into novel advanced applications. Owing to cellulose chain rigidity, some cellulose derivatives can form thermotropic or lyotropic mesophases (in suitable solvents). Among cellulose ethers, hydroxypropylcellulose (HPC) have encouraged the scientific community due to its cholesteric liquid crystalline organization at high concentration [30]. These liquid crystalline phases, with an internal periodic modulation of the refractive index, exhibit many remarkable optical properties as a result of their photonic band structure, which have applications such as polarized light sources, information displays, and storage devices [31]. These phases may also mimic the structural organization of type I collagen and are good analogues of the extracellular matrix, with a structure close to that of biological tissues. These materials can be used either in tissue repair or as models for the culture of cells in 3D, the study of their migration and signaling activities, in a manner close to physiological conditions [32].

In the next section, the functionalization of cellulose will be addressed in detail. Novel functionalized cellulose-based materials have been developed for biosensors and energy storage devices. Some approaches for enzyme immobilization methods including covalent attachment of enzymes by reaction with chemically modified cellulose as well as by adsorption of proteins will be described.

4. Cellulose-based bioelectronic devices

4.1. Cellulose-based matrices for biological immobilization

Both cellulose and cellulose derivatives, such as cellulose nitrate, cellulose acetate and carboxymethyl cellulose, exhibit an excellent biocompatibility which makes them appropriate for immobilization of biological compounds [33, 34]. As is known, the ideal support for enzymes should be inert, stable and mechanically resistant making the use of cellulose matrices ideal for adsorption and covalent bond immobilization.

The modification of cellulose with dendritic structures is a novel and interesting path to synthesize functional and unconventional cellulose-based supports for the immobilization of enzymes. Moreover, the introduction of reactive groups into the cellulose structure may allow a covalent nonreversible attachment of biomolecules. Maria Montanez [35] and her team suggested the hybridization of cellulose surface with branched dendritic entities that improves the sensitivity toward biomolecules. The described methodology delivers a new toolbox for the design of sophisticated biosensors with advantages such as low detection limit, versatility and suppression of nonspecific interactions providing highly sophisticated cellulose surfaces with unprecedented tunability. Dendrimers are synthetic macromolecules with highly branched structure and globular shape. They possess unique properties such as high density of active groups, good structural homogeneity, internal porosity, and good biocompatibility [36]. When addressed to biosensor applications, the well defined dendritic structures generate surfaces with increased reproducibility and high affinity for biomolecular immobilization. This is due to the extraordinary control over the architecture coupled to the possibility of designing a large number of accessible active sites at the periphery of the dendritic scaffolds.

A further approach is the modification of cellulose-based structures with ionic liquids (ILs). Ionic liquids are often used in the preparation of functional materials by its covalent attachment to the support surface forming a stable composite. Moccelini [37] have reported the development of a novel polymeric support based on cellulose acetate and 1-n-butyl-3-methylimidazolium bis(trifluoromethylsulfonyl)imide-based IL, BMI.N(Tf)2 IL, for enzyme immobilization. The introduction of the IL probably causes an increase in the distance between the cellulose chains due to the interactions of the anion of the IL and the hydrogen bond networks of the cellulose acetate. Thus, the enzyme can be entrapped within the interstitial space of the formed composite, which results in a considerable stabilization of the enzyme structure, and consequently increases its activity. The study performed demonstrates that this material was able to immobilize Laccase, leading to high efficient and robust biocatalysts thus improving the electrochemical performance of the biosensor.

The use of ILs is an alternative either for cellulose dissolution or to facilitate the dispersion of carbon nanotubes. For that reason, Xuee Wu [38] describes a method to immobilize enzymes in a cellulose-multiwalled carbon nanotube (MWCNT) matrix via the IL reconstitution process. This method consists in the dissolution of cellulose in the IL, followed by dispersion of MWCNT in the solution and enzyme addition. Subsequently, the IL is removed by dissolution, leaving the cellulose-MWCNT matrix with the enzyme encapsulated on the surface. The cellulose–MWCNT matrix possesses a porous structure which allows the immobilization of a large amount of enzyme close to the electrode surface, where direct electron communication between active site of enzyme and the electrode is enabled. The –OH groups of cellulose can also provide a good environment for the encapsulation of the enzyme. The authors have employed the resulting porous matrix in the immobilization of Glucose oxidase (GOx). The encapsulated GOx showed good bioelectrochemical activity, enhanced biological affinity as well as good stability.

The simple electrode fabrication methodology and the biocompatibility of the cellulose–MWCNT matrix mean that the immobilization matrix can be extended to diverse proteins, thus providing a promising platform for further research and development of biosensors and other bioelectronics devices.

The use of ILs as an intermediary solvent to facilitate the combination of cellulose and CNTs has been suggested by Jun Wan [39]. A cellulose and single wall carbon nanotube (SWNTs) composite was utilized to immobilize leukemia K562 cells on a gold electrode to form a cell impedance sensor.

Envisaging the immobilization of other biomolecules, Alpat and Telefoncu [40] describes the development of a novel biosensor based on the co-immobilization of TBO (Toluidine Blue O), NADH (Nicotinamide adenine dinucleotide) and ADH (alcohol dehydrogenase) on a cellulose acetate coated glassy carbon electrode for ethanol identification. In fermentation and distillation processes, ethanol can reach toxic concentrations that may cause inflammation and conjunctiva of the nasal mucous membrane and irritation of the skin. Therefore proper detection and quantification of ethanol is of extreme importance. The detector is made by simply deposition on the surface of a glassy carbon electrode and an active layer was prepared by covalent linkage between the mediator TBO and a cellulose acetate membrane. This mediator is commonly used for the oxidation and determination of NADH. Then, a NADH solution and the ADH were added to the cellulose acetate-TBO-modified glassy carbon electrode and tested. The developed biosensor exhibited good thermal stability and long-term storage stability.

The immobilization of proteins on solid surfaces is a key step for the development of medical diagnostic systems. An alternative approach for the immobilization of specific proteins is the chemical modification of cellulose. Stephan Diekmann [41] and his colleagues have described a targeted chemical modification of cellulose to be used as substrate for proteins and biocatalysts bonding. A new cellulose derivative obtained by modification of cellulose with nitrilotriacetic acid (NTA) was used for the complexation of nickel (II). The complex formed was used to immobilize labeled molecules. In that way, the Ni-cellulose derivative allows the development of specific and sensitive molecular diagnostic systems. Another approach is proposed by Jianguo Juang [42] using protein-functionalized cellulose sheets. The surface of the individual cellulose nanofibers was coated with an ultrathin titania gel. The titania coated surfaces were then biotinylated creating a biotin monolayer on each nanofiber by the coordination of carboxyl group. Subsequently, bovine serum albumin (BSA) was added to the functionalized surface to prevent nonspecific adsorption of streptavin. The immobilization of streptavin molecules on its surface was made through biotin-streptavin interaction. Streptavidin has two pairs of binding sites for biotin on opposite's faces of molecule. When immobilized on the cellulose nanofiber with one pair, the other pair is available for further attachment of biotinylated species. The cellulose sheet, composed by numerous nanofibers modified with titania/biotin/BSA layers with anchored streptavidin molecules, gives a large surface area to detect biotin-tagged biomolecules. Thus, biofunctionalized cellulose is a promising substract for specific biomolecular detection.

As previously described, the immobilization of biological compounds can be an important parameter for implantable biosensors due to the fact that it dictates the sensitivity, selectivity and long-term stability of the device. Thus, cellulose appears as an easy functionalized material and an ideal support for adsorption and covalent bond immobilization of biomolecules.

4.2. Cellulose-based energy storage devices

There is currently a strong demand for the development of new inexpensive, flexible, lightweight and environmentally friendly energy storage devices. As a result of these needs, research is currently carried out to develop new versatile and flexible electrode materials as alternatives to the materials used in batteries and fuel cells.

Bacterial cellulose membranes have been widely used as an active layer for the construction of electrodes for fuel cells. Barbara Evans [43] and her colleagues describe the ability of bacterial cellulose to catalyze the precipitation of palladium within its structure. Since bacterial cellulose fibrils are extruded by bacteria and then self-assemble to form a three-dimensional network configuration, a structure with a high surface area with catalytic potential is generated. Bacterial cellulose has reducing groups able to promote the precipitation of palladium, and others metals such as gold, and silver from aqueous solution. Then, the metalized bacterial cellulose can be used as anode or cathode in biofuel cells and in biosensors. The possibility of bacterial cellulose to be used for the anodic oxidation of H_2 envisaging an energy conversion device has been proved. Another combination of bacterial cellulose and carbon based electrodes was suggested by Yan Liang [44]. He proposes the fabrication of a novel composite based on the combination of carbonized bacterial cellulose nanonofibers and carbon paste electrode. Due to its nano-dimension, lower cost and prominent electrochemical properties, bacterial cellulose-based carbonaceous materials would be an ideal candidate for the preparation of novel carbon paste electrodes. A conductive polyaniline (PANI)/bacterial cellulose nanocomposite membrane was reported by Weili Hu [45]. The author reports on the oxidative polymerization of aniline using the tridimensional structure of the bacterial cellulose as a template. The resulting PANI-coated bacterial cellulose composite formed a uniform and flexible membrane with a high conductivity and good mechanical properties which could be applied in sensors and flexible electrodes.

A different approach is proposed by Xueyan Zhao [46]. He reports on the use of cellulose materials for the preparation of hierarchical carbon materials. A novel method of fabrication of CNT-carbon fibers was developed through carbonization of cellulose fibers being the growth of CNT in the presence of a metal catalyst. A single CNT modified carbon fiber was used as a microelectrode, and then tested for the efficiency of oxidation reaction of NADH (Nicotinamide adenine dinucleotide) generated from the glycerol oxidation reaction. The single fiber microelectrode is promising for applications such as enzyme, glycerol, and NADH biosensors. Also, Sungryul Yun [47] suggests the fabrication of MWCNTs/cellulose composites. In this work, MWCNTs were covalently grafted to cellulose. The covalently grafted MWCNTs improve the mechanical properties of cellulose due to their homogeneous distribution in the composite. Moreover, if MWCNTs can be aligned by the cellulose chains the mechanical properties will be greatly enhanced. Thus, homogeneous distribution of MWCNTs covalently grafted to a cellulose matrix allows the construction of stable electron pathways for cellulose-based electronics and mechanical reinforcing.

Recently, cellulose paper has been (re)discovered as a smart material that can be utilized for sensors and actuators. Celulose-based energy storage devices have significant inherent advantages in comparison with many currently employed batteries and supercapacitors regarding environmental friendliness, flexibility, cost and versatility. The development of cellulose-based flexible energy storage devices is particularly interesting due to the simple procedures for manufacturing these cellulosic composites being, consequently, relatively inexpensive. Various types of devices, such as thin film transistors [48], active matrix displays, sensors, batteries [49] and capacitors [50] have been fabricated on paper substrate [51]. Liangbing Hu [50] and his colleagues have demonstrated that the application of paper can be expanded to energy storage devices by coating it with a simple solution of CNTs. Because paper absorbs solvents easily and binds with CNTs strongly, the fabrication process for the conductive paper is much simpler than that for other substrates, such as glass or plastics. CNTs deposited on porous paper are more accessible to ions in the electrolyte than those on flat substrates which can result in high power density. Because of the high conductivity and the large surface area, the conductive paper was studied in supercapacitors applications as active electrodes and current collectors.

A new design and fabrication method for a supercapacitor based on a flexible CNT-cellulose-IL nanocomposite sheets was developed by Victor Pushparaj [52]. They used unmodified plant cellulose dissolved in an IL and subsequently embedded in the MWCNTs. The nanocomposite paper formed, which has a few tens of microns thickness, contains MWCNTs as the working electrode and the cellulose surrounding individual MWNTs, as well as the IL in cellulose, as the self-sustaining electrolyte. In addition to using the IL electrolyte, the authors propose the use of a suite of electrolytes based on body fluids, suggesting the possibility of the device being useful as a dry-body implant. Indeed, the use of biological fluids as an electrolyte for energy applications became an ideal alternative for implantable medical devices and disposable diagnostic kits. The earliest urine-activated paper batteries have been developed and reported by Ki Bang Lee [53]. This device consists in a copper chloride (CuCl)-doped filter paper between a copper layer and a magnesium one. Then, the whole assembly is sandwiched between two plastic layers and later laminated by passing it through heated rollers at 120ºC. Magnesium and copper chloride are used as the anode and the cathode of the device respectively, and the Cu layer acts as an electron-collecting layer. When a droplet of human urine is added to the battery, the urine soaks through the paper between the Mg and Cu layers, and after that the chemicals dissolve and react to produce electricity. The chemical composition of urine is widely used as a way of testing various diseases and also as an indicator of a general state of health. For instance, the concentration of glucose in urine can be a useful diagnostic tool for diabetics. Thus, the described work has demonstrated the viability of a urine-activated paper battery for biological application devices including home based health test kits.

Undeniably paper substrates are widely used for flexible electronics not only for being by far the cheapest but also for being one of the most flexible and lightweight material for that purpose. Since paper is manly composed by cellulosic fibers it also exhibits a high surface area which is an advantage for energy applications.

Recently, the electrospinning technique has attracted attention for the preparation of functional materials. Electrospinning is a broadly used technology for electrostatic fiber formation which utilizes electrical forces to produce polymer fibers with diameters ranging from 2 nm to several micrometers using polymer solutions of both natural and synthetic polymers (Figure 3). This technique allows the production of nanofibers, nanotubes, nanobelts and highly porous membranes. Electrospun nanofibers offer several advantages such as, an extremely high surface-to-volume ratio, tunable porosity, and exhibit a wide variety of cross-sectional shapes [31]. Because of these advantages, electrospun nanomaterials have unique properties applicable to a wide range of fields, including the fabrication of nanomaterials for use in energy conversion devices.

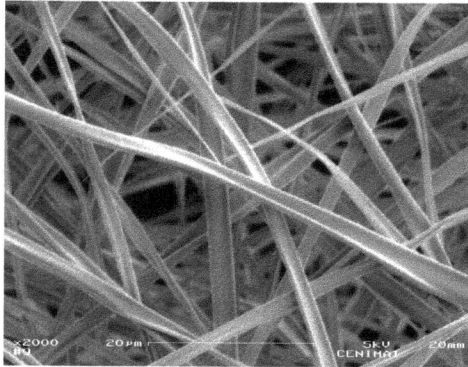

Figure 3. Scanning electron microscopy image of an electrospun cellulose acetate membrane.

Thus, the electrospinning of cellulose and derivatives has been actively studied [31, 54]. Due to their extraordinary properties, such as porosity and large specific surface area, electrospun polysaccharide fibers have been used in biomedical applications such as tissue engineering [55], drug delivery [56], antimicrobial medical implants [57] and biosensors [58, 59].

Liu Shuiping [59] describes the fabrication of photochromic nanofibrous mats through the electrospinning technique. The spiropyrans (SP) are a well-known class of materials that have reversible photochromic properties. On this work a blend solution of cellulose acetate and NO_2SP (1, 3', 3'-trimethyl-6-nitrospiro (2H-1-benzopyran-2, 2'-indoline) was electrospun forming a homogeneous and highly porous membrane. The photochromic and fluorescent properties of the functionalized nanofibers were determined, showing that the nanofibers exhibited an excellent photosensitivity. These nanofibers have a great potential for application in optical devices and biosensors. Another approach is described by Nafiseh Sharifi [60] selecting the electrospinning technique to develop a nanostructure with electrocatalytic properties. This study focuses on a new, simpler and low cost fabrication method of silver nanostructures by using cellulose as a template. Silver nanoparticles were deposited onto electrospun cellulose fibers followed by thermal removal of the cellulose template. The self-standing silver nanostructure formed is highly porous and exhibited a specific surface area which is in fact appropriate for applications in high surface area electrodes in electrochemistry such as fuel cells.

In fact, the use of electrospun fibers in the development of functionalized materials opens a new path for the creation of novel, lightweight and flexible nanostructures. Our research team is currently working on the development of a bio-battery based on an electrospun cellulose acetate membrane [54]. The bio-battery reported by us is composed by an ultrathin monolithic structure in which the separator and the electrodes are physically integrated into a thin and flexible polymeric structure. A highly porous structure is produced by electrospinning to work as a bio-battery after the deposition of metallic layers (electrodes) in each one of the faces (Figure 4). In order to power electronic medical implants, power-supply systems must be able to operate independently over a prolonged period of time, without the need of external recharging or refueling. This cellulose-based structure demonstrated the ability to generate electrical energy from physiological fluids showing a power density of $3\mu W.cm^{-2}$ [54]. This is a really promising achievement since a typical power required for a pacemaker operation is around $1\mu W$. Besides the supplying of low power consumption devices, biochemical monitoring systems and artificial human muscles stimulation mechanisms can also be foreseen as potential field of applications where it is desirable this kind of implantable micro power sources.

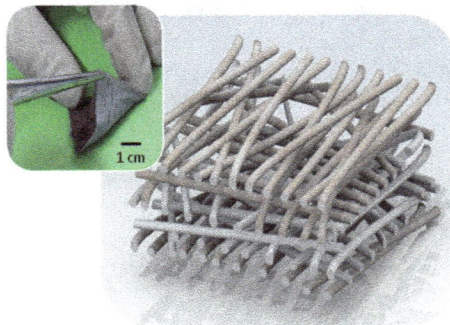

Figure 4. Schematic and macroscopic image of the bio-battery developed by our group. It consists in a cellulose acetate membrane, produced by electrospinning, covered with metallic layers to form the electrodes.

The inspiring advances in the development of innovative cellulose-based bioelectronic devices and its promising perspectives make it a challenging field of study. Electronics can be made lightweight, flexible, and capable of intimate, non-invasive integration with the soft, curvilinear surfaces of biological tissues offering important opportunities for diagnosing and energy harvesting.

5. Conclusion

Cellulose and its derivatives have demonstrated to be a versatile material with a unique chemical structure which provides a good platform for the construction of new biomaterials and biodevices. Indeed, the high density of free hydroxyl groups in the cellulose structure makes it a helpful solid substrate that can undergo functionalization allowing the production of new materials for novel advanced applications. From biological immobilization to energy storage devices, the progresses in cellulose functionalization are described as innovative and challenging. Future advances in cellulose-based devices can envisage the development of essential medical implantable devices and healthcare systems.

Author details

Ana Baptista, Isabel Ferreira and João Borges*
CENIMAT/I3N and Materials Science Department,
Faculty of Science and Technology of New University of Lisbon (FCT/UNL), Portugal

Acknowledgement

The authors work was partially supported by Portuguese Science and Technology Foundation (FCT-MCTES) through the Strategic Project PEst-C/CTM/LA0025/2011. Ana Baptista also acknowledges FCT-MCTES for the doctoral grant SFRH/BD/69306/2010.

6. References

[1] Katz E, Buckmann A.F, and Willner I (2001) Self-powered enzyme-based biosensors. J. Am. Chem. Soc. 123: 10752-10753

[2] Tersch C, and Lisdat F (2011) Label-free detection of protein-DNA interactions using electrochemical impedance spectroscopy. Electrochimica Acta 56: 7673-7679

[3] Fang X, Tan O.K, Tse M.S, and Ooi E (2010) A label-free immunosensor for diagnosis of dengue infection with simple electrical measurements. Biosensors and Bioelectronics 25: 1137-1142

[4] Osman M.H, Shah A.A, and Walsh F.C (2011) Recent progress and continuing challenges in bio-fuel cells. Part I: Enzymatic cells. Biosensors and Bioelectronics 26: 3087-3102

[5] Malhotra B.D, and Chaubey A. (2003) Biosensors for clinical diagnostics industry. Sensors and Actuators B. 91: 117-127

[6] Massafera M.P, and Torresi S.I.C (2009) Urea amperometric biosensors based on a multifunctional bipolymeric layer: comparing enzyme immobilization methods. Sensors and Actuators B. 137: 476-482

[7] Zhang B, Weng Y, Xu H, and Mao Z (2012) Enzyme immobilization for biodiesel production. Appl Microbiol Biotechnol. 93: 61-70

[8] Li X, Wang X, Ye G, Xia W, and Wang X (2010) Polystyrene-based diazonium salt as adhesive: A new approach for enzyme immobilization on polymeric supports. Polymer 51: 860-867

[9] Frasconi M, Mazzei F, and Ferri T (2010) Protein immobilization at gold - thiol surfaces and potential for biosensing. Anal Bioanal Chem. 398: 1545-1564

[10] Hua M-Y, Chen H-C, Tsai R-Y, Lin Y-C, and Wang L (2011) A novel biosensing mechanism based on a poly(N-butyl benzimidazole)-modi?ed gold electrode for the detection of hydrogen peroxide. Analytica Chimica Acta 693: 114-120

[11] Lu L-M, Li H-B, Qu F, Zhang X-B, Shen G-L, and Yu R.Q (2011) In situ synthesis of palladium nanoparticle-graphene nanohybrids and their application in nonenzymatic glucose biosensors. Biosensors and Bioelectronics 26: 3500-3504

* Correspondig Author

[12] Yang D-S, Jung D-J, and Choi S-H (2010) One-step functionalization of multi-walled carbon nanotubes by radiation-induced graft polymerization and their application as enzyme-free biosensors. Radiation Physics and Chemistry 79: 434-440

[13] Grayson A.C.R, Shawgo R.S, Johnson A.M, Flynn N.T, Li Y, Cima M.J, and Langer R (2004) A BioMEMS Review: MEMS technology for physiologically integrated devices. Proceedings of the IEEE 92 (1): 6-21

[14] Lueke J, and Moussa W.A (2011) MEMS-based power generation techniques for implantable biosensing applications. Sensors 11: 1433-1460

[15] Harb A, (2011) Energy harvesting: State-of-the-art. Renewable Energy 36: 2641-2654

[16] Bouendeu E (2011) A low-cost electromagnetic generator for vibration energy harvesting. IEEE Sensors Journal 11 (1): 107-113

[17] Xu S, Qin Y, Xu C, Wei Y, Yang R and Wang Z.L (2010) Self-powered nanowire devices. Nature Nanotechnology 5: 366-373

[18] Sun C, Shi J, Bayerl D.J, and Wang X (2011) PVDF microbelts for harvesting energy from respiration. Energy Environ. Sci., 4: 4508–4512

[19] Bhatia D, Bairagi S, Goel S, and Jangra M (2010) Pacemakers charging using body energy. J Pharm Bioallied Sci. 2(1): 51–54.

[20] Rincón R.A, Lau C, Luckarift H.R, Garcia K.E, Adkins E, Johnson G.R, and Atanassov P (2011) Enzymatic fuel cells: Integrating flow-through anode and air-breathing cathode into a membrane-less biofuel cell design. Biosensors and Bioelectronics 27: 132–136

[21] Wang H-Y, Bernarda A, Huang C-Y, Lee D-J, and Chang J-S (2011) Micro-sized microbial fuel cell: A mini-review. Bioresource Technology 102: 235–243

[22] Yang L, and Shijie Liu S (2005) Kinetic Model for Kraft Pulping Process. Ind. Eng. Chem. Res. 44: 7078-7085

[23] Wang X, Li H, Cao Y, and Tang Q (2011) Cellulose extraction from wood chip in an ionic liquid 1-allyl-3-methylimidazolium chloride (AmimCl). Bioresource Technology 102: 7959–7965

[24] Dufresne A, Cavaille J-Y, and Vignon M.R (1997) Mechanical behavior of sheets prepared from sugar beet cellulose microfibrils. Journal of Applied Polymer Science 64(6):1185-1194

[25] Zuluaga R, Putaux J-L, Restrepo A, Mondragon I, and Gañán P (2007) Cellulose microfibrils from banana farming residues: isolation and characterization. Cellulose 14: 585–592

[26] Tsekos I (1999) The sites of cellulose synthesis in algae: diversity and evolution of cellulose synthesizing enzyme complexes. J. Phycol. 35: 635–655

[27] Szymańska-Chargot M, Cybulska J, and Zdunek A (2011) Sensing the structural differences in cellulose from apple and bacterial cell wall materials by Raman and FT-IR spectroscopy. Sensors 11: 5543-5560

[28] Kondo T (2005) Hydrogen bonds in cellulose and cellulose derivatives In Severian Dumitriu Editor. Polysaccharides, structural diversity and functional versatility, NY, USA: Marcel Dekker, pp 69-95

[29] Moon R.J, Martini A, Nairn J, Simonsen J, and Youngblood J (2011) Cellulose nanomaterials review: structure, properties and nanocomposites. Chem. Soc. Rev. 40: 3941–3994

[30] Godinho M.H, Filip D, Costa I, Carvalho A.L, Figueirinhas J.L, and Terentjev E.M (2009) Liquid crystalline cellulose derivative elastomer films under uniaxial strain. Cellulose 16:199-205

[31] Canejo J.P, Borges J.P, Godinho M.H, Brogueira P, Teixeira P.I.C, and Terentjev E.M (2008) Helical Twisting of Electrospun Liquid Crystalline Cellulose Micro- and Nanofibers. Advanced Materials 20(24): 4821–5

[32] Giraud-Guille M.M, Balamie E, Mosser G, Helary C, Gobeaux F, and Vegier S (2008) Liquid crystalline properties of type I collagen: Perspectives in tissue morphogenesis. C. R. Chimie. 11: 245–252.

[33] Frey M.W (2008) Electrospinning Cellulose and Cellulose Derivatives. Polymer Reviews. 48(2): 378–91

[34] Klemm D, Heublein B, Fink H, and Bohn A (2005) Cellulose: Fascinating Biopolymer and Sustainable Raw Material. Angew. Chem. Int. Ed 44(22): 3358–93.

[35] Montanez M.I, Hed Y, Utsel S, Ropponen J, Malmstrom E, Wagberg L, Hult A, and Malkoch M (2011) Bifunctional dendronized cellulose surfaces as biosensors. Biomacromolecules 12(6):2114–25.

[36] Pohl M, Michaelis N, Meister F, and Heinze T (2009) Biofunctional surfaces based on dendronized cellulose. Biomolecules 10: 382-389

[37] Moccelini S.K, Franzoi A.C, Vieira I.C, Dupont J, and Scheeren C.W (2011) A novel support for laccase immobilization: Cellulose acetate modified with ionic liquid and application in biosensor for methyldopa detection. Biosensors and Bioelectronics 26(8):3549–54.

[38] Wu X, Zhao F, Varcoe J.R, Thumser A.E, Avignone-Rossa C, and Slade R.C.T (2009) Direct electron transfer of glucose oxidase immobilized in an ionic liquid reconstituted cellulose–carbon nanotube matrix. Bioelectrochemistry 77(1):64–8.

[39] Wan J, Yan X, Ding J, and Ren R (2010) A simple method for preparing biocompatible composite of cellulose and carbon nanotubes for the cell sensor. Sensors and Actuators B 146: 221–225

[40] Alpat S, and Telefoncu (2010) Development of an alcohol dehydrogenase biosensor for ethanol determination with Toluidine Blue O covalently attached to a cellulose acetate modified electrode. Sensors 10: 748- 764

[41] Stephan Diekmann S, Siegmund G, Roecker A, and Klemm D.O (2003) Regioselective nitrilotriacetic acid–cellulose–nickel-complexes for immobilisation of His 666 -tag proteins. Cellulose 10: 53–63

[42] Huang J, Ichinose I, and Kunitake T (2006) Biomolecular modification of hierarchical cellulose fibers through tinania nanocoating. Angew. Chem. Int. Ed., 45: 2883-2886

[43] Evans B.R, O'Neill H.M, Malyvanh V.P, Lee I, and Woodward J (2003) Palladium-bacterial cellulose membranes for fuel cells. Biosensors and Bioelectronics 18(7):917–23.

[44] Liang Y, He P, Ma Y, Zhou Y, Pei C, and Li X (2009) A novel bacterial cellulose-based carbon paste electrode and its polyoxometalate-modified properties. Electrochemistry Communications 11: 1018–1021

[45] Hu W, Chen S, Yang Z, Liu L, and Wang H (2011) Flexible electrically conductive nanocomposite membrane based on bacterial cellulose and polyaniline. J. Phys. Chem. 115: 8453 – 8457

[46] Zhao X, Lu X, Tze W.T.Y and Wang P (2010) A single carbon fiber microelectrode with branching carbon nanotubes for bioelectrochemical processes. Biosensors and Bioelectronics 25: 2343–2350

[47] Sungryul Yun S, and Kim J (2011) Mechanical, electrical, piezoelectric and electro-active behavior of aligned multi-walled carbon nanotube/cellulose composites. Carbon 49: 518-527

[48] Martins R, Barquinha P, Pereira L, Correia N, Gonçalves G, Ferreira I, and Fortunato E (2009) Selective floating gate non-volatile paper memory transístor. Phys. Status Solidi RRL 3 (9): 308– 310

[49] Ferreira I, Brás B, Correia N, Barquinha P, Fortunato E, and Martins R (2010) Self-Rechargeable Paper Thin-Film Batteries: Performance and Applications. J. Display Technol. 6: 332-335

[50] Liangbing Hu L, Choi J.W, Yang Y, Jeong S, Mantia F, Cui L-F, and Cui Y (2009) Highly conductive paper for energy-storage devices. PNAS 106(51): 21490–21494

[51] Tobjörk D, and Ronald Österbacka R (2011) Paper Electronics. Adv. Mater. 23: 1935–1961

[52] Pushparaj V.L, Shaijumon M.M, Kumar A, Murugesan S, Ci L, Robert Vajtai R, Linhardt R.J, Nalamasu O, and Ajayan P.M (2007) Flexible energy storage devices based on nanocomposite paper. PNAS 104 (34): 13574–13577

[53] Lee K.B (2005) Urine-activated paper batteries for biosystems. J. Micromech. Microeng. 15: S210–S214

[54] Baptista A.C, Martins J.I, Fortunato E, Martins R, Borges J.P, and Ferreira I (2011) Thin and flexible bio-batteries made of electrospun cellulose-based membranes. Biosensors and Bioelectronics 26(5): 2742–5

[55] Lee KY, Jeong L, Kang YO, Lee S.J, and Park W.H (2009) Electrospinning of polysaccharides for regenerative medicine. Advanced Drug Delivery Review 61(12):1020–32

[56] Sill T.J, and von Recum H.A (2008) Electrospinning: Applications in drug delivery and tissue engineering. Biomaterials 29(13): 1989–2006

[57] Penchev H, Paneva D, Manolova N, and Rashkov I (2010) Hybrid nanofibrous yarns based on N-carboxyethylchitosan and silver nanoparticles with antibacterial activity prepared by self-bundling electrospinning. Carbohydrate Research 345(16):2374–80

[58] Li J, Vadahanambi S, Kee C-D, and Oh I-K (2011) Electrospun fullerenol-cellulose biocompatible actuators. Biomacromolecules 12(6):2048–54

[59] Shuiping L, Lianjiang T, Weili H, Xiaoqiang L, and Yanmo C (2010) Cellulose acetate nanofibers with photochromic property: Fabrication and characterization. Materials Letters 64 (22): 2427 – 2430

[60] Sharifi N, Tajabadi F, and Taghavinia N (2010) Nanostructured silver fibers: Facile synthesis based on natural cellulose and application to graphite composite electrode for oxygen reduction. International Journal of hydrogen energy 35: 3258-3262

Chapter 3

Cellulose - A Biomaterial with Cell-Guiding Property

Miretta Tommila, Anne Jokilammi, Risto Penttinen and Erika Ekholm

Additional information is available at the end of the chapter

1. Introduction

A biomaterial is defined as a material, either man-made or natural, intended to interact with biological systems. It does not have a chemical effect in the organism, nor thus it need to be metabolised to be active like for example drugs [1]. When inserted into the body, a local tissue inflammatory reaction called foreign body reaction is induced [2]. This reaction may either favour or adversely affect the tissue repair process.

Cellulose and its derivatives are well tolerated by most tissues and cells [3-5]. These non-toxic materials have good biocompability, therefore, they offer several possibilities in medical applications. Regenerated cellulose sponges have also been used in experimental surgery for decades as it does not affect the healing process, but acts as a chemoattractant inducing cells involved in the repair process to migrate towards it [6-8].

We have studied different biomaterials including cellulose in search for an optimal bone substitute. In bone defects, regenerated cellulose supported with cotton fibres was shown to allow new bone in-growth to some degree [9-11]. Oxidation with periodate and hydrogen peroxide, or carbamination further improved its biocompability but not enough to be used as bone substitutes. We also expected to increase the osteostimulating property of regenerated cellulose by coating it with a silica-rich hydroxyapatite (HA) as it resembles the mineral composition of bone. To our disappointment, the HA-coated cellulose did not promote bone formation but favoured instead inflammation and fibroplasia. Since the bone implant study revealed unexpectedly an enormous ability of the HA-implants to induce granulation tissue, the coated cellulose was tested subcutaneously as well. These studies showed that the HA-coated cellulose not only attracted inflammatory cells but also bone marrow-derived progenitor cells of both haematopoietic and mesenchymal origin (see box 1). In this chapter, we will discuss cellulose as implant material with emphasis on the cell guiding properties of regenerated cellulose coated with silica-rich HA.

2. Cellulose for medical applications and as a tissue engineering matrix

Cellulose, the most common organic compound on Earth, is degraded by microbial enzymes. Animal cells cannot cleave the $\beta(1\rightarrow4)$-bond between the two glucose moieties in cellulose. Thus, cellulose degradation in tissues takes place by a slow non-enzymatic hydrolysis of the $\beta(1\rightarrow4)$-bond and therefore cellulose can be regarded as an almost stable matrix. Despite this, cellulose and its derivatives are well tolerated by cells and tissues and induce a moderately strong foreign body reaction in the tissue [3-8].

BOX 1. ADULT BONE MARROW-DERIVED STEM CELLS

Adult stem cells are immature cells, dispersed in tissues throughout the body after development. Like all stem cells, they are capable of either making identical copies of themselves or to differentiate depending on their local environment into mature cell types with characteristic morphology and function. Stem cells usually generate an intermediate, partly differentiated, cell type, called precursor or progenitor cell, before they achieve their fully differentiated state. Adult stem cells are rare, however. Their primary functions are to replenish dying cells, and with limitations, to regenerate damaged tissues.

The best characterised adult stem cells are those found in the bone marrow, which provides a unique niche for haematopoietic stem cells (HSCs) and the mesenchymal stem or stromal cells (MSCs). HSCs are responsible for the production and replacement of all blood cells during the entire lifetime [13]. The earliest haematopoietic precursor, the haemangioblast, is not only a precursor of haematopoietic cell lineages but also of cells that line all blood vessels and lymphatics, namely the endothelial cells [14].

Mesenchymal stromal cells are a heterogenous population of stem/progenitor cells able to differentiate into several cell types such as chondrocytes, osteocytes, fibroblasts, myocytes, adipocytes, epithelial and neuron-like cells. When stimulated by specific signals, these cells can be released from their niche in the bone marrow into circulation and recruited to the target tissues where they undergo *in situ* differentiation and contribute to tissue homeostasis and repair [15]. MSCs also secrete factors that promote survival and differentiation of endogenous cells as well as angiogenetic factors essential for blood vessel formation. MSCs possess remarkable immunosuppressive properties and can inhibit the proliferation and function of the major immune cell population [16, 17] as well as antimicrobial properties [18]. Furthermore, these multipotential stromal stem and progenitor cells at different stages of maturation contribute to the formation of HSC stem cell niche and play a critical role in haematopoiesis [19].The characteristic of MSCs makes these cells exceptionally suitable for various therapeutic possibilities such as supporting tissue regeneration, correcting inherited disorders, dampening chronic inflammation, and delivering biological agents [15].

Cellulose is non-toxic and has good biocompability, therefore, it offers several possibilities in medical applications. Cellulose and its derivatives are used, among other things, as coating materials for drugs, additives of pharmaceutical products, blood coagulant, supports for immobilized enzymes, artificial kidney membranes, stationary phases for optical resolution, in wound care and as implant material and scaffolds in tissue engineering [3, 12].

2.1. Regenerated cellulose

Cellulose sponges can be manufactured by adding supportive strengthening fibres (8-10 mm long cotton fibres; about 20% of the weight of the cellulose) and sodium sulphate crystals as pore forming material to a cellulose viscose (sodium xantogenate) solution (4-6 g cellulose/100 g viscose). The cellulose is regenerated by heating the solution in a water bath after which the sponge is washed with hot water, treated with a dilute acid and sodium hypochlorite bleaching solution, and finally washed repetitively in distilled water before drying and sterilisation [20, 21]. When inserted subcutaneously, a vital and well vascularised repair tissue, called granulation tissue, grows rapidly into this cellulose sponge. Due to this good granulation tissue formation ability, cellulose sponges have been used in experimental surgery for decades [6, 7, 22] and the subcutaneous implantation of the cellulose sponge is widely accepted method for wound healing (see box 2) studies [8, 23]. Several cellulose products for wound healing purposes (e.g. Cellospon®, Cellstick®, Sponcal®, Visella®, and Absorpal®) are commercially available. These products are made from the sponge form viscose cellulose and have homogenous porous structure, characterized by thin pore walls with one or more inter-pore openings. They are elastic and can be compressed and expanded repeatedly with no damage to their internal structure, hence providing free entrance for the invading cells to the inner parts of the sponge [24].

Host reactions following implantation of biomaterials include injury, blood-material interactions, provisional matrix formation, inflammation, granulation tissue development, foreign body reaction, and fibrosis/fibrous capsule development [25]. When implanted subcutaneously, a blood-material interaction occurs with protein adsorption to the cellulose sponge and a blood-based transient provisional matrix, a blood clot; is formed on and around the sponge. The platelets, originated from the injured blood vessels, not only participate to haemostasis but also liberate bioactive agents like cytokines and growth factors that will attract inflammatory and phagocytosing cells. The first cells to arrive are polymorphonuclear leucocytes, i.e. neutrophils, which are characteristic for the acute inflammatory response. These cells secrete pro-inflammatory cytokines that, in turn, attract circulating monocytes, which are activated and converted in the tissue to macrophages that kill bacterial pathogens, scavenge tissue debris and destroy remaining neutrophils. Biomaterial surface adherent macrophages can also fuse to form multinucleated foreign body giant cells. In their attempt to phagocytose the biomaterial, adherent macrophages become active [25]. By releasing a variety of chemotactic, neovasculogenic and growth factors that stimulate cell migration, proliferation and formation of new blood vessels and tissue matrix, macrophages mediate the transition from the inflammatory phase to the proliferative phase. During the proliferative phase, the provisional extracellular matrix in the cellulose sponge is gradually replaced with granulation tissue, which is formed from infiltrated mature fibroblasts and rapidly proliferating mesenchymal stromal cells (MSCs) differentiating to fibroblasts *in situ*. The newly formed extracellular matrix is rich in blood vessels, which carry oxygen and nutrients to maintain the metabolic processes. The sponge is surrounded by a well-vascularised fibrous capsule, which becomes somewhat thinner during the final remodelling phase [38].

Similar biocompatible regenerated cellulose developed for wound healing studies has also been tested as a scaffold for cartilage tissue engineering. Although the cellulose sponge provided a non-toxic environment for cartilage cells, the construct remained soft and lacked the extracellular matrix composition typical for normal articular cartilage [26]. When implanted into bone defects, regenerated cellulose strengthened by cotton fibres allowed new bone in-growth to some extent [9-11].

2.1.2. Hydroxyapatite-coating of regenerated cellulose

The number of cells and tissue in-growth are affected to a certain limit by the porosity, size of pores, and the thickness of the pore walls of the cellulose sponge [8]. We hypothesised that coating the regenerated cellulose with hydroxyapatite (HA) that resembles the mineral composition of bone, would improve its bone forming properties. The mineral originated from a specific bioactive glass, S53P4 (23% Na_2O, 20% CaO, 4% P_2O_5, 53% SiO_2) that has a good osteoconductivity and is in clinical use [27-32]. However, glass as such, is difficult to trim to the desired size and form. Furthermore, it is brittle and fragile, and therefore, not suited in sites subjected to load like in femoral and tibial bone defects.

In our studies, the calcium phosphate layer was precipitated on cellulose sponges (10 x 100 x 100 mm) with average pore sizes between 50 and 350 µm by the biomimetic method of Kokubo et al [33]. Mineralisation was initiated in 500 ml of sterile simulated body fluid (SBF) supplemented with a 2.0 g of the bioactive glass at 37°C for 24h and was then grown in 500 ml sterile 1.5 x SBF for 14 days at the same temperature under continuous shaking. The SBF solution was changed every second day. The formed calcium phosphate layer rich in silica was verified by scanning electron microscope (figure 1) and characterised with Fourier transform infrared spectroscopy [11]. (1 x SBF = 136.8 mM NaCl, 4.2 mM $NaHCO_3$, 3.0 mM KCl, 1.0 mM K_2HPO_4 x $3H_2O$, 1.5 mM $MgCl_2$ x $6H_2O$, 2.5 mM $CaCl_2$ and 0.5 mM Na_2SO_4, pH 7.4; ion concentration close to that of human plasma)

Sterile HA-cellulose and untreated cellulose sponges, sized 2.3 x 3 x 8 mm, were implanted into femoral bone defects of male rats aged 10-13 weeks (for further details see [11]) and were followed up for 52 weeks. The implants were analysed histologically and with biochemical and molecular biologic methods. The HA layer did not improve the bone in-growth into the cellulose sponge. In fact, the new bone was instead mainly formed beneath the implant at the bottom of the defect leaving the implant filled with a well vascularised fibrous tissue rich in inflammatory cells (figure 2). The inflammatory reaction was much stronger than in the uncoated cellulose indicated by the larger number of inflammatory cells

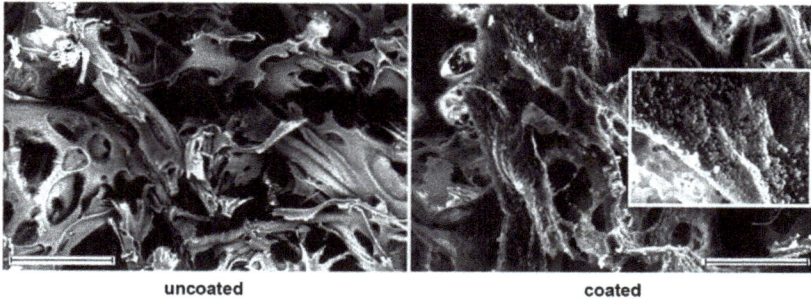

uncoated   **coated**

Figure 1. SEM micrograph of regenerated uncoated and HA-coated cellulose sponges (bar = 50 μm). The hydroxyapatite layer was initiated in sterile 1 x SBF with bioactive glass at 37 °C for 24 h and was then grown in sterile 1.5 x SBF at the same temperature for 14 days under continuous shaking.

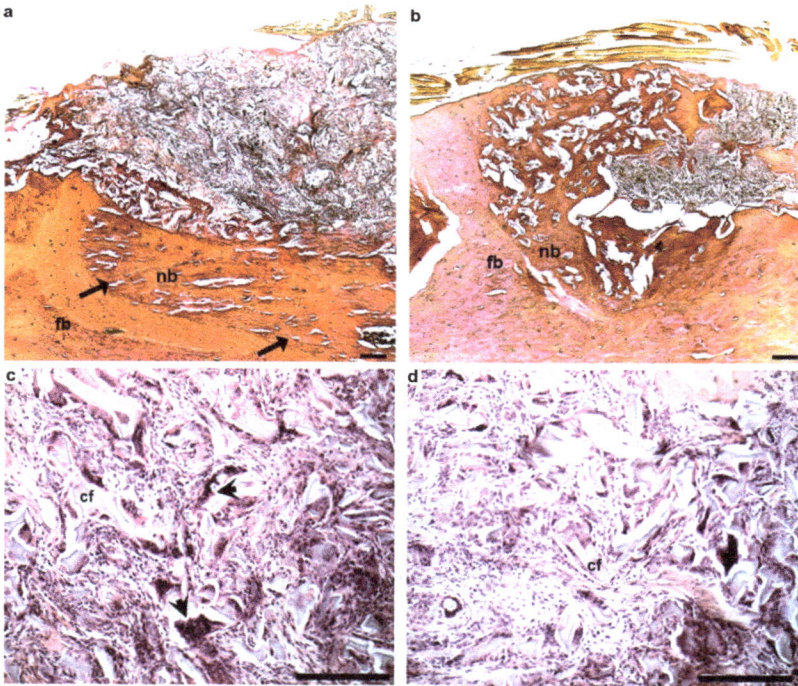

Figure 2. HA-coating of cellulose prevents bone in-growth. One year after implantation into rat femoral bone defect, new bone (nb) growth is mainly observed beneath (arrows) the HA-implant (a), which has been pushed out from the defect area. The HA-implant itself (b) is mostly filled with soft connective tissue containing abundant giant cells (arrow heads). Uncoated cellulose implant (c) allows new bone in-growth and the non-ossified parts contain less inflammatory cells. (a and c; van Gieson stain; b and d haematoxylin-eosin stain; cf = cellulose fragment; scale bar = 100 μm, modified from [11]).

including macrophages and foreign body cells, which also is a sign of chronic inflammation. Activated inflammatory cells produce many pro-inflammatory bioactive agents, such as tumour necrosis factor-alpha (TNF-α), which is known to interfere with the bone specific transcription factor Cbfa1 and to depress the function of differentiated osteoblasts [34,35]. Continuous exposure to these agents may, thus, inhibit differentiation of the progenitor cells into bone forming osteoblasts explaining, as least partly, the less osteoid tissue in HA-coated cellulose implants. Furthermore, the HA layer did intensify the attachment of transforming growth factor beta 1 (TGFβ1) [11], a growth factor involved in fibroplasia. Hence, the HA surface did not offer any advantages in comparison with untreated cellulose in cortical bone defect healing.

2.2. The effect of increased biodegradability of cellulose

Another approach to improve the biocompatibility of cellulose was to alter its chemical structure in order to increase its biodegradability. The mild bleaching and oxidation of regenerated cellulose with sodium hypochlorite carried out during the preparation of cellulose sponge does not cleave the glucose ring and the resultant cellulose is not biodegradable, which probably prevented complete ossification of the implanted sponge. Therefore, in the search for suitable bone defect fillers, we extended the material development with a two sequential oxidation steps. Firstly, the cellulose was oxidated by periodate for 1-3 hours. This treatment opens some glucose molecules and should theoretically make them more susceptible to glucosidases and other enzymes capable for carbohydrate degradation. Excess periodate was washed by ascorbate or thiosulphate and water before the second oxidation by hydrogen peroxide (H_2O_2) for 3 or 4 hours. As the oxidation reactions were not complete, the resultant materials are combinations of 2,3-dialdehyde and 2,3-dicarboxyl celluloses. The biogradability of the celluloses was tested in SBF for 7, 15 and 30 days. Oxidations for 3 h in periodate followed by 4 h in H_2O_2 turned out to be the best combination as 70% of the material was dissolved. Therefore, this material was used for further testing. No cytotoxicity was observed in fibroblast cultures. The material has to be sterilised by 70-95 % ethanol or ethylene oxide because autoclaving destroys the porous structure of the scaffolds.

The results of the bone implantation experiments (figure 3 a, b) showed that oxidised scaffolds were flattened, their pores had disappeared and the material was completely replaced by cells so that no visible cellulose fibrils were observed in the implantation sites. The degradation was not complete as the phagocytosing cells were full of homogenous material. It is conspicuous, however, that no giant cells were observed in the oxidised samples, whereas normal cellulose always induces a number of foreign body giant cells. If the sponges were oxidised more extensively their structures collapsed and the material could not be used for implantation. The implanted scaffolds did not show, on the other hand, any significant bone in-growth. Instead they consisted of cell masses that histologically were strikingly homogeneous. New bone had been grown on the opposite site of the implant strengthening the defect site. Despite improved biodegradability, oxidised cellulose was considered to have no value as a bone substitute.

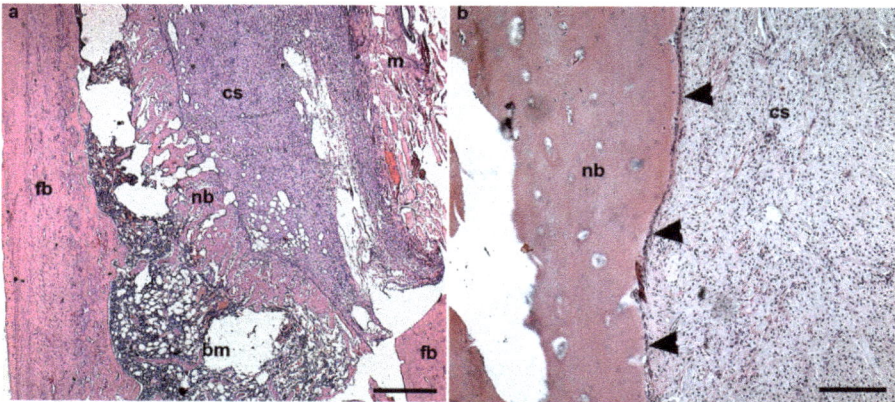

Figure 3. Oxidations with periodate and H_2O_2 increase the biocompability and degradation of cellulose. Oxidised cellulose (a, b) allows new bone (nb) formation when implanted into femoral bone (fb) defects of rat. (cs = cellulose scaffold, bm = bone marrow, m=muscle overlaying the implant site, arrow heads point at osteoblasts lining the new bone; haematoxylin-eosin stain; scale bars = 100 µm (a), and 25 µm (b)).

Biodegradation of cellulose can also be improved by treating it with urea. The resultant carbamino cellulose showed increased solubility that could be regulated by the duration of treatment. The fundamental aim was to develop material that could be used as a vehicle for drugs in tablets, or perhaps for subcutaneous long-lasting administration of drugs. Small, round or oval cellulose pearls with 50-500 µm diameters can be manufactured from regular or carbamino cellulose by dropping viscose into a solution containing 100 g H_2SO_4 and 200 g Na_2SO_4/l at 20°C followed by centrifugation [36]. Four and six per cent viscose solutions were used to make the 0.5 mm diameter cellulose pearls. The material was collected, washed with distilled water and 5g H_2SO_4/l and dried for 24 hours at 40°C. Sterilisation was carried out by autoclaving or with 70% ethanol.

For implantation studies, several pearls were glued together with alginate [37] in moulds. The results from the subcutaneous implantation experiment (Figure 4 a, b) were encouraging as implanted 4% cellulose pearls were infiltrated with new granulation tissue and most of the pearls showed signs of nearly complete degradation where as 6%-pearls were more resistant during the observation period of two weeks. Intramedullary implantation into rat femoral bone (figure 3 c-e) showed similar behaviour: many of the 4 %-pearls were infiltrated by new granulation tissue and some were surrounded by new osteoid tissue. There was some variation in the degree of degradation; while some pearls had been digested completely, some remained almost intact. No foreign body giant cells were observed, however. We do not know whether alginate surroundings affected the degradation of pearls in the bony environment, but to make the carbamino cellulose more useful in medical applications, the structure should be further altered to become even more vulnerable to hydrolytic enzyme attacks, especially if used for subcutaneous administration of drugs.

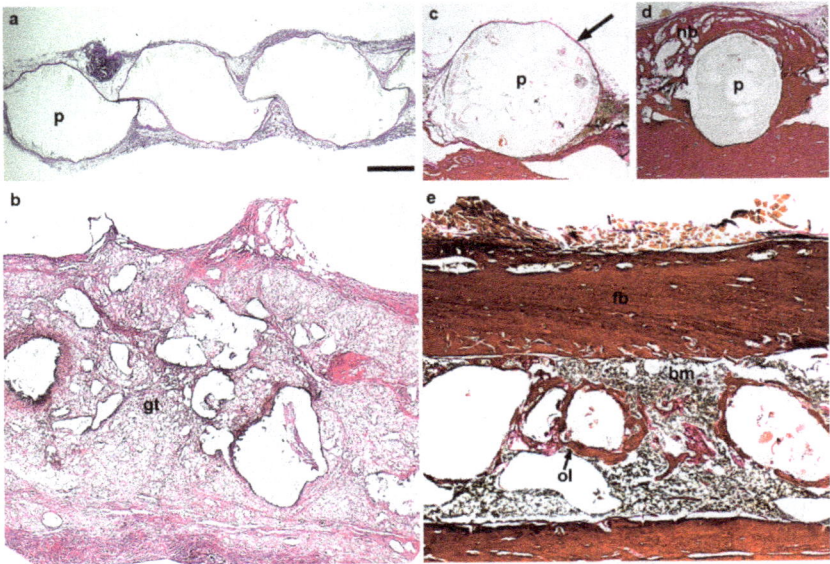

Figure 4. Tissue reactions of carbamino cellulose two weeks after implantation. Subcutaneously implanted 6%-cellulose pearls (p) stayed intact and showed only modest degradation (a), whereas b) 4%-cellulose pearls were degraded and infiltrated with new granulation tissue (gf). Similar behaviour was observed in bone implants: c) 6%-cellulose pearls were surrounded by a thin connective tissue capsule (arrow) whereas about half of the b) 4%-cellulose pearls were partially degraded and surrounded by bone (nb) or a thin osteoid layer (ol) even in the bone marrow (bm) area. (van Gieson stain; equal magnifications; scale bar 200 μm).

2.3. The biological effect of subcutaneously implanted hydroxyapatite-coated cellulose

The bone defect study showed that HA-coated cellulose favoured rapid fibrous tissue proliferation instead of bone formation [11]. Therefore, it was considered to have no value as a bone replacement material but might be useful in other applications in which accelerated granulation tissue formation is needed. Subcutaneously (figure 5 a, b) implanted silica rich HA-implants showed a massive inflammatory reaction with an intense foreign body reaction and increased invasion of fibrovascular tissue already 1-3 days after implantation. Such strong tissue reaction was not seen with any other subcutaneously implanted cellulose sponge. Tissue growth into uncoated regenerated cellulose was much slower and took place mainly on their surface (figure 6). [38]

Subcutaneously implanted HA-sponges activate the inflammatory response and the secretion of cytokines and growth factors important to wound healing, such as TGF-β1, TNF-α, vascular endothelial growth factor (VEGF) and platelet derived growth factor A (PDGF-A) The long-term study revealed, however, that the excessive connective tissue

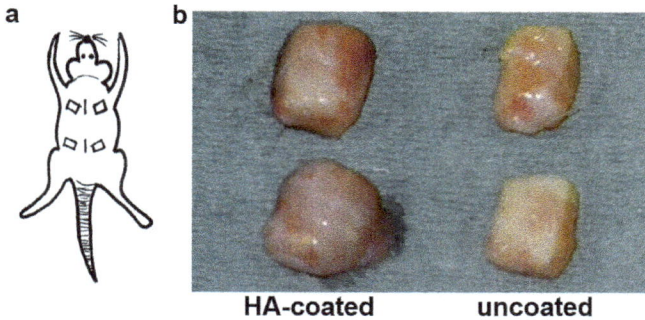

HA-coated **uncoated**

Figure 5. a). A schematic presentation of the subcutaneous implantation model used in our studies. Two midline incisions were made on the back of the rats, and sterilised, moistened sponge implants (10 x 5 mm) were inserted bilaterally into subcutaneous pockets under general anaesthesia. b). Subcutaneously implanted cellulose sponges 7 days after implantation. HA-coated implants are darker in colour as a sign of high cellularity and rich neovascularisation, whereas the uncoated implants are pale.

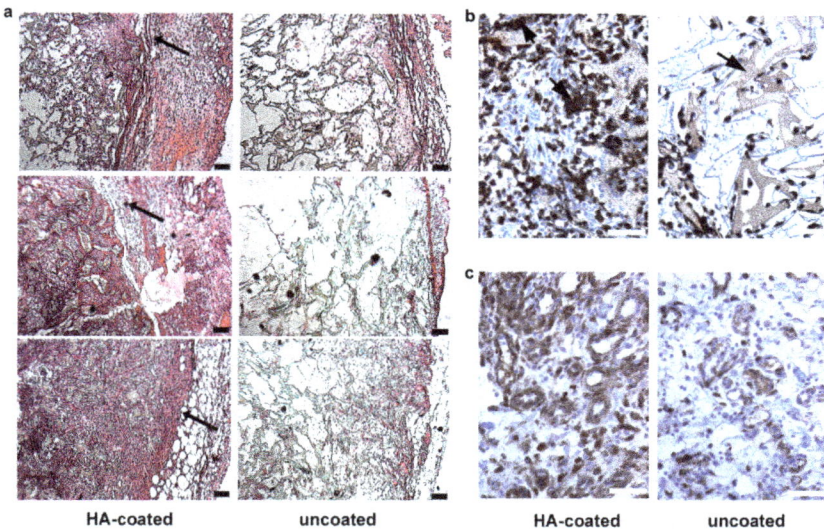

HA-coated **uncoated** **HA-coated** **uncoated**

Figure 6. The HA-coating accelerated tissue growth into subcutaneously implanted cellulose sponges as well as the inflammatory response and blood vessel formation. a) Haematoxylin-eosin–stained sections 1 (upper), 3 (middle), and 7 (lower) days after implantation. The arrows in HA-coated sponges point at the border between the implant and the surrounding capsule (scale bar = 100 μm). b) HA-coated sponges contain large clusters (arrows) of accumulated macrophages (brownish coloured cells). Macrophages favour gathering near to cellulose fibres (arrow head) (day 5; scale bar = 50 μm). c) More blood vessels, as indicated by CD31-staining, can bee seen in 5-day-old HA-coated sponge compared to uncoated one (scale bar = 50 μm).

formation, which is histologically normal, does not disturb the animals in any way. After 14 days postoperatively, the foreign body reaction in HA-coated sponges starts to diminish. At one month, the difference between the HA-coated and uncoated cellulose had levelled off and at the end of the study, at one year no obvious histological difference between the coated and uncoated were detected (figure 7). [38]

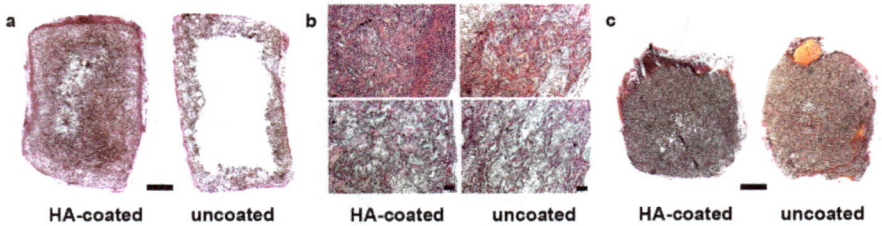

Figure 7. Histology of subcutaneous cellulose implants. a) At 14 days HA-coated sponge is filled with granulation tissue (van Gieson-stained whole implants, scale bar = 1000 μm). b) Haematoxylin-eosin-stained sections one and three months after implantation, scale bar 100 μm. c) At one year no significant difference can be observed between HA-coated and uncoated sponges (van Gieson-stained whole implants, scale bar = 1000 μm. Modified from [38]).

2.3.1. Cell trafficking and homing to regenerated cellulose

Cellular movement and re-localisation are essential for many fundamental physiologic properties, not only during embryonic development, but also during wound healing and organ repair. At the wound site, local and infiltrated cells release chemokines that recruit blood-circulating stem and progenitor cells. These bioactive agents also increase bone marrow cell mobility, thus facilitating cell mobilisation into the peripheral blood and consequently into the sites of wound healing [39]. Stromal-derived factor-1 (SDF-1) is one powerful chemokine in stem cell trafficking that regulates both haematopoietic, endothelial and mesenchymal progenitor cells. The biological effects of SDF-1 are mediated by the chemokine receptor CXCR4 [40-43]. During the early stages of wound healing, SDF-1 seems to be up-regulated by the influence of pro-inflammatory factors like TNF-α, which creates a SDF-1 concentration gradient that triggers the recruitment of CXCR4-expressing cells from the blood stream to the site of injury, where these cells further differentiate into other functional repair cells [44].

Mineralised cellulose implant not only attracts more inflammatory cells than uncoated cellulose but also circulating bone marrow-derived stem cells of both haematopoietic and mesenchymal origin [45]. SDF-1 expression (GEO series accession no. GSE19748 and GSE19749; http://www.ncbi.nlm.nih.gov/geo/query/acc.cgi?acc=GSExxx) is upregulated in HA-sponges together with its receptor CXCR4 (figure 8). This strongly indicates that the HA-coated implant has a better homing capacity of circulating bone marrow-derived stem cells than the uncoated one.

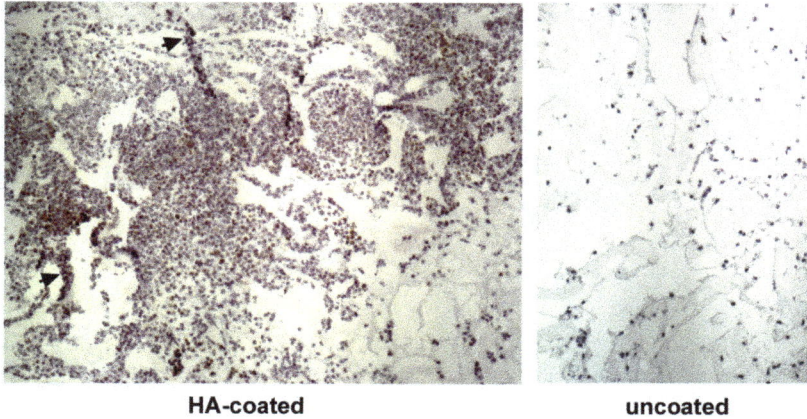

HA-coated   **uncoated**

Figure 8. HA-coated cellulose contain large amount of CXCR4-positive cells. Numerous clusters (arrow heads) of and individual CXCR4-positive (brownish coloured) cells are detected throughout the HA-coated sponge at day 7.

Haematopoietic stem cells seem to be the first stem cells to invade the empty centres of the HA-coated cellulose implants (figure 9 a-c). The more abundant occurrence of HSCs is most probably responsible for the augmented blood vessel formation in HA-coated cellulose. The earliest haematopoietic precursor, the haemangioblast, is namely the precursor for both haematopoietic and endothelial cell lineages, not only during embryogenesis but also in adults [14, 46]. The haematopoietic progenitors, especially in the HA-coated implants, were located in close contact with the cellulose fragments (figure 10 a). Hence, the coating of cellulose with HA creates an environment that facilitates stem cell homing more efficiently than uncoated cellulose. In the bone marrow, undifferentiated HSCs are detected near the inner surface of the medullary cavity, i.e. the endosteum, in the so-called endosteal stem cell niche. At this site, the bone is in constant turnover: bone is formed by the osteoblasts and removed by specific macrophages, the osteoclasts. Due to bone degradation, soluble calcium ions (Ca^{2+}) are released into the bone marrow fluid. Various cells, including primitive HSCs, respond to extracellular ionic calcium concentrations through a calcium sensing receptor, CaSR. This receptor seems to have a function of holding HSCs in close physical nearness to the endosteal surface [47]. The mineral layer on the cellulose resembles that of bone. When the numerous foreign body giant cells/macrophages gathered around the mineralised cellulose try to get rid of the foreign material, Ca^{2+} is released generating a beneficial milieu for the primitive HSCs as it resembles the endosteal stem cell niche in the bone marrow. This theory is supported by the numerous CaSR-positive cells near the mineralised cellulose fibres, in the same areas as cells positive for CD34, a common marker for endothelial cells, are observed. These cells are not only found in the granulation tissue but also in the central parts of the implant. Similar cells are seen in uncoated cells, but in remarkably less quantity (figure 9 d-e).

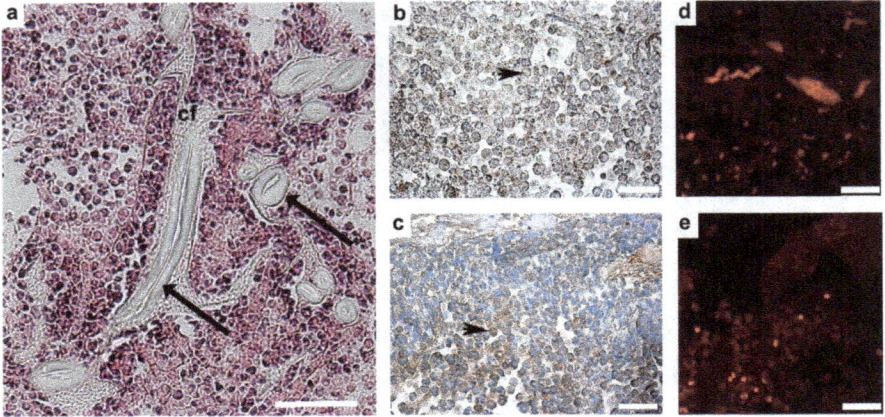

Figure 9. Stem cells are located near the cellulose fragments. a) General histology showing cells gathering around the cellulose fragments (cf) at day 7. The arrows point at pores (scale bar = 50 µm). b). HA-coated implants contains numerous cells (arrowhead) positive for c-kit, a marker for premature cells (day 7; scale bar = 25 µm). c) Small rounded cells (arrowhead) positive for CD34, a commonly used marker for HSCs (day 7; scale bar = 25 µm). More CaSR-positive cells (red fluorescence) are observed in HA-coated implants at day 7 (d) than in uncoated (e) sample (scale bar = 25 µm).

In the cellulose implants, mesenchymal stem cells are mainly found in the forming granulation tissue [45] in line with the fact that these primitive cells home to the wound site and differentiate into connective tissue cells that produce the extracellular matrix of the granulation tissue [48]. In addition, MSCs secrete signals that limit systemic and local inflammation, decrease apoptosis in the threatened tissue, stimulate neovascularisation, activate local stem cells, modulate the immune cells, and exhibit direct antimicrobial activity [18, 49, 50]. Therefore, the more abundant occurrence of MSCs in the HA-coated cellulose sponge most probably contribute to the enhanced blood vessel formation compared to uncoated cellulose and to the declining of the foreign body reaction during the second week of implantation. MSCs also secrete many cytokines that stimulate haematopoiesis, mainly the myeoloid cell lineage, but MCSs seems to have a supportive effect on erythropoiesis, the process of red blood cell formation, as well [51].

2.3.2. *Experimental granulation tissue expresses haemoglobin*

An unexpected finding was that the granulation tissue induced by cellulose sponge contains haemoglobin producing glycophorin A-positive cells (figure 10 a-d) indicating that the haematopoietic precursor cells are also able to differentiate into the erythropoietic lineage [52]. This, in turn, suggests that this repair tissue is capable of making blood. In healthy adults, globin has been considered to be expressed only in the bone marrow area by immature erythropoietic precursors. When the mature red blood cell or erythrocyte emerges from the bone marrow, it has lost its nucleus, ribosomes and mitochondria, which means that the cell is no longer capable of gene expression. As in bone marrow, where erythroid progenitors mature in association with macrophages [53, 54], the plentiful macrophages, especially in the HA-coated implants, might further back up the erythropoietic differentiation of HSCs in the granulation tissue. Microarray data (GEO series accession no. GSE19748 and GSE19749; http://www.ncbi.nlm.nih.gov/geo/query/acc.cgi?acc=GSExxx) revealed many genes related to erythropoiesis like erythropoietin and its receptor EpoR, the transciption factors Hif-1α, gata-1 and -2, and particularly Alas2, which is exclusively expressed in developing red blood cells called erythroblasts and is required for the expression of β-globin [55].

Haemoglobin has traditionally been thought to serve as the main oxygen transporter in erythrocytes. Many studies, including ours [56-65], show, however, that haemoglobin expression is much more versatile than previously has been assumed. During granulation tissue formation in the cellulose sponges, the haemoglobin expression pattern showed a biphasic pattern [52]. The first peak appeared during the most intense inflammatory response in the initiation of the healing process before invasion of HSCs, indicating that also another cell type is participating in the haemoglobin expression. Since active macrophages are known to express globin [64], these cells (figure 10 e-g) are most likely responsible for the early globin expression in the granulation tissue. In macrophages, the globins are most probably involved in processes different from oxygen transport and delivery to tissues. There is accumulating evidence that haemoglobin also binds, stores and transports nitric oxide. Nitric oxide is an important gaseous signalling molecule in wound healing [66] involved, among other things, in the formation of granulation tissue and new blood vessels [67-69]. While nitric oxide is a prerequisite for successful wound healing, an excess of this signalling molecule may be as harmful as its underproduction [67]. The fact that an intense expression of inducible nitric oxidase synthase (iNOS), an enzyme that catalyses the

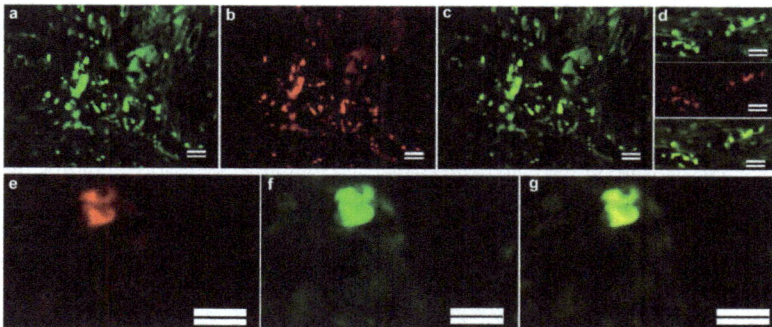

Figure 10. Double staining confirmed different haemoglobin positive cell types in cellulose implants. The granulation tissue in cellulose sponges contains haemoglobin (a) -producing glycophorin A-positive cells (b) implying that haematopoietic precursor cells are able to differentiate into red blood cells. c) Merged image of haemoglobin- and glycophorin A-positive cells. d). Red blood cells in a blood vessel in the capsule area of HA-implant; haemoglobin (upper) positive, glycophorin A (middle) and merge image (lower). e) CD-68 positive cells indicating macrophages. The same cells are also positive for haemoglobin (f). (g) Merged image of CD-68- and glycophorin A-positive cells (scale bar 20 μm, modified from [52]).

formation of nitric oxide, which reflects the production of nitric oxide observed in 3-day-old HA-implant, but not at day 10 [52], coincides with the strong inflammatory reaction that starts to decline during the second week of implantation [38]. The production of haemoglobin during this phase might eliminate the excess nitric oxide and prevent its negative effect on matrix deposition, neovascularisation and apoptosis. In uncoated cellulose implants, iNOS is detected at day 10, which supports the observation of slower sequence of events in the granulation tissue formation in these uncoated implants

3. Conclusions and future perspectives

Regenerative medicine involves tissue formation and healing in order to restore the functionality of damaged organs or tissues. As tissue repair and regeneration after injury involve the selective recruitment of circulating or resident stem cell populations, stem cell therapy is often employed as one mean to promote tissue regeneration. Its success might, nevertheless, be complicated by strong immune-rejection of transplanted cells or shortage of autologous cell supply. Furthermore, if a scaffold, with or without bioactive agents, is used to administrate the stem cells, poor integration between the scaffold/implant and the host tissue might affect the outcome.

An interesting tissue engineering concept is cell guidance aimed at total *in vivo* tissue engineering without the need of adding bioactive agents or cells. Numerous studies have shown that cellulose itself functions as a chemoattracant and is able to stimulate granulation tissue formation. Uncoated cellulose sponge has been tested in treatment of chronic leg ulcers (Pajarre, unpublished data) and in severe burn injuries (Lagus, unpublished data) in the 1990's with good results. The cellulose sponge adsorbs debris and bacteria from the wound site and attracts inflammatory cells. In these cases, a short-term, powerful inflammatory response is actually necessary. After cleaning the wound bed, the cellulose induces vital granulation tissue formation, and smoothens and prepares the wound bed for successful skin transplantation.

The fascinating property of HA-coated cellulose sponge is its ability to even further amplify the healing mechanisms of the body. The HA-coated cellulose acts as a cell-guiding material, attracting stem cell reserves. The novel finding of haemoglobin expression during wound healing brought into daylight new data concerning blood formation and development of neovascularisation. The clinical relevance of this is the production of more vascularised granulation tissue in the critical early phases of wound healing.

We hypothesise that the cell guiding property of the HA-coated cellulose is due to the combination of silica and calcium phosphate. Preliminary results (Stark et al, unpublished) from our on-going study show that a mineral layer induced by dipping the cellulose sponge in a calcium phosphate solution has not the same beneficial feature on granulation tissue formation (not shown) than the mineral layer induced by the bioactive glass in simulated body fluid. Although there was a somewhat stronger inflammatory reaction when compared to uncoated cellulose it was not as intense as in HA-coated cellulose. Our deduction is that the silica elevates the inflammatory reaction with enhanced level of bioactive agent production that attracts more circulating bone marrow-derived progenitor cells whereas the calcium phosphate layer contributes to hastened stem cell homing to the cellulose sponge.

Due to the cell-guiding property of silica rich HA-coated regenerated oxidised cellulose in combination with the capacity to promote proliferation of richly vascularised connective tissue, this material might have potential in clinical situations when rapid granulation tissue growth is needed as in treatment of poorly healing wounds. The contact with the HA-cellulose sponge would be local and temporary, therefore minimizing any possible disadvantages. In addition to safety issues, the manufacturing process of coating cellulose with HA is relatively simple and cheap, and the HA-coated cellulose sponge is easy to handle, form and sterilise.

BOX 2. BIOLOGY OF WOUND HEALING

Wound healing is a complex and dynamic process of restoring cellular structures and tissue layers in the body. The physiological and coordinated response to injury is practically similar in all tissues and involves three distinct but overlapping phases that can be divided into inflammation, new tissue formation and remodelling [70]. In turn, these three phases comprehend coordinated series of events that includes chemotaxis, phagocytosis, neocollagenesis, collagen degradation, and collagen remodelling. Furthermore, neovascularisation, epithelisation, and the production of new glycosaminoglycans (GAGs) and proteoglycans are vital during wound healing process.

The key initiators of the healing process are the platelets, which within minutes after injury aggregate and form fibrin clot in aim to control bleeding. In addition to their important role in hemostasis, platelets also liberate growth factors that will attract inflammatory and phagocytosing cells. The first cells to arrive are polymorphonuclear leucocytes, i.e. neutrophils that secrete proinflammatory cytokines. Shortly thereafter circulating monocytes will appear, are activated and converted in the tissue to macrophages that kill bacterial pathogens, scavenge tissue debris and destroy remaining neutrophils. Macrophages also mediate the transition from the inflammatory phase to the proliferative phase by releasing a variety of chemotactic agents and growth factors that stimulate cell migration, proliferation and formation of tissue matrix.

The second phase of wound healing is often called the proliferative phase or the granulation tissue formation phase. This stage starts normally two to three days after injury and lasts approximately two to three weeks. During this phase the provisional extracellular matrix is gradually filled with granulation tissue. The phenomenal feature is to diminish the area of tissue loss by contraction and fibroplasia. The infiltrated cells produce a new extracellular matrix, rich in blood vessels, which carry oxygen and nutrients to maintain the metabolic processes. Although new collagen and other extracellular matrix proteins are continuously actively synthesised, the earlier formed fibrin clot is enzymatically degraded. This process allows the proceeding of re-epithelisation that is needed to control the growth of the repair tissue and wound closure. The proteolytic activity is also a prerequisite of the neovascularisation.

Usually by three weeks after injury, new tissue formation starts to decrease, and the emphasis of wound healing process turns to the remodelling and maturation. The main objective of this phase is to achieve maximum tensile strength by reorganisation, degradation and re-synthesis of the extracellular matrix. This final process may last even several years, before the new granulation tissue rich in cells and vascular capillaries has matured into a relatively acellular and avascular scar that lacks appendages, including hair follicles, sebaceous glands, and sweat glands [70].

apoptosis	programmed cell death
biocompability	the ability of a material to perform with an appropriate host response in a specific application
chemoattracant	a chemical (chemotactic) agent that induces an organism or a cell to migrate toward it
chemokine	small chemotactic pro-inflammatory cytokine
chemotaxis	directional movement in response to the influence of chemical stimulation
cytokine	a small cell-signalling protein molecule secreted by numerous cells; involved in intercellular communication
endosteum	a thin layer of connective tissue lining the medullary cavity of bone
fibroplasia	the process of forming fibrous tissue
growth factor	a naturally occurring substance capable of stimulating cell growth, proliferation and cell differentiation
haemostasis	the process that causes bleeding to stop
haematopoiesis	production of all types of blood cells including formation, development, and differentiation of blood cells
phagocytosis	an important defence mechanism against infection by microorganisms (e.g. bacteria) and the process of removing cell debris (e.g. dead tissue cells) and other foreign bodies
receptor	a structure on the surface of or inside a cell that selectively receives and binds a specific substance
stem cell niche	a local tissue microenvironment that maintains and regulates stem cells
transcription factors	molecules, usually proteins, which are involved in regulating gene expression

Table 1.

Author details

Miretta Tommila, Anne Jokilammi, Risto Penttinen and Erika Ekholm*
Department of Medical Biochemistry and Genetics, University of Turku, Turku, Finland

Miretta Tommila
Department of Anaesthesiology, Intensive Care, Emergency and Pain Medicine, Turku University Hospital, Turku, Finland

Acknowledgement

Johanna Holmbom, Christoffer Stark and Ville Peltonen are acknowledged for technical assistance. We thank Bruno Lönnberg, Kurt Lönnqvist and the Cellomeda Ltd. for various celluloses and the Swedish Cultural Foundation for supporting our ongoing cellulose study.

4. References

[1] Buddy D Ratner, Allan S Hoffman, Frederick J Schoen, Jack E Lemons (2004) Biomaterials Science: An Introduction to Materials in Medicine. San Diego, London. Elsivier Academic Press. 864 p.

[2] Anderson JM, Jones JA (2007) Phenotypic dichotomies in the foreign body reaction. Biomaterials. 28: 5114-5120.

[3] Miyamoto T, Takahashi S, Ito H, Inagaki H (1989 Tissue compability to cellulose and its derivatives. J. biomed. mat. res. 23: 125-133.

[4] Barbié C, Chaveaux D, Barthe X, Baquey C, Poustis j (1990) Biological behavior of cellulosic material after bone implants: Preliminary results. Clin. mat. 5: 251-258.

[5] Kino Y, Sawa M, kasai S, Mito M (1998) Multiporous cellulose microcarrier for the development of hybrid artificial liver using isolated hepatocytes. J. surg. res. 79 71-76.

[6] Viljanto J, Kulonen E (1962) Correlation of tensile strength and chemical composition in experimental granuloma. Acta pathol. microbial. scand. 52: 120-126.

[7] Raekallio J, Viljanto J (1975) Regeneration of subcutaneous connective tissue in children. A histological study with application of the CELLSTICK device. J. cutan. pathol. 2: 191-197.

[8] Pajulo Q, Viljanto J, Lönnberg B, Hurme T, Lönnqvist K, Saukko P (1996) Viscose cellulose sponge as an implantable matrix: changes in the structure increase the production of granulation tissue. J. biomed. mater. res. 32: 439-446.

[9] Märtson M, Viljanto V, Laippala P, Saukko P (1997) Bone formation in implant is a milieu dependent phenomenon. An experimental study with cellulose sponge in rat. Surg. childh. intern. 5: 251-255.

---

* Corresponding Author

[10] Märtson M, Viljanto J, Hurme T, Saukko P (1998) Biocompatibility of cellulose sponge with bone. Eur. surg. res. 30: 426-432.

[11] Ekholm E, Tommila M, Forsback AP, Märtson M, Holmbom J, Ääritalo V, Finnberg C, Kuusilehto A, Salonen J, Yli-Urpo A, Penttinen R (2005) Hydroxyapatite coating of cellulose sponge does not improve its osteogenic potency in rat bone. Acta biomater. 1: 535-544.

[12] Hoenich N 2006 Cellulose for medical applications: past, present, and future: Available: http://ojs.cnr.ncsu.edu/index.php/BioRes/article/viewFile/BioRes_01_2_270_280_Hoenich_Cellulose_Medical_Review/26.

[13] Boisset JC, Robin C (2012) On the origin of hematopoietic stem cells: progress and controversy. Stem cell res. 8: 1-13.

[14] Cao N, Yao ZX (2011) The hemangioblast: from concept to authentication. Anat. rec. (Hoboken). 294: 580-588. doi: 10.1002/ar.21360

[15] Liu ZJ, Zhuge Y, Velazquez OC (2009) Trafficking and differentiation of mesenchymal stem cells. J. cell biochem. 106: 984-991.

[16] Uccelli A, Moretta L, Pistoia V (2008) Mesenchymal stem cells in health and disease. Nat. rev. immunol. 8: 726-736.

[17] Shi M, Liu ZW, Wang FS (2011) Immunomodulatory properties and therapeutic application of mesenchymal stem cells. Clin. exp. immunol. 164:1-8. DOI:10.1111/j.1365-2249.2011.04327.x

[18] Krasnodembskaya A, Song Y, Fang X, Gupta N, Serikov V, Lee JW, Matthay MA (2010) Antibacterial effect of human mesenchymal stem cells is mediated in part from secretion of the antimicrobial peptide LL-37. Stem cells. 28: 2229-2238.

[19] Lévesque JP, Winkler IG (2011) Hierarchy of immature hematopoietic cells related to blood flow and niche. Curr. opin. hematol. 18: 220-225.

[20] Viljanto J (1964) Biochemical basis of tensile strength of wound healin. An experimental study with viscose cellulose sponges in rat. Acta chir. scand. Suppl. 333.

[21] Matis Märtson (1999) Cellulose sponge as tissue engineering matrix: The effect of cellulose content, implant size and location on connective tissue and bone formation in rat. Annales Universitatis Turkuensis D 349. ISBN 951-29-1452-2. Doctoral thesis

[22] Viljanto J, Jääskeläinen A (1973) Stimulation of granulation tissue growth in burns. Ann. chir. gyn. fenniae. 62: 18-24.

[23] Inkinen K, Wolff H, Lindroos P, Ahonen J (2003) Connective tissue growth facor and its correlation to other growth factors in experimental granulation tissue. Connect. tissue res. 44: 19-29.

[24] Viljanto J (1995) Cellstick device for wound healing research. In: Altmeyer P, Hoffmann K, el Gammal S, Hutchinson J, editors. Wound healing ad skin physiology. Springer-Verlag. pp. 513-522.

[25] Anderson JM, Rodriguez A, Chang DT (2008) Foreign body reaction to biomaterials. Semin. immunol. 20: 86-100.

[26] Pulkkinen H, Tiitu V, Lammentausta E, Laasanen MS, Hämäläinen ER, Kiviranta I, Lammi MJ (2006) Cellulose sponge as a scaffold for cartilage tissue engineering. Biomed. mater. eng. 16: S29-35.

[27] Aitasalo K, Kinnunen I, Palmgren J, Varpula M (2001) Repair of orbital floor fractures with bioactive glass implants. J. oral maxillofac. surg, 59: 1390-1396.

[28] Peltola M, Aitasalo K, Suonpää J, Varpula M, Yli-Urpo A (2006) Bioactive glass S53P4 in frontal sinus obliteration: a long-term clinica lexperience. Head neck. 28: 834-841.

[29] Peltola M, Suonpää J, Aitasalo K, Määttänen H, Andersson Ö, Yli-Urpo A, Laippala P (2000) Experimental follow-up model for clinical frontal sinus obliteration with bioactive glass (S53P4). Acta otolaryngol. suppl. 543: 167-169.

[30] Stoor P, Pulkkinen J, Grénman R (2010) Bioactive glass S53P4 in the filling of cavities in the mastoid cell area in surgery for chronic otitis media. Ann. otol. rhinol. laryngol. 119: 377-382.

[31] Lindfors NC, Hyvonen P, Nyyssönen M, Kirjavainen M, Kankare J, Gullichsen E, Salo J. (2010) Bioactive glass S53P4 as bone graftsubstitute in treatment of osteomyelitis. Bone. 47: 212-218.

[32] Lindfors NC, Koski I, Heikkilä JT, Mattila K, Aho AJ. (2010) A prospective randomized 14-yearfollow-up study of bioactive glass andautogenous bone as bone graft substitutes in benign bone tumors. J. biomed. mater. res. B appl. biomater. 94: 157-164.

[33] Kokubo T, Hata K, Nakamura T, Yamamuro T (1991) Apatite formation on ceramics, metals and polymers induced by CaO SiO_2 based glass in a simulated body fluid. Bioceramics. 4: 113-120.

[34] Gilbert L, He X, Farmer P, Rubin J, Drissi H, van Wijnen AJ, Lian JB, Stein GS, Nanes MS (2002) Expression of the osteoblast differentiation factor RUNX2 (Cbfa1/AML3/Pebp2alpha A) is inhibited by tumor necrosis factor-alpha. J. biol. chem. 277: 2695-2701.

[35] Nanes MS (2003) Tumor necrosis factor ⊛: molecular and cellular mechanisms in skeletal pathology. Genes. 231: 1-15.

[36] Glen Österås (2002) Utveckling av metod för framställning av cellulosapärlor. (Development of the method to make cellulose pearls). Åbo Academy University, Finland. MSc thesis.

[37] Kalakovich KW, Boynton RE, Murphy JM, Barry F (2002) Chondrogenic differentiation of human mesenchymal cells within alginate culture system. In vitro cell dev. biol. anim. 38: 457-466.

[38] Tommila M, Jokinen J, Wilson T, Forsback AP, Saukko P, Penttinen R, Ekholm E (2008) Bioactive glass-derived hydroxyapatite-coating promotes granulation tissue growth in subcutaneous cellulose implants in rats. Acta biomater. 4: 354-361.

[39] Jo DY, Rafii S, Hamada T, Moore MA (2000) Chemotaxis of primitive hematopoietic cells in response to stromal cell-derived factor-1. J. clin. invest. 105: 101-111.

[40] Dar A, Kollet O, Lapidot T (2006) Mutual, reciprocal SDF-1/CXCR4 interactions between hematopoietic and bone marrow stromal cells regulate human stem cell migration and development in NOD/SCID chimeric mice. Exp. hematol. 34: 967-975.

[41] Kaplan RN, Psaila B, Lyden D (2007) Niche-to-niche migration of bone-marrow-derived cells. Trends mol. med. 13: 72-81.

[42] Schantz JT, Chim H, Whiteman M (2007) Cell guidance in tissue engineering: SDF-1 mediates site-directed homing of mesenchymal stem cells within three-dimensional polycaprolactone scaffolds. Tissue eng.13: 2615-2624.

[43] Sasaki M, Abe R, Fujita Y, Ando S, Inokuma D, Shimizu H (2008) Mesenchymal stem cells are recruited into wounded skin and contribute to wound repair by transdifferentiation into multiple skin cell type. J. immunol. 180: 2581-2587.

[44] Ding J, Hori K, Zhang R, Marcoux Y, Honardoust D, Shankowsky HA, Tredget EE (2011) Stromal cell-derived factor 1 (SDF-1) and its receptor CXCR4 in the formation of postburn hypertrophic scar (HTS). Wound repair regen. 19: 568-578. doi: 10.1111/j.1524-475X.2011.00724.x.

[45] Tommila M, Jokilammi A, Terho P, Wilson T, Penttinen R, Ekholm E (2009). Hydroxyapatite coating of cellulose sponges attract bone-marrrow-derived stem cells in rat subcutaneous tissue. J. r. soc. interface. 6: 873-880.

[46] Lancrin C, Sroczynska P, Serrano AG, Gandillet A, Ferreras C, Kouskoff V, Lacaud G (2010). Blood cell generation from hemangioblast. J mol. med. 88: 167-172.

[47] Adams GB, Chabner KT, Alley IR, Olson DP, Szczepiorkowski ZM, Poznansky MC, Kos CH, Pollak MR, Brown EM, Scadden DT (2006). Stem cell engraftment at the endosteal niche is specified by the calcium-sensing receptor. Nature. 439: 599-603.

[48] Wu Y, Wang J, Scott PG, Tredget EE (2007) Bone marrow-derived stem cells in wound healing: a review. Wound repair regen. 15: S18-26.

[49] Yagi H, Soto-Gutierrez A, Parekkadan B, Kitagawa Y, Tompkins RG, Kobayashi N, Yarmush ML (2010) Mesenchymal stem cells: Mechanisms of immunomodulation and homing. Cell transplant. 19: 667-679.

[50] Wannemuehler TJ, Manukyan MC, Brewster BD, Rouch J, Poynter JA, Wang Y, Meldrum DR (2012). Advances in mesenchymal stem cell research in sepsis. J. surg. res. 17: 113-126.

[51] Lazar-Karsten P, Dorn I, Meyer G, Lindner U, Driller B, Schlenke P (2011) The influence of extracellular matrix proteins and mesenchymal stem cells on erythropoietic cell maturation. Vox sang. 101: 65-76. doi: 10.1111/j.1423-0410.2010.01453.x

[52] Tommila M, Stark C, Jokilammi A, Peltonen V, Penttinen R, Ekholm E (2011) Hemoglobin expression in rat experimental granulation tissue. J. mol. cell biol. 3:190-196.

[53] Palis P (2009) Molecular biology of erythropoiesis. In: Wickrema A, Kee B, editors. Molecular basis of hematopoieis, New York: Springer. pp. 73-93.

[54] Chasis JA, Mohandas N (2008) Erythroblastic islands: nisches for erythropoiesis. Blood. 112: 470-478.

[55] Sadlon TJ, Dell'Oso T, Surinya KH, May BK (1999) Regulation of erythroid 5-aminolevulinate synthase expression during erythropoiesis. Int. j. biochem. cell biol. 31: 1153-1167.

[56] Newton DA, Rao, KM, Dluhy RA, Baatz (2006) Hemoglobin is expressed by alveolar epithelial cell. J. biol. chem. 281: 5668-5676.

[57] Grek CL, Newton DA, Spyropoulus DD, Baatz JE (2011 Hypoxia upregulates expression of haemoglobin in alveolar epithelial cells. Am. j. respir. mol. biol. 44: 439-447.

[58] Nishi H, Inagi R, Kato H, Tanemoto M, Kojima I, Son D, Fujita T, Nangaku M (2008) Hemoglobin is expressed by mesangial cells and reduces oxidant stress. J. am. soc. nephrol. 2008 19: 1500-1508.

[59] Dassen H, Kamps R, Punyadeera C, Dijcks F, de Goeij A, Ederveen A, Dunselman G, Groothuis P (2008). Haemoglobin expression in human endometrium. Hum. reprod. 23: 635-641.

[60] Wride MA, Mansergh FC, Adams S, Everitt R, Minnema SE, Rancourt DE, Evans MJ (2003) Expression profiling and gene discovery in the mouse lens. Mol. vis. 9: 360-396.

[61] Biagioli M, Pinto M, Cesselli D, Zaninello M, Lazarevis D, Roncalgia P, Simone R, Vlachouli C, Plessy C, Bertin N, Beltrami A, Kobayashi K, Gallo V, Santoro C, Ferrer I, Rivella S, Beltrami CA, Carninci P, Raviola E, Gustincich S (2009) Unexcpected expression of alpha- and beta-globin in mesencephalic dopaminergic neurons and glial cells. Proc. natl. acad. sci. USA. 106: 15454-15459.

[62] Schelshorn DW, Schneider A, Kuscinsky W, Weber D, Kruger C, Dittigen T, Burgers HF, Sabouri F, Gassler N, Bach A, Mauer MH (2009) Expression of haemoglobin in rodents neurons. J. cereb. blood flow metab. 29: 585-595.

[63] Richter F, Meurers BH, Zhu C, Medvedeva VP, Chesslet MF (2009) Neurons express haemoglobin alpha- and beta-chains in rat and humans brains. J. comp. neurol. 515: 538-547.

[64] Setton-Avruj CP, Musolino PL, Salis S, Allo M, Bizzozero O, Villar MJ, Soto EF, Pasquini JM (2007) Prescence of alpha-globin mRNA and migration of bone marrow cells after sciatic nerve injury suggest their participation in the degeneration/regeneration process. Exp. neurol. 203: 568-578.

[65] Liu L, Zeng M, Stamler JS (1999) Hemoglobin induction in mouse macrophages. Proc natl. acad. sci. USA 96: 6643-6647.

[66] Weller R (2003) Nitric oxide: a key mediator in cutaneous physiology. Clin. exp. dermatol. 28: 511-514.

[67] Schwentker A, Vodovotz Y, Weller R, Billiar TR (2002) Nitric oxide and wound repair: role of cytokines? Nitric oxide. 7: 1-10.

[68] Soneja A,Drews M, Malinski T (2005). Role of nitric oxide, nitroxidative and oxidative stress in wound healing. Pharmacol. rep. 57 S108-109.

[69] Luo J, Chen A (2005) Nitric oxide: a newly discovered function on wound healing. Acta pharmacol. sin. 26: 259-264.

[70] Gurtner GC, Werner S, Barrandon Y, Longaker MT (2008) Wound repair and regeneration. Nature. 453: 314-321.

Chapter 4

Effect of Polymorphism on the Particle and Compaction Properties of Microcrystalline Cellulose

John Rojas

Additional information is available at the end of the chapter

1. Introduction

Cellulose is the most abundant natural linear polymer. It consists of 1,4-linked-β-D-glucose units and is known to exist in the following distinct allomorphs: I_α (from algae and bacteria), I_β (from superior plants), II (the most stable form produced by mercerization), III_I and III_{II} (prepared from ammonia at -30 ºC), and IV_I and IV_{II} (produced at 260 ºC in glycerol). Each allomorph differs in its physicochemical properties [1,2]. Cellulose III is formed when native cellulose is treated with liquid ammonia at low temperatures, whereas cellulose IV is obtained by treatment of regenerated cellulose at high temperatures (Figure 1) [3]. However, the last two forms have no pharmaceutical applications.

Of these, the cellulose I (MCCI) allomorph is the most prevalent form and cellulose II is the most stable form [4]. MCCI can be converted to MCCII, but not vice versa [5,6]. As shown in Figure 2, in cellulose I (MCCI), the chain orientation is exclusively parallel [3], whereas in cellulose II (MCCII) the chains are arranged in an anti-parallel orientation.

Commercial microcrystalline cellulose (MCCI) contains the cellulose I lattice. It is obtained from wood pulp by treatment with dilute strong mineral acids (HCl, H_2SO_4, HNO_3) at boiling temperatures until the degree of polymerization levels-off [7,8]. The acid hydrolyzes the less ordered regions of the polymer chains, leaving the crystalline regions intact. This MCCI is also called hydrolyzed cellulose or hydrocellulose.

Since the 1970s, microcrystalline cellulose I (MCCI) has been the dominant excipient used for direct compression due to its good diluent and binding properties and low moisture content. The strong binding properties of MCCI are due to hydrogen bonding among the plastically deforming cellulose particles. However, it suffers from sensitivity to lubricants

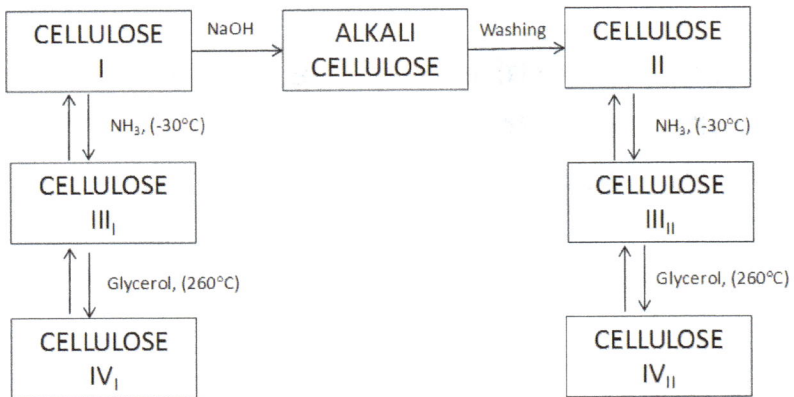

Figure 1. Scheme for the formation of cellulose allomorphs

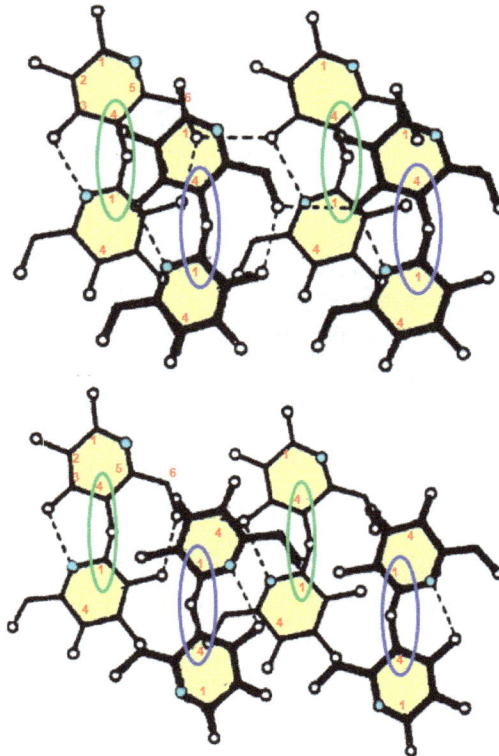

Figure 2. Conformations of MCCI (A) and MCCII (B)

and poor flow [9,10]. Because of its strong binding properties, it requires the addition of a disintegrant for an effective drug release, making formulations more costly. The compactibility of MCCI is also adversely affected when processed by high shear wet granulation since upon drying part of the water interacts with cellulose through hydrogen bonding and as a result, these hydrogen bonds are not available for further particle bonding [11].

Recently, microcrystalline cellulose II (MCCII) was introduced as a new direct compression excipient [12]. It can be produced by soaking MCCI in an aqueous sodium hydroxide solution (> 5 N) at a 1:6 weight-to-volume ratio for 14 h at room temperature, with occasional stirring. The resulting MCCII gel is then precipitated (regenerated) with a 50-60% aqueous ethanolic solution, filtered, washed with distilled water to neutrality by decantation, filtered again, and dried at 40 °C until reaching a moisture content of less than 5 % [13]. During this process, the amorphous regions of the microfibrils are partially eliminated leaving the most crystalline parts intact. The resulting product is usually washed and spray-dried to get a powder [14].

In most cases, a polymorphic transformation could modify some of the particle properties of a material. One of those properties is related to the water uptake capacity of the powder. In fact, this uptake depends on the crystalline structure of the polymer. Further, the mechanical and disintegrating properties of these materials are related to the degree of crystallinity and water uptake capacity, respectively. Thus, knowledge of the water sorption behavior of the cellulose allomorphs is essential to understand and predict their stability, especially during storage, alone or combined with other materials in a dosage form under variable ambient conditions. In this study, the effect of polymorphic transformation on the microcrystalline cellulose functionality was evaluated. The particle and mechanical properties of MCCII were assessed and compared with those of commercial MCCI (Novacel® PH-101).

2. Experimental

2.1. Materials

MCCI (Novacel PH-101, lot 6N608C) was donated by FMC Biopolymers, Philadelphia, PA, USA. Sodium hydroxide (lot 58051305C) and concentrated hydrochloric acid (37%, lot k40039517) were obtained from Carlo Erba, and Merck, respectively. Magnesium stearate (lot 2256KXDS) was purchased from Mallinckrodt Baker and acetaminophen (lot GOH0A01) was obtained from Sigma-Aldrich.

2.2. Methods

2.2.1. Preparation of Microcrystalline Cellulose II (MCCII)

Approximately, 500 g of MCCI was soaked in 3 L of 7.5 N NaOH for 72 h at room temperature. The cellulose II thus obtained was washed with distilled water until it reached neutral pH. The slurry was sequentially passed through 6 (3350 μm), 10 (2000 μm), 24 (711 μm), 40 (425 μm) and 100 (150 μm) mesh screens using an oscillating granulator (Riddhi

Pharma Machinery, Gulabnagar, India) when the moisture content was ~60, 50, 40, 30 and 20%, respectively. The final material was dried in a convection oven at 60°C (Model STM 80, Rigger Scientific Inc, Chicago, IL) to a moisture content of less than 5%.

2.2.2. Fourier-Transform Infrared Spectroscopy (FT-IR) Characterization

Approximately, 1.5 mg of sample was mixed with about 300 mg of dry potassium bromide (previously dried at 110 °C for 4 h) with an agate mortar and pestle. The powdered sample was compressed into a pellet using a 13 mm flat-faced punch a die tooling fitted on a portable press at a dwell time of five minutes. A Perkin Elmer spectrophotometer (Spectrum BX, Perkin Elmer, CA, USA) equipped with the Spectrum software (Perkin Elmer, Inc, CA, USA) was used to obtain the spectrum between 650 to 4000 cm^{-1}. The resolution, interval length and number of scans employed were 16, 2.0 and 16 cm^{-1}, respectively.

2.2.3. Powder X-Rays (P-XRD) Characterization

Powder X-ray diffraction (P-XRD) measurements were conducted over a 5 to 45º 2θ range using a Rigaku Bench top, diffractometer (Miniflex II, Rigaku Americas, The Woodlands, TX, USA) at 40 kV and 30 mA equipped with monochromatic CuKα (α_1=1.5460 Å, α_2= 1.54438 Å) X-ray radiation. The sweep speed and step width were 0.5º 2θ/min and 0.008°, respectively. The PeakFit software, version 4.12 (SeaSolve Inc, Framingham, MA) was used for the calculation of the areas. The degree of crystallinity was found from the equation [15]:

$$DC = \frac{I_C}{I_T} * 100\% \qquad (1)$$

Where I$_C$ is the sum of the areas of all crystalline peaks and I$_T$ is the area of the amorphous and crystalline regions.

2.3. Powder properties

The microphotographs were taken on an optical microscope (BM-180, Boeco, Germany) coupled with a digital camera (S8000fd, Fujifilm Corp., Japan) at 700X magnification. The true density was determined on a helium picnometer (AccuPyc II 1340, Micromeritics, USA) with ~2 g of sample. Bulk density was determined by the ratio of 20 g of sample divided the measured volume. Tap density was measured directly from the final volume of the tapped sample obtained from the AUTO-TAP analyzer (AT-2, Quantachrome instruments, USA). Flow rate was obtained by measuring the time for ~20 g of sample to pass through a glass funnel (13 mm diam). Porosity (ε) of the powder was determined from the equation:

$$\varepsilon = 1 - \left(\frac{\rho_{bulk}}{\rho_{true}} \right) \qquad (2)$$

Where, ε, ρ$_{bulk}$, and ρ$_{true}$ are the porosity, bulk density and true density of the powder, respectively. The degree of polymerization (DP) was obtained by the intrinsic viscosity method [η] at 25 ±0.5 °C using a Canon-Fenske capillary viscometer (cell size 50) and cupriethylendiamine hydroxide (CUEN) as solvent [16]. The DP was found by the relationship:

$$DP = 190 * [\eta] \tag{3}$$

The compressibility of the powder was obtained by applying the Kawakita model [17]:

$$N / \left[(V_i - V_n) / V_i \right] = N / a + 1 / ab \tag{4}$$

Where, N is the tap number, V_i the initial volume and V_n the volume at the respective tap number. The constant "a" is related to the total volume reduction for the powder bed (compressibility index) and the constant "b" is related to the resistant forces (friction/cohesion) to compression [18].

2.4. Particle size

Samples were fractionated on a RO-TAP sieve shaker (RX29, W.S. Tyler Co., Mentor, OH, USA) using stainless steel 420, 250, 180, 125, 105, 75 μm sieves, stacked together in that order (Fisher Scientific Co., Pittsburgh, PA, USA). Approximately, 50 g of the sample was shaken for 30 min followed by weighing the fractions retained in each sieve. The particle size distributions and the geometric means were found from the log-Normal distributions using the Minitab software (v.16, Minitab®, Inc., State College, PA).

2.5. Swelling studies

The swelling value is expressed as the ratio of the expanded volume of the powder upon water addition and the initial sample weight. Approximately, 500 mg of the powder was vigorously dispersed in a 10 ml graduate cylinder filled with 10 ml of distilled water at room temperature and the increase in volume of the powder was measured with time [19].

2.6. Water sorption isotherms

Water sorption isotherms were conducted on a VTI Symmetric Gravimetric Analyzer (Model SGA-100, VTI Corporation, Hialeah, FL), equipped with a chilled mirror dew point analyzer (Model Dewprime IF, Edgetecth ford, MA) at 25 °C. The water activity employed ranged from 0 to 0.9. Water uptake was considered at equilibrium when a sample weight change of no more than 0.01% was reached. Samples were analyzed in triplicate. The non-linear curve fitting and the resulting parameters were obtained using the Statgraphic software vs. 5 (Warrenton, VA). The Young-Nelson Model (YN) was used for data fitting. This model distinguishes between the tightly bound monolayer, normally condensed externally adsorbed water, and internally absorbed water [20]. In this model, water uptake is given by equations 5-9:

$$m = A(\theta + \beta) + B\Psi \tag{5}$$

$$\theta = \frac{a_w}{a_w + (1 - a_w)E} \tag{6}$$

$$\Psi = a_w \theta \tag{7}$$

$$\beta = -\frac{Ea_w}{E - (E-1)a_w} + \left(\frac{E^2}{E-1}\right)\ln\frac{E - (E-1)a_w}{E} - (E+1)\ln(1 - a_w) \tag{8}$$

$$E = e^{-(H_1 - H_l)/RT} \tag{9}$$

Where, m, θ, Ψ, and B correspond to the total fractional moisture content, the fraction of molecules cover by monolayer, the fraction covered by a layer 2 or more molecules thick, and the amount of absorbed water in the multilayer. H_1 is the heat of adsorption of water bound to the surface, H_L the heat of condensation, R is the gas constant (8.31 J/Kmol), and T the temperature. A and B are dimensionless constants related to the fraction of adsorbed and absorbed water on the polymer, respectively. E is the equilibrium constant between the monolayer and liquid water. The product $A\theta$ is related to the amount of water in the monolayer and $A(\theta+B)$ is the externally adsorbed moisture during the sorption phase. $B\Psi$ is the amount of moisture absorbed during the sorption phase [21].

2.7. Tableting properties

Compacts of ~500 mg each were made on a single punch tablet press (Compac 060804, Indemec Ltd, Itagui, Colombia) coupled with a load cell (Model LCGD-10K, Omega Engineering, Inc., Stamford, CT) using flat-faced 13 mm punches and die tooling for 1 and 30 s. Pressures ranged from ~35 to ~190 MPa. Forces were measured on a strain gauge meter (Model DPiS8-EI, Omega Engineering, Inc., Stamford, CT). Compact heights were measured immediately after production and after 5 days of storage to measure the elastic recovery of the material.

2.8. Compressibility analysis

The natural logarithm of the inverse of compact porosity, $[-\ln(\varepsilon)]$, was plotted against compression pressure (P) to construct the Heckel plot [22, 23]. The slope (m) of the linear region of this curve is inversely related to the material yield pressure (P_y), which is a measurement of its plasticity [24]. Thus, a low P_y (usually values <100 MPa) indicates a high ductile deformation mechanism upon compression. The Heckel model is given by:

$$-\ln\varepsilon = mP + A \tag{11}$$

Where, A is the intercept obtained by extrapolating the linear region to zero pressure. Other parameters useful in assessing compressibility are D_0, D_a, and D_b, which are related to initial powder packing/densification, total compact densification, and particle rearrangement/fragmentation at the initial compaction stage, respectively [25]. D_0 was calculated by dividing the bulk density with the true density [26]. The strain rate sensitivity (SRS) was found by the percentage change of the P_y resulted from 1 and 0.03 compact/s speeds, respectively.

2.9. Compact tensile strength

It was determined on a VanKel hardness tester (UK 200, VanKel, Manasquan, NJ, USA). Each compact was placed between the platens and the crushing force was then measured. The radial tensile strength (TS) values were obtained according to the Fell and Newton equation [27]:

$$TS = \frac{2\sigma}{\pi DH} \tag{12}$$

Where, F is the crushing force (N) needed to break the compact into two halves, D is the diameter of the compact (mm), and t is the compact thickness (mm). The crosshead speed of the left moving platen was 3.5 mm/s.

2.10. Dilution potential

Tablets containing different levels of acetaminophen (25, 50, 75, 85 or 95%) and a poorly compressible drug, were prepared and their crushing strength was determined. Acetaminophen and the test excipient were mixed in a V-Blender for 30 min and then compressed on a single punch tablet press at 120 MPa and a dwell time of 30 s. Samples were analyzed in triplicate.

2.11. Lubricant Sensitivity (LSR)

Lubricant sensitivity was assessed by mixing powders with magnesium stearate at the 99:1 weight ratio in a V-blender (Riddhi Pharma Machinery, Gulabnagar, India) for 30 min. Tablets were prepared using a single punch tablet press at 120 MPa and a dwell time of 30 s. The lubricant sensitivity was expressed as the lubricant sensitivity ratio (LSR) according to the equation:

$$LSR = \frac{H_0 - H_{lub}}{H_0} \tag{13}$$

Where, H_0 and H_{lub} are the crushing strengths of tablets prepared without and with lubricant, respectively. Samples were analyzed in triplicate.

2.12. Compact friability

The friability test was performed on a friabilator (FAB-25, Logan Instruments Corp., NJ, USA) at 25 rpm for 4 min. An amount of ~6.5 g of compacts made at 150 MPa, each weighing ~500 mg, was tested in a friabilator. Compacts were then dusted and reweighed. The percentage weight loss was taken as friability.

2.13. Compact disintegration

Tablets, each weighing ~500 mg, were made on a single punch tablet press (060804, Indumec, Itagui, Columbia) using a 13 mm round, flat-faced punches and die set. Five replicates were tested in distilled water at 37 °C employing a Hanson disintegrator (39-133-115, Hanson Research Corporation, Northridge, CA, USA) operating at 30 strokes/min.

3. Results and discussion

3.1. FT-IR characterization

As seen in Figure 3 the cellulose allomorphs showed no major differences in the infrared bands, except for the vibration peak at 3423 cm^{-1} corresponding to intramolecular OH stretching, including hydrogen bonds and at 2893 cm^{-1} due to CH and CH_2 stretching. It is reasonable that these two bands appear displaced due to the rearrangement of the cellulose chains and hydrogen bonding pattern, which is parallel and antiparallel for MCCI and MCCII, respectively. Other main vibrational peaks with virtually no change due to the polymorphic transformation are: 1642 cm^{-1} corresponding to OH from absorbed water, 1427 cm^{-1} due to CH_2 symmetric bending, 1375 cm^{-1} due to CH bending, 1330 cm^{-1} due to OH in plane bending, 1255 cm^{-1} corresponding to C-O-H bending, 1161 cm^{-1} due to C-O-C asymmetric stretching (β-glucosidic linkage), 1063 cm^{-1} due to C-O/C-C stretching and 895 cm^{-1} due to the asymmetric (rocking) C-1 (β-glucosidic linkage) out-of-plane stretching vibrations. This band is associated to the cellulose II lattice [28, 29].

3.2. P-XRD characterization

It is well known that the acid hydrolysis of powdered α-cellulose I reduces the degree of polymerization and in turn increases slightly the degree of crystallinity, true density, compact tensile strength and fragmentation tendency [30, 31]. In this study, the change in the above properties is only attributed to the polymorphic form, which is evidenced in the powder x-Rays difractograms (Figure 4). MCCI displayed the characteristic diffraction peaks of the cellulose I lattice at 14.8, 16.3 and 22.4° 2θ, corresponding to the 11 $\bar{0}$, 110 and 200 reflections, respectively. A shoulder at 20.4° 2θ has also been also identified in some MCCI excipients [32]. MCCII materials showed crystalline peaks at 12, 20 and 22° 2θ corresponding to the 11 $\bar{0}$, 110 and 200 reflections, respectively [2]. The degree of crystallinity of MCCI was larger than that of MCCII. It is plausible that the antiparallel arrangement along with the monoclinic unit cell renders a loose molecular packing and hence causes a lower degree of crystallinity. During the soaking step, the alkali could partially attack the glycosidic bonds located on the less ordered surface of the microfibrils since they are more accessible than the glycosidic bonds located in the ordered regions of the microfibrils of cellulose. This might explain why MCCII has a slightly lower degree of crystallinity than MCCI.

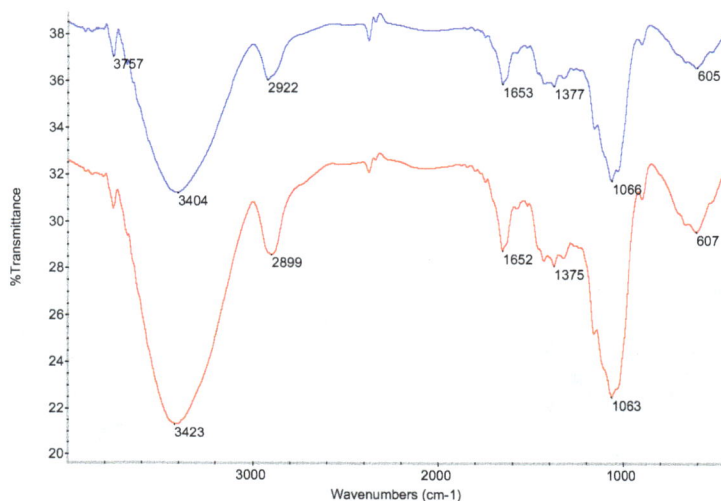

Figure 3. FT-IR of microcrystalline cellulose allomorphs

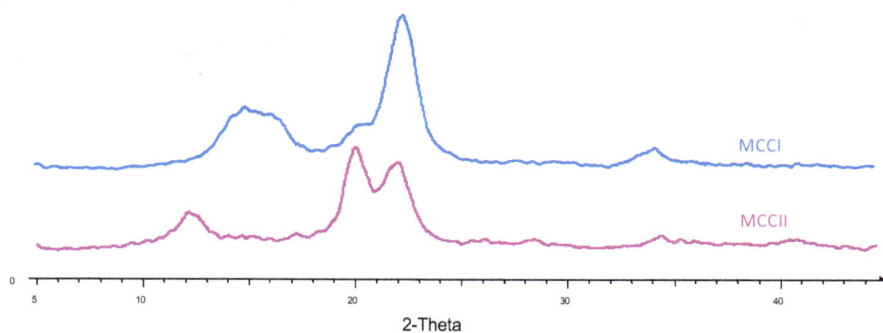

Figure 4. Powder XRD of microcrystalline cellulose allomorphs

3.3. Powder properties

Figure 5 shows micrographs depicting particle morphologies. It seems to be that the polymorphic transformation had little effect on the morphology of these particles. Both, MCCI and MCCII consisted of aggregated and irregularly-shaped particles with rough surfaces and sharp edges. However, elongated particles were more predominant for MCCI. Table 1 lists the powder properties of these materials. Since the polymorphic transformation did not cause major morphological changes in the particles, the mean particle size of the two polymorphs was comparable. The particle size distribution is depicted in Figure 6. In this case, both materials showed a positively skewed distribution, but MCCII had a slightly larger tendency to have high frequencies in the low particle size region.

MCCI had a larger true density than MCCII. On the other hand, MCCII presented larger bulk and tap densities and consequently lower total powder porosity as compared to MCCI. The great ability of MCCII for particle packing could be due to morphological factors including its lower proportion of elongated particles, lower particle porosity and lower roughness. As a result, upon tapping or application of compression forces, MCCII is more likely to undergo a major volume reduction as indicated by its larger compressibility (52%). All these factors also contributed to an improvement in flow for MCCII. Thus, flowability was 3-fold larger for MCCII than for MCCI.

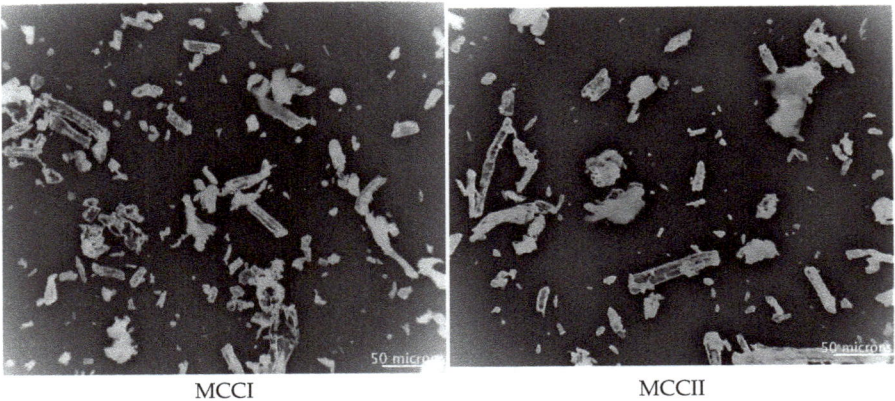

MCCI   MCCII

Figure 5. Optical micropictures of microcrystalline allomorphs

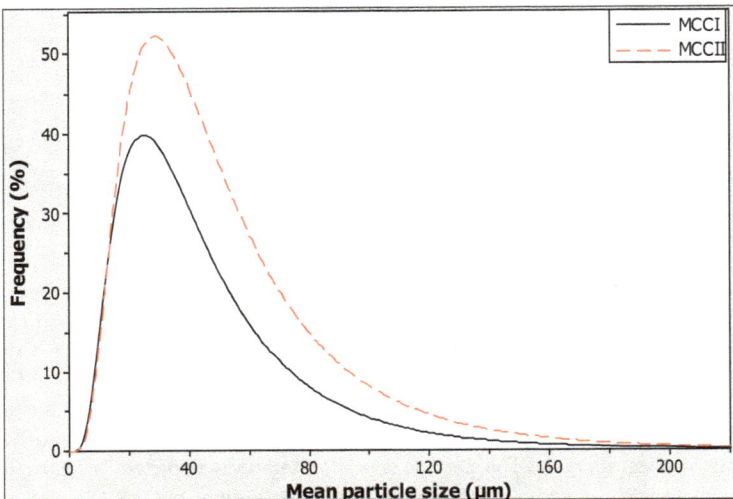

Figure 6. Particle size distribution of microcrystalline celluloses

The voluminosity of the powder was found by taking the reciprocal of the bulk density and thus, MCCII, due to its large packing ability, presented a lower voluminosity or bulkiness than MCCI.

The degree of polymerization (DP) is an indication of the average polymer length. The lower values of DP obtained for MCCII are probably due to the partial hydrolysis occurring during the NaOH treatment.

Property	n	MCCI	MCCII
Geometric mean (μm)	1	50.4 ± 7.4	56.5 ± 7.1
Compressibility (%)	1	33	52
Degree of polymerization	3	233.3 ± 0.6	183.6 ± 3.2
Molecular weight (g/mol)	1	37794.6	29743.2
Degree of crystallinity (%)	1	74.3	65.2
Porosity	3	0.78 ± 0.01	0.64 ± 0.01
Bulk density (g/cm³)	3	0.35 ± 0.01	0.55 ± 0.00
Tap density (g/cm³)	3	0.51 ± 0.00	0.76 ± 0.00
Voluminosity (ml/g)	3	2.9 ± 0.1	1.8 ± 0.0
True density (g/cm³)	3	1.578±0.05	1.540 ± 0.03
Flow rate (g/s)	3	1.61 ± 0.43	4.93 ± 0.96
Swelling value (ml/g)	3	0.2 ±0.1	0.8 ±0.0
Specific surface area (m²/g)	3	1.49 ± 0.05	0.53±0.05
Moisture content (%)	1	1.8	2.2

n=replicate

Table 1. Powder properties of microcrystalline celluloses

Figure 7 shows the isotherms curves fitted according to the Young-Nelson (YN) model and Table 2 shows the parameters derived from this model. All cellulosic materials exhibited a typical sigmoidal type II isotherm and the sorption and desorption curves showed hysteresis. Hysteresis is defined as the difference between the amount of water desorbed and sorbed. This difference creates a loop in the isotherm and is very common in hydrophilic materials. The amplitude of the loop is observed in Figure 7. This hysteresis was high between 0.2 and 0.7 water activities. Results indicate that hysteresis occurred throughout the sorption range and not just in the capillary condensation region as reported previously [33]. It is well known that the higher moisture content obtained when a polymer is desorbing from a saturated state is due to microcapillary deformation accompanied by the creation of more permanent hydrogen bonds which are no longer attainable in subsequent re-wetting processes. This phenomenon is known in cellulose as hornification [34].

Once cellulose sorbs water it swells slightly because microcapillaries expand due to the thermal motion of incoming water molecules forming new internal surfaces. Once water is removed, relaxation of the matrix to the original state is prevented. As a result, microcapillaries become greater on desorption compared to the adsorption step. In other words, this hysteresis is caused by structural changes due to the disruption of the hydrogen bonding network of the polymer while interacting with water molecules. Therefore, the extra sorbed water showed by hysteresis is related to the structural or conformational changes of cellulose chains, which expose previously inaccessible high affinity sorption sites [35].

Since hysteresis took place mostly at a water activity of 0.2-0.7, it cannot be completely attributed to pore effects (ink bottle pores), but to swelling due to specific interactions of water sorbed in the bulk. The lower hysteresis of MCCI might indicate weaker specific interactions of its hydrophilic sites with water, and consequently, smaller structural reorganization of the chains due to cellulose swelling.

Figure 8 shows the deconvoluted curves for the monolayer and multilayer formation, respectively according to the YN model. It assumes that the monolayer and multilayers of water molecules are formed simultaneously at very low water activities. Furthermore, it considers the absorption of water into the core of particles to be the first step, followed by the layering process on the surface. MCCI showed the lowest fraction of water molecules that formed a monolayer and multilayer, respectively. Further, the amount of water absorbed in the core was considered as negligible. Thus, most of the cellulose water uptake can be attributed to adsorption due to layering.

The YN model also demonstrated that the fraction of water absorbed in the particles core was smaller than the fraction adsorbed as a monolayer and multilayer. Also, the H_1-H_L value for MCCII was very high, indicating the prevalence for layering formation, which is in line with the high hydrophilicity of this material. MCCII always showed higher affinity for water than MCCI as seen by the higher monolayer and multilayer formation at different water activities. This trend proved that the polymorphic transformation of MCCI into MCCII enhanced the water sorption capacity.

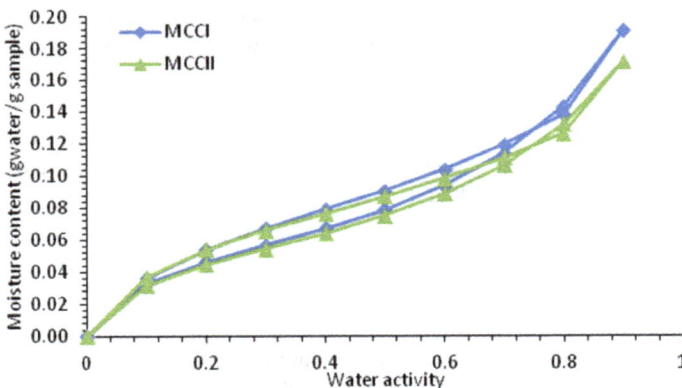

Figure 7. Water sorption and desorption isotherms according to the Young-Nelson model.

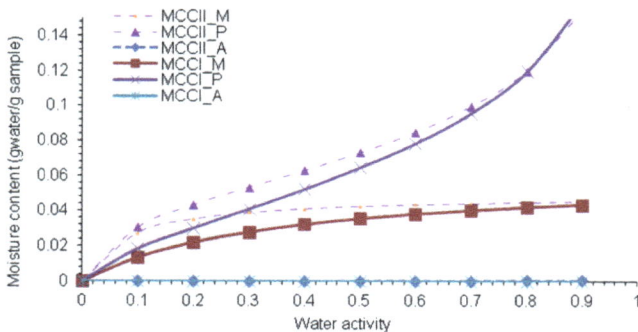

Figure 8. Deconvoluted Young-Nelson model for water sorption of microcrystalline celluloses. M, indicates monolayer formation, P, multilayer formation and A, the fraction attributed to the absorption process in the core of the particles.

In the YN model, the absorbed water has diffused through pores into the core of the particle from the adsorbed monolayer. The monolayer water is then the result of the balance between surface binding forces of the multilayer and water in the core. It is possible that water molecules bind as succeeding layers rather than to empty sites on the surface of the solid. Thus, the formation of a second layer probably starts at lower concentrations than those corresponding to the monolayer formation because the completion of the perfect monolayer would lead to a substantial decrease in entropy, which is very unlikely for natural polymers [36]. During the desorption phase vapor pressure is reduced and water molecules at the surface are removed before diffusion forces pull moisture out of the core of the material [37].

In the sorption phase, the curves of the monolayer formation presented a type I Langmuir isotherm, whereas the curves of the multilayer sorption showed a type II isotherm. The sorption of monolayer water is almost complete at a water activity of 0.1 for MCCII, whereas for MCCI it increased steadily up to 0.9 water activity. On the other hand, both polymers showed a constant increase of the multilayer formation throughout the whole water activity range. This proves that not all the adsorption sites of the first layer were filled when the formation of multilayers started. For this reason, the external moisture component isotherm had a type II shape and its contribution to the total amount of sorbed moisture increased with increasing water activities.

The parameters derived from the YN model are shown in Table 2. All samples had a good fit to the model ($r^2 > 0.9990$). The A (0.04-0.05) parameter confirmed that layering of water molecules either as a monolayer or multilayers was more prevalent than absorption into the core of the particles (B <0.01). Further, this absorption was more predominant for MCCII than for MCCI. The E and H_1-H_L values indicate that the formation of a monolayer for MCCII was energetically more favorable than the formation of multilayers. Further, H_1-H_L was positive, indicating the execution of an endothermic process. The E parameter was low for MCCII indicating that the driving force for water sorption was higher than that of MCCI during the sorption phase. The first water molecule could bind to the 6-hydroxyl group because it is the most exposed hydroxyl group in cellulose [38]. Then, more incoming water molecules could bind to hydroxyl groups located in carbons 2 and 3 of the cellulose monomers.

Sample	A Mean ± SD	B Mean ± SD	E Mean ± SD	H_1-H_L kJ/mol	r^2
MCCI	0.04 ± 0.00	0.00 ± 0.00	0.26 ± 0.02	3.3	0.9990
MCCII	0.05 ± 0.01	0.01 ± 0.00	0.07 ± 0.05	7.4	0.9997

SD, standard deviation, A, fraction of adsorbed water; B, fraction of absorbed water;
E, equilibrium constant between the monolayer and liquid water; H_1-H_L, heat difference between absorption and condensation of water; r^2 coefficient of determination.

Table 2. Parameters obtained from the Young-Nelson model

The degree of crystallinity of MCCI was higher than that of MCCII. Therefore, as expected the less crystalline MCCII presented more water sorption sites due to its large amorphous component as compared to MCCI.

4. Compression studies

Although initially developed for metals, the Heckel analysis is widely used to assess the compressibility of pharmaceutical powders. Table 3 lists the Heckel analysis results. The yield pressure value, P_y, which is obtained from the inverse of the slope of the linear portion of the Heckel curve, refers to the pressure at which the material begins to deform plastically. A plastic deformation implies deformation and sliding of the crystals planes that consolidate the particles. Usually, a plastic deformation causes a minimum change in the surface area available for particle binding. On the other hand, brittle materials require extensive fragmentation for the formation of available surfaces for particle binding. In this case, large particles brake down into smaller particles upon compression. In general, the lower the P_y value, the higher the ductility of the material. In the present study, MCCI showed a low value, whereas MCCII exhibited a high P_y value (~93 MPa and 144 MPa, respectively). Therefore, MCCII is considered as less ductile than MCCI.

The D_0, D_a and D_b parameters, calculated from the Heckel plots, represent the initial packing of the material upon die filling, total packing at low pressures, and the degree of powder bed arrangement due to fragmentation at low pressures, respectively. The D_0 values follow the same trend than the bulk density values suggesting that the polymorphic transformation of MCCI into MCCII increased the ability of this material for packing. On the contrary, the total powder packing at low compression pressures remained unchanged. The D_b value as expected was higher for MCCI due to the less tendency to pack in the powder bed and thus, its particles were more able to rearrange at low compression forces.

Another widely used model for assessing compressibility of powders is the one proposed by Kawakita [17,18]. The "a" parameter indicates that MCCII was more compressible than MCCI. In other words, this material had a larger ability to reduce in volume upon compression. Likewise, the large "b" value for MCCII indicates that a tight packing arrangement in the powder bed generates a large resistant force for volume reduction. However, once consolidation starts, it is easier for MCCII to have a larger compressibility than MCCI.

5. Compactibility studies

Figure 9 shows the relationship between compact tensile strength and compression pressure. MCCI formed stronger compacts than MCCII. The magnitude of this difference appears to increase with increasing compression forces. The area under the curve of the tensile strength is an indication of the material compactibility. This compactibility was 4 times larger for MCCI than for MCCII suggesting this polymorphic transformation had a negative effect on the tensile strength of MCCI compacts. The small compact tensile strength values of MCCII are due to its low plastic deforming ability. These results indicate that the tight molecular arrangement of the chains and the higher plastic deforming ability of MCCI facilitate the formation of hydrogen bonding upon compression resulting in the formation of strong compacts.

Model	Parameter	MCCI	MCCII
	P_y (MPa)	92.6	144
	D_0	0.22	0.36
Heckel	D_a	0.48	0.45
	D_b	0.22	0.10
	r^2	0.9910	0.9850
	A	0.33	0.52
Kawakita	B	0.03	0.12
	r^2	0.9980	0.9999
	AUCTS	885.1	180.6
	Friability (%)	0.11	0.23
Test	SRS (%)	7.3	15.6
	D. Pot. (%)	24	20
	LSR	0.38	0.36

P_y, powder yield pressure; D_0, initial rearrangement as a result of die filling; D_a, total packing at low pressures; Particle rearrangement/fragmentation at early compression stages; a, Compressibility parameter; b, Indicates ease of compression; LSR, lubricant sensitivity ratio; SRS, strain rate sensitivity; D. Pot, dilution potential for acetaminophen; AUCTS, area under the tensile strength curve.

Table 3. Parameters obtained from the Heckel and Kawakita models and other tableting tests

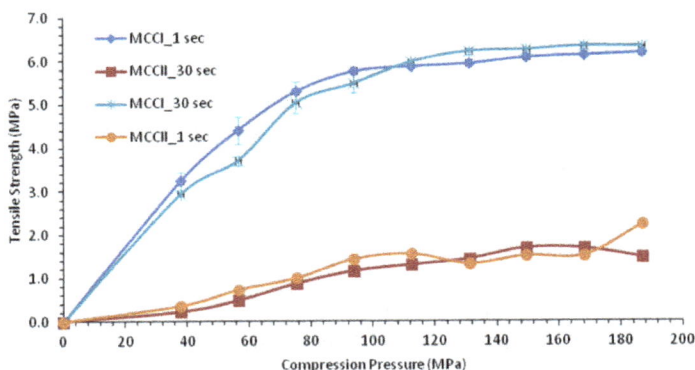

Figure 9. Compact tensile strength of the cellulose allomorphs

5.1. Dilution potential

To assess the effect of a poorly compressible substance on the compactibility of cellulose, compacts containing different weight ratios of the test material and acetaminophen were prepared and their crushing strength was determined. In this case, the crushing strength of the powder mixtures were plotted against the mass fraction of excipient and the resulting straight lines were interpolated to the X-axis to find the dilution potential. The dilution potential of MCCII (20%) was comparable to that of MCCI (24%). These results clearly suggest that MCCII and MCCI serve as effective binders and offer potential to produce tablets with poorly compressible drugs by direct compression. In fact, MCCI showed the highest compactibility but presented a comparable dilution potential for acetaminophen. This indicates that the presence of a poorly compressible drug affected the compactibility of MCCI much more than that of MCCII since a larger dilution potential was expected for MCCI.

5.2. Compact friability

Tablets prepared using 150 MPa compression pressures were tested for friability. All tablets had less than 1% friability indicating an optimum strength for handling and shipping. These results correlate well with the tensile strength results shown in Fig. 7. These findings clearly suggest that MCCI and MCCII serve as effective binders and offer potential to produce compacts with poorly compressible drugs by direct compression.

5.3. Lubricant sensitivity

Lubricant sensitivity was tested with magnesium stearate at a 1.0% w/w level in compacts made at 60 MPa of compression pressure. The presence of a surrounding layer of this hydrophobic lubricant had a large detrimental effect on MCCI since this material was the most highly deforming material. Therefore, MCCI compactibility was reduced to a value comparable to that of MCCII. Moreover, the more fragmenting character of MCCII makes it able to create a large number of binding surfaces which counteract the coating effect exerted by magnesium stearate. In other words, magnesium stearate coats easily the surface of MCCI particles and thereby restricts the contact points between particles resulting in compacts of low strength. Thus, the lubricant film around particles in more fragmenting materials such as MCCII is not complete easing the formation of hard compacts.

5.4. Compact disintegration

Independent of compact porosity, compact disintegration time was less than 20 s for MCCII (Fig. 10). Only compacts of MCCI made at ~40 MPa passed the test requiring disintegration times of less than 30 min. The rapid disintegration times of MCCII are due to its larger swelling value and larger affinity for water as compared to MCCI. Further, it is possible that the low degree of crystallinity of MCCII played an important role in its rapid disintegration. The effect of compact porosity was considered as negligible since all compacts were made at 10-20% porosity. These results indicate that when MCCI is employed as excipient for making tablets, it always requires the addition of a superdisintegrant, which is a material with a large affinity for water that enhances compact disintegration. On the contrary, MCCII does not require that ingredient due to its intrinsic disintegrating properties.

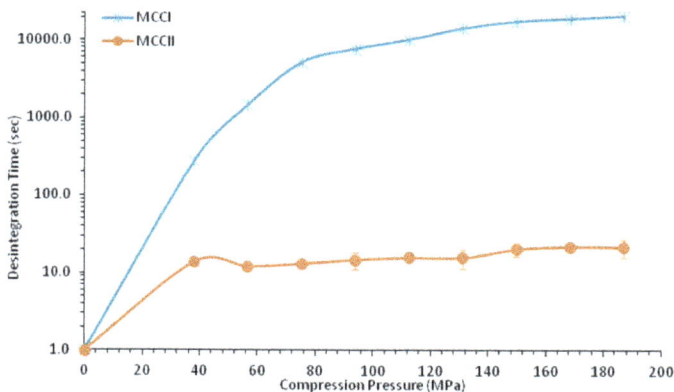

Figure 10. Compact disintegration time of the cellulosic materials.

6. Conclusions

The polymorphic transformation had a major effect on the degree of polymerization, degree of crystallinity, packing tendency, bulk and tap densities, voluminosity, swelling ability and water sorption rate of microcrystalline cellulose. MCCII exhibited a faster disintegration ability, but lower compactibility than MCCI due to its large amorphous component. However, when MCCII was mixed and compressed with acetaminophen it was as compactable as MCCI and possessed a comparable acetaminophen loading capacity and lubricant sensitivity as those of MCCI.

Particle morphology and particle size were not affected, but compact tensile strength was highly affected by the polymorphic transformation. Most of the resulting particle and tableting properties depended on the polymorphic form of MCC. For this reason, it is important to select the right crystalline form of MCC before formulating a drug in a solid dosage form since it could affect the overall particle and tableting properties of the mixture.

Cellulosic excipients are hydrophilic materials and the polymorphic transformation caused differences in the hydrophilic properties of cellulose. Most of the water sorption isotherms exhibited a type II sigmoid shape and MCCII presented the largest water uptake by absorption into the core of the particles.

Cellulosic materials showed a hysteresis loops which were caused by capillary shrinking during the desorption step. The YN model assume sorption and desorption as a dynamic process, in which the primary sorption sites are filled up throughout the whole water activity range and do not require the formation of a complete monolayer to form a multilayer.

Author details

John Rojas

Department of Pharmacy, School of Pharmaceutical Chemistry, The University of Antioquia; Medellín, Colombia

Acknowledgement

The author truly appreciates the sponsorship of the committee for the development of research (CODI) of the University of Antioquia.

7. References

[1] Klemm D, Philipp B, Heinze T, Heinze U (1998a). Comprehensive Cellulose Chemistry: Functionalization of Cellulose. New York, USA: John Wiley. 389 p.

[2] Klemm D, Philipp T, Heinze U, Wagenknecht W, editors (1998b). Comprehensive Cellulose Chemistry: Fundamental and Analytical Methods. John Willey. pp.107-249.

[3] Krassig H (1996). Cellulose, Structure, Accessibility and Reactivity. Gordon and Breach Science Publishers, Amsterdam, Holland. 376 p.

[4] Kroon-Batenburg LMJ & Kroon J (1997). The crystal and molecular structure of cellulose I and II. *Glycocon J.* 14: 677-690.

[5] Blackwell J & Kolpak FJ (1975). The Structure of Regenerated Cellulose. *Macromolecules,* 8: 563-564.

[6] Battista O (1965). Colloidal macromolecular phenomena. *J Polym Sci Pol Sym.* 9:135-155.

[7]   Kolpak FJ & Blackwell, J (1976). Determination of the structure of cellulose II. *Macromolecules* 9, 273-278.

[8]   Battista O & Smith P (1961). Level-off degree of polymerization cellulose products. US Patent No. 2978446.

[9]   Lerk C & Vromans H (1988). Densification properties and compactibility of mixtures of pharmaceutical excipients with and without magnesium stearate. *Int J Pharm.* 46: 183–192.

[10] Moreton R (1996). Tablet excipients to the year 2001: A look into the crystal ball. *Drug Dev Ind Pharm.* 22:11-23.

[11] Westermarck S, Juppo AM, Kervinen L, Yliruusi J (1999). Microcrystalline cellulose and its microstructure in pharmaceutical processing. *Eur J Pharm Biopharm.* 48:199-206.

[12] Kumar V, Reus M, Yang D (2002). Preparation, characterization, and tableting properties of a new cellulose-based pharmaceutical aid. *Int J Pharm.* 235: 129–140.

[13] Kumar V & Reus M (2006). Evaluation of cellulose II powders as a potential multifunctional excipient. *Int J Pharm.* 322: 31–35

[14] Rojas J & Kumar V 2012. Effect of polymorphic form on the functional properties of cellulose: A comparative study. *Carbohyd Polym.* 87: 2223-2230.

[15] Rabek JF 1980. Experimental Methods in Polymer Chemistry. Bristol, UK: John Wiley. 861 p.

[16] Meyer V, editor (2006). Annual Book of American Society of Testing and Materials. Section 6. Paints Related Coatings and Aromatics. West Conshhocken, PA: ASTM International. 403 p.

[17] Kawakita K & Ludde KH (1971). Some considerations on powder compression equations. *Adv Powder Tech.* 4: 61-68.

[18] Yamashiro M, Yuasa Y, Kawakita K (1983). An experimental study on the relationships between compressibility, fluidity and cohesion of powder solids at small tapping numbers. *Powder Technol.* 34:225-231.

[19] Edge S, Steele DF, Staniforth JN, Chen A, Woodcock P (2002a). Powder compaction properties of sodium starch glycolate disintegrants. *Drug Dev Ind Pharm.* 28: 989-999.

[20] Young JH & Nelson GH (1967). Theory of hysteresis between sorption and desorption isotherms in biological materials. *Trans Am. Soc. Agric. Eng.* 10: 260-263.

[21] Nokhodchi A, Ford JL, Rubinstein HL (1997). Studies on the interaction between water and (hydroxypropyl) methyl cellulose. *J Pharm Sci.* 86: 608-615.

[22] Heckel RW (1961a). An analysis of powder compaction phenomena. *Trans Metal Sci AIME* 221:1001-1008.

[23] Heckel RW (1961b). Density-pressure relationship in powder compaction. *Trans Metal Sci AIME* 221:671-675.

[24] Alderborn G & Nyström C (1996). Pharmaceutical powder compaction technology. New York: Marcel Dekker Inc. 615p.

[25] York P (1992). Crystal engineering and particle design for the powder compaction process. *Drug Dev Ind Pharm* 18: 677-721.

[26] Chow YP & Chowhan CT (1980). Compression behavior of pharmaceutical powders. *Int J Pharm.* 5: 139-148.

[27] Fell JT & Newton JM (1968) The tensile strength of lactose tablets. *J Pharm Pharmacol.* 20: 657-758.

[28] Carrillo F, Colom X, Saurina J, Sunol J (2004). Structural FTIR analysis and thermal characterization of lyocel and viscose-type fibers. *Eur Polym J.* 40: 229- 234.

[29] Zhbankov RH (1966). Infrared spectra of cellulose and its derivatives. New York: Consultants Bureau, 333 p.

[30] Iida K, Aoki K, Danjo K, Otsuka A, Chen CY (1997). A comparative evaluation of the mechanical properties of various celluloses. *Chem Pharm Bull.* 45: 217-220.

[31] Suzuki T & Nakagami H (1999). Effect of microcrystalline cellulose on the compactibility and dissolution of tablets. *Eur J Pharm Biopharm.* 47: 225-230.

[32] Kothari SH, Kumar V, Banker GS (2002). Comparative evaluation of powder and mechanical properties of low crystallinity celluloses, microcrystalline celluloses, and powdered celluloses. Int J Pharm. 232:69-80.

[33] Parker ME, Bronlund JE, Mawson AJ (2006). Moisture sorption isotherms for paper and paperboard in food chain conditions. *Packag Technol Sci.*19: 193-209.

[34] Pizzi A, Eaton NJ, Barista M (1987). Theoretical water sorption energies by conformational analysis. *Wood Sci Technol.* 2: 235-48.

[35] Kmetes V & Koskar R (2005). Evaluation of the moisture sorption of several excipients by BET, GAB and microcalorimetric approaches. *Chem Pharm Bull.* 53: 662-665.

[36] Khan F & Pipel N (1986). Water solid interactions II. Effect of moisture sorption and glass transition temperature on compactibility of microcrystalline cellulose alone or in binary mixtures with polyvinyl pyrrolidone *Powder Tecnol.* 48: 145-50.

[37] Kochervitov V, Ulvenlund S, Kober M, Jarring K, Arnebrant T (2008). Hydration of microcrystalline cellulose and milled cellulose studied by sorption calorimetry, J Phys Chem B. 112: 3728-3734.

[38] Hartley DI, Kamke FA, Peemoeller H (1992). Cluster theory for water sorption in wood *Wood Sci Technol.* 26: 83-99.

Chapter 5

Application of Cellulose and Cellulose Derivatives in Pharmaceutical Industries

Javad Shokri and Khosro Adibkia

Additional information is available at the end of the chapter

1. Introduction

Cellulose probably is the most abundant organic compound in the world which mostly produced by plants. It is the most structural component in herbal cells and tissues. Cellulose is a natural long chain polymer that plays an important role in human food cycle indirectly. This polymer has versatile uses in many industries such as veterinary foods, wood and paper, fibers and clothes, cosmetic and pharmaceutical industries as excipient. Cellulose has very semi-synthetic derivatives which is extensively used in pharmaceutical and cosmetic industries. Cellulose ethers and cellulose esters are two main groups of cellulose derivatives with different physicochemical and mechanical properties. These polymers are broadly used in the formulation of dosage forms and healthcare products. These compounds are playing important roles in different types of pharmaceuticals such as extended and delayed release coated dosage forms, extended and controlled release matrices, osmotic drug delivery systems, bioadhesives and mucoadhesives, compression tablets as compressibility enhancers, liquid dosage forms as thickening agents and stabilizers, granules and tablets as binders, semisolid preparations as gelling agents and many other applications. These polymeric materials have also been used as filler, taste masker, free-flowing agents and pressure sensitive adhesives in transdermal patches. Nowadays cellulose and cellulose based polymers have gained agreat popularity in pharmaceutical industries and become more and more important in this field owing to production of the new derivatives and finding new applications for existed compounds by pharmaceutical researchers.

2. Classification of cellulose-based polymers

2.1. Cellulose

Pure cellulose is available in different forms in the market with very different mechanical and pharmaceutical properties. The difference between various forms of cellulose is related to the shape, size and degree of crystallinity of their particles (fibrous or agglomerated). Microcrystalline cellulose (MCC) is the most known cellulose which extensively used in pharmaceutical industries. MCC grades are multifunctional pharmaceutical excipients which can be used as compressibility enhancer, binder in wet and dry granulation processes, thickener and viscosity builder in liquid dosage forms and free-flowing agents in solid dosage forms. Mechanical properties of MCC grades are greatly influenced by their particle size and degree of crystallization. In recent years the new grades of MCC are prepared with improved pharmaceutical characteristics such as silisified MCC (SMCC) and second generation MCC grades or MCC type II (MCC-II). These grades are prepared by co-processing of cellulose with other substances such as colloidal silicon dioxide or by special chemical procedures. Other types of available pure cellulose are powdered cellulose (PC) and low crystallinity powdered cellulose (LCPC).

Regenerated cellulose is one of the other forms of processed cellulose which produced by chemical processing on natural cellulose. In the first step, cellulose dissolves in alkali and carbon disulfide to make a solution called "viscose". Viscose reconverted to cellulose by passing through a bath of dilute sulfuric acid and sodium sulphate. Reconverted cellulose passed through several more baths for sulfur removing, bleaching and adding a plasticizer (glycerin) to form a transparent film called cellophane. Cellophane has several applications in pharmaceutical packaging due to its suitable characteristics such as good compatibility, durability, transparency and elasticity.

2.2. Cellulose ether derivatives

Cellulose ethers are high molecular weight compounds produced by replacing the hydrogen atoms of hydroxyl groups in the anhydroglucose units of cellulose with alkyl or substituted alkyl groups. The commercially important properties of cellulose ethers are determined by their molecular weights, chemical structure and distribution of the substituent groups, degree of substitution and molar substitution (where applicable). These properties generally include solubility, viscosity in solution, surface activity, thermoplastic film characteristics and stability against biodegradation, heat, hydrolysis and oxidation. Viscosity of cellulose ether solutions is directly related with their molecular weights. Examples of mostly used cellulose ethers are: Methyl cellulose (MC), Ethyl cellulose (EC), Hydroxyethyl cellulose (HEC), Hydroxypropyl cellulose (HPC), hydroxypropylmethyl cellulose (HPMC), carboxymethyl cellulose (CMC) and sodium carboxymethyl cellulose (NaCMC).

2.3. Cellulose ester derivatives

Cellulose esters are generally water insoluble polymers with good film forming characteristics. Cellulose esters are widely used in pharmaceutical controlled release preparations such as osmotic and enteric coated drug delivery systems. These polymers are often used with cellulose ethers concurrently for preparation of micro-porous delivery membranes. Cellulose esters categorized in organic and inorganic groups. Organic cellulose esters are more important in pharmaceutical industries. Various types of organic cellulose esters have been used in commercial products or in pharmaceutical investigations such as cellulose acetate (CA), cellulose acetate phthalate (CAP), Cellulose acetate butyrate (CAB), Cellulose acetate trimelitate (CAT), hydroxupropylmethyl cellulose phthalate (HPMCP) and so on (Heinämäki et al., 1994). The most available formulations in market which made by these polymers are enteric coated dosage forms which are usually produced applying acid resistant polymeric coats containing phthalate derivatives of cellulose esters especially cellulose acetate phthalate (Lecomte et al., 2003; Liu & Williams III, 2002). Inorganic cellulose esters such as cellulose nitrate and cellulose sulphate are less important than organic cellulose esters in pharmaceutical industries. Cellulose nitrate or pyroxylin is a transparent compound with good film forming ability but rarely applied alone in pharmaceutical formulations due to its very low solubility in currently used pharmaceutical solvents as well as their very high flammability. The use of pure cellulose nitrate in drug formulations only limited to one topical anti-wart solution named collodion that made with 4%w/v concentration in diethyl ether/ethanol mixture as solvent. Cellulose nitrate/cellulose acetate mixture are also exploited to prepare micro-porous membrane filters used in pharmaceutical industries.

3. Applications of cellulose and its derivatives in pharmaceutical industries

3.1. Application in bioadhesive and mucoadhesive drug delivery systems

Bioadhesives and mucoadhesives are drug containing polymeric films with ability of adhering to biological membranes after combining with moisture or mucus compounds. Bioadhesives were developed in mid 1980s as a new idea in drug delivery and nowadays they have been accepted as a promising strategies to prolong the residence time and to improve specific localization of drug delivery systems on various biological membranes (Lehr, 2000; Grabovac et al., 2005; Movassaghian et al., 2011).

In compared with tablets, these dosage forms have higher patient compliance due to their small size and thickness. Other advantage of these drug delivery systems is their potential to prolong residence time at the site of drug absorption and thus they can reduce the dosing frequency in controlled release drug formulations. These dosage forms can also intensify the contact of their drug contents with underlying mucosal barrier and improve the epithelial transport of drugs across mucus membranes especially in the case if poorly absorbed drugs (Ludwig, 2005; Lehr, 2000). Some special polymers can be used in these formulations with epithelial permeability modulation ability by loosening the tight intercellular junctions. Some of these polymers also can act as proteolytic enzymes inhibitor in orally used adhesive formulations of sensitive drugs (Lehr, 2000).

Bioadhesives considered as novel drug delivery systems. These dosage forms are formulated to use on the skin and mucus membranes of gastrointestinal, ear, nose, eye, rectum and vagina. The main excipients of these formulations are adhesive and film-former polymer(s). Adhesive polymers are synthetic, semi synthetic or natural macromolecules with capablility of attaching to skin or mucosal surfaces. Very different types of polymers have been used as bioadhesive polymers. Synthetic polymers such as acrylic derivatives, carbopols and polycarbophil, natural polymers such as carageenan, pectin, acacia and alginates and semi-synthetic polymers like chitosan and cellulose derivatives are used in bioadhesive formulations (Deshpande et al., 2009; Grabovac et al., 2005). Cellulose derivatives especially cellulose ethers are widely used in bioadhesives. There are used in various types of these formulations such as buccal, ocular, vaginal, nasal and transdermal formulations alone or with combination of other polymers. More recently used cellulose ethers in bioadhesives include nonionic cellulose ethers such as ethyl cellulose (EC), hydroxyethyl cellulose, hydoxypropyl cellulose (HPC), methyl cellulose (MC), carboxymethyl cellulose (CMC) or hydroxylpropylmethyl cellulose (HPMC) and anionic ether derivatives like sodium carboxymethyl cellulose (NaCMC). Ability of polymer to take up water from mucus and pH of target place are important factors determining the adhesive power of polymers. Some bioadhesive polymers such as polyacrylates show very different adhesion ability in various pH values thus the selection of adhesive polymer should be made based on the type of bioadhesive preparation. One advantage of cellulose ethers such as NaCMC and HPC is lesser dependency of adhesion time and adhesion force of them to pH of medium in compared with polyacrylate and thiolated bioadhesive polymers (Grabovac et al., 2005). Cellulose ethers, alone or their mixtures with other polymers, have been studied in oral (Deshpande et al., 2009; Venkatesan et al., 2006), buccal (Perioli et al 2004), ocular (Ludwig, 2005), vaginal (Karasulu et al., 2004) and transdermal (Sensoy et al., 2009) bioadhesives. In some studies, other groups of adhesive polymers or polysaccharides are used with cellulose ethers to improve their adhesion characteristics such as adhesion time and adhesion force. Concurrent use of polyvinyl pyrrolidone (PVP), hydroxypropyl beta cyclodextrin, polycarbophil, carbopol(s), pectin, dextran and mannitol with HPMC, HEC or NaCMC have been reported in the literatures. (Karavas et al., 2006).

3.2. Application in pharmaceutical coating processes

Solid dosage forms such as tablets, pellets, pills, beads, spherules, granules and microcapsules are often coated for different reasons such as protection of sensitive drugs from humidity, oxygen and all of inappropriate environmental conditions, protection against acidic or enzymatic degradation of drugs, odor or taste masking or making site or time specific release characteristics in pharmaceuticals to prepare various modified release drug delivery systems such as sustained release, delayed release, extended release, immediate release, pulsatile release or step-by-step release dosage forms (Barzegar jalali et al., 2007; Gafourian et al., 2007). Both ether and ester derivatives of cellulose are widely used as coating of solid pharmaceuticals. Cellulose ethers are generally hydrophil and convert to hydrogel after exposing to water. Although, some of the cellulose ethers e.g. ethyl cellulose are insoluble in water but majority of them such as methyl, hydroxypropyl and hydroxylpropylmethyl cellulose are water soluble. Both of soluble and insoluble cellulose ethers can absorb water and form a gel. After exposing of these coated dosage forms with water, the coating polymers form to hyrogel and gradually dissolve in water until disappear but the insoluble cellulose ether coatings remain as a viscose gel around tablets and drug release is performed by diffusion of drug molecules within this layer. These two types of dosage forms called dissolution-controlled and diffusion-controlled drug delivery systems, respectively. Despite cellulose ethers, the cellulose esters are generally water insoluble or water soluble in a distinct pH range. These polymers like cellulose acetate (CA), cellulose acetate phthalate (CAP) and cellulose acetate butyrate (CAB) do not form gel in presence of water and they are widely used for preparing of pH sensitive and semi-permeable micro-porous membranes. These membranes are employed for wide variety of controlled release coating of pharmaceuticals especially in enteric or osmotic drug delivery devices. These polymers are benefited to make different cellulosic membrane filters applied in pharmaceutical industries.

3.3. Application in extended release (ER) solid dosage forms

3.3.1. In coated extended release formulations

Extended release pharmaceuticals refer to dosage forms that allow a twofold or greater reduction in frequency of the drug administration in comparison with conventional dosage forms. These formulations can be made as coated or matrix type. Coated ER formulations are generally made with water insoluble polymeric film coating with or without gel-forming ability. The dominant mechanism of drug release in coated ERs is diffusion whereas in matrix type of ERs, erosion of matrix is the main mechanism of drug release. The most used cellulosic polymer in these modified release dosage forms is ethyl cellulose. Ethyl cellulose is completely insoluble in water, glycerin and propylene glycol and soluble in some organic solvents such as ethanol, methanol, toluene, chloroform and methyl acetate. Aqueous dispersions of ethyl cellulose such as Surelease® (Colorcon) or Aquacoat® (FMC BioPolymer) or its organic solutions can be used for coating of extended release formulations. After ingestion of these formulations, an insoluble viscose gel is forming around the tablet which doesn't allow to drug to freely release from dosage form. Drug molecules should pass across this barrier by diffusion mechanism to enter the bulk dissolution medium and thusthe release duration is extended much more than the same uncoated conventional formulation. Larger solid pharmaceuticals like tablets can be coated with rotating pan coaters whereas the smaller types as pills, beads or granules are coated with fluidized bed or air-suspension coater equipments. Because of water insolubility of EC, it is often used in conjunction with water soluble polymers such as MC and HPMC in aqueous coating liquids (Frohoff-Hülsmann et al., 1999a, 1999b). EC solutions in organic solvents such as ethanol can be thickened by HPMC or HPC (Rowe, 1986; Larsson et al., 2010). Water soluble cellulosic polymers with higher amounts can be used as pore former in micro-porous types of extended release and enteric systems. Using of plasticizers is necessary for achieving acceptable coating of pharmaceuticals by these polymers. EC is compatible with commonly used plasticizers such as dibutyl phthalate, diethyl phthalate, dicyclohexyl phthalate, butyl phthalyl butyl glycolate, benzyl phthalate, butyl stearate and castor oil. Other plasticizers such as triacetin, cholecalciferol and α-tocopherol also have been used in EC film coats (Arwidson et al., 1990; Kangarlou et al., 2008). The molecular weights of ECs are in a wide range and different grades of them are existed from 4 to 350 (Colorcon official website). Concentration of 5%w/v from these EC grades in toluene/ethanol mixture at 25°C can produce about 3 to 380 cp viscosity.

3.3.2. In extended release polymeric matrices

Matrices are very simple and efficient systems for controlling drug release from dosage forms. Production of these systems is less time consuming and no needs to special or sophisticated equipments. Majority of ER matrices are made by a simple mixing of drug, polymer(s) and filler followed by one or two stage compaction process. Polymeric matrices as drug delivery systems are very important in developing of modified release dosage forms. In these devices, the drug is dispersed either molecularly or in particulate form within a polymeric network. The main types of drug delivery matrices included swellable and hydrophilic monolithic, erosion controlled and non-erodible matrices (Roy et al., 2002). The use of hydrophilic matrices has become extremely popular in controlling the release rate of drugs from solid dosage forms due to their attractiveness in the case of economic and process development points of view (Conti et al., 2007). During the last two decades, hydrophilic swellable polymers have been widely used for preparation of controlled release matrix tablet formulations. Although various types of rate controlling polymers have been used in hydrophilic matrices, cellulose derivatives especially cellulose ethers are probably the most frequently encountered in pharmaceutical literatures and the most popular polymers in formulation of commercially available oral controlled release matrices. They good compressibility characteristics so they are easily converted to matrices by direct compression technique. In contact with an aqueous liquid, i.e., dissolution medium or gastrointestinal fluid, the hydrophilic polymers present in the matrix swell and a viscose gelatinous layer formed in outer surface of matrix. This layer controls the drug release from matrix. Drug molecules can release out of system by diffusion across this layer. Viscosity of the gel layer is a critical rate-controlling factor in drug release rate from matrices. Erosion of polymeric matrices also can influence the release of the drug from system. Increasing viscosity of the gel, gives rise to increase the resistance against polymer erosion and drug diffusion resulting in reduction of the drug release rate. Various types of cellulose derivatives have been used in formulation of hydrophilic polymeric matrices such as HPMC, NaCMC, CMC, HEC, HPC and EC with different molecular weights (Barzegar-jalali et al., 2010; Javadzadeh et al., 2010; adibkia et al., 2011; Asnaashari et al., 2011). Both of soluble and insoluble cellulose ethers can be used in hydrophilic polymeric matrices due to their hydrophilic nature and ability of them to forming gel in aqueous media. The highest swelling power and hydration rate among cellulose ethers is related to HEC (Saša et al., 2006) but the mostly used cellulose ether is hydrophilic matrices is HPMC due to its excellent swelling properties, good compressibility and fast hydration in contact with water (Ferrero et al., 2008, 2010; Nerurcar et al., 2005). For achieving the good release characteristics, mixtures of various cellulose ethers or mixtures of different grades of a distinct polymer with different ratios can be used based on the intended release rate of controlled release system (Chopra et al., 2007). Some specialized hydrophilic matrices can be made with cellulose ethers for special purposes for example, HPMC matrices with alkalizing buffers like sodium citrate for protection of acid labile drugs have been investigated (Pygall et al., 2009).

3.4. Application in osmotic drug delivery systems

In recent years, considerable attention has been focused on development of novel drug delivery systems (NDDS). Among various NDDS available in the market, oral controlled release (CR) systems hold the major market share because of their advantages over others. Majority of oral CR systems fall in the category of matrices, reservoirs and osmotic devices. Among various types of CR systems, osmotic devices are considered as novel CR systems (J. Shokri et al., 2008a). These formulations utilize osmotic pressure as energy source and driving force for delivery of drugs. Some physiological factors such as pH, presence of food and gastrointestinal motility may affect drug release from conventional CR systems (matrices and reservoirs), whereas, drug release from oral osmotic systems is independent of these factors to a large extent. A classic osmotic device basically consists of an osmotically active core surrounded by a semi-permeable membrane (SPM) and a small orifice drilled through SPM using LASER or mechanical drills. In fact, this system is really a coated tablet with an aperture which acts as drug delivery port (figure 1). This type of devices is called

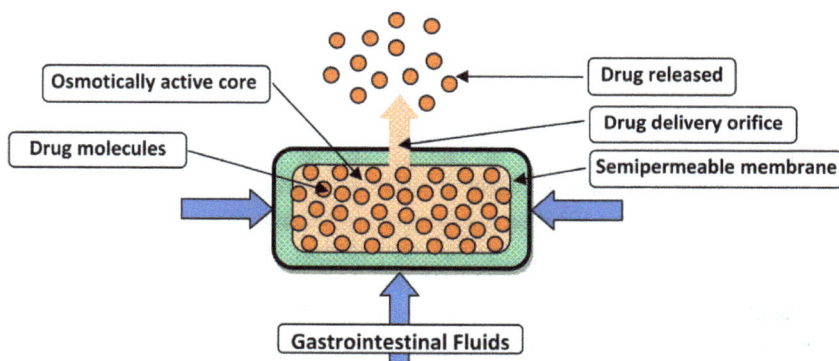

Figure 1. Schematic diagram of an EOP osmotic system.

monolithic or elementary osmotic pumps (EOPs). The more sophisticated osmotic devices have bi-layer (push-pull systems) or tri-layer (sandwich osmotic pumps) cores consisted of an osmotically active drug layer and polymeric layer(s) in one or two sides. Some of osmotic systems called asymmetric membrane or controlled porosity osmotic pumps have not any orifice in their SPM (wang et al., 2005). In these devices, water soluble polymers are used in their SPM as pore formers. Pore formers dissolve after exposing of dosage form to aqueous media and numerous micro pores are created in SPM for drug delivery reason. When an osmotic tablet exposed to an aqueous environment, water pumps from outside into the system due to the great osmotic pressure difference between two sides of SPM. Pumping of water into the system increases the inner hydrostatic pressure leading the saturated drug solution to flow through the small drug delivery orifice or micro pores (in the case of asymmetric membrane devices). Because of high difference of osmotic pressure between two sides of SPM, the osmotic pressure gradient remain constant and thus, the release rate of drug from these devices is almost constant and independent to environmental conditions. EOPs are the most commercially important osmotic devices so that more than 240 patents have been devoted. Procardia XL® and Adalat CR (nifedipine), Acutrium® (phenylpropanolamine), Minipress XL® (prazocine) and Volmax® (salbutamol) are examples of EOPs available in the market (J. Shokri et al., 2008a; Nokhodchi et al., 2008).

3.4.1. In SPM formulation of osmotic systems

As noted earlier, each osmotic delivery system is consisted of two main components included osmotically active core and semi-permeable membrane (SPM). Cellulose acetate (CA) is the mostly used polymer in formulation of SPM in all types of osmotic drug delivery devices. This polymer is the most important cellulose ester derivative with good film forming ability and mechanical characteristics for using in osmotic systems. CA is insoluble in water in both acidic and alkaline conditions. The CA films are only permeable to small molecules such as water while larger molecules like organic drugs can not pass through them. Plasticizers are used in SPM composition for improving the flexibility and mechanical properties of membrane. Various types of plasticizers have been used in formulation of osmotic pharmaceuticals such as castor oil, low and medium molecular weights polyethylenglycols (PEGs), sorbitol, glycerin, propylene glycol, triacetine, ethylene glycol monoacetate, diethyl phthalate, diethyl tartrate and trimethyl phosphate (J. Shokri et al., 2008a, 2008b; Prabakaran et al., 2004; Makhija & Vavia, 2003; Liu et al., 2000a, 2000b; Okimoto et al., 1999). Generally, the mixture of hydrophilic and hydrophobic plasticizers is used for producing the intended drug release characteristics. In controlled porosity osmotic pumps (CPOPs), the additional components such as pore formers are needed. The most efficient pore formers are hydrophilic polymers with high water solubility properties. Water soluble cellulose ether derivatives can be used as pore former in SPM of these devices. Low molecular weight grades of these polymers are suitable for this purpose due to their faster dissolution rate and lower viscosities. Low molecular weight MCs and HPMCs have been used as pore former in CPOP formulations. Central cores are coated with a coating formulation containing SPM components such as film former (CA), pore former(s) and plasticizer(s) dissolved or dispersed in a suitable liquid base. Acetone/ethanol mixtures are generally used as solvent system to dissolve cellulose acetate in coating liquid (J. Shokri et al., 2008a; Nokhodchi et al., 2008; M.H. Shokri et al., 2011). In some studies, cellulose acetate is used as fine particles suspended in an aqueous medium for coating of osmotic cores (Liu et al., 2000b). Ethyl cellulose (EC) and ethylhydroxyl propyl cellulose also have been used as SPM of osmotic devices in some studies but permeability of these membranes is lower than CA membranes. In these formulations, hydrophilic cellulose ether derivatives such as

HPMC have been used for improving SPM permeability (Marucci et al., 2010; Wang et al., 2005; Hjärtstam et al., 1990).

3.4.2. In central core of osmotic systems

Central core of an osmotic pump is generally a simple compressed tablet basically consisted of the active drug(s), osmotically active agent(s), hydrophilic polymer(s) and other commonly used ingredients such as filler, compressibility enhancer, free flowing agent and lubricant. In one compartment devices (EOPs and controlled porosity OP), these polymers mixed with other ingredients and compressed to a tablet whereas in two layered (Push-Pull systems), or tri layered (Sandwich systems) cores, these polymers compressed in one or two separated layer in one or both sides of drug layer (J. Shokri et al., 2008b; Kumaravelrajan et al., 2010). These polymers should have high water uptake and swelling capacity. Cellulose derivatives play an important role in core formulations of osmotic devices. Water soluble cellulose ethers commonly used as core polymers due to their hydrophilicity and good swelling properties. Most currently used polymers for this purpose are MC, HEC, HPC and HPMC with various molecular weights. After exposing of system to water, water move into the system due to great osmotic pressure difference between outer and inner part of device. This water is imbibed to polymer(s) and causes swelling of them. Swelling of core polymer(s) produce the driving force for ejecting the drug solution from drug release orifice with constant rate (J. Shokri et al., 2008a, 2008b; Prabakaran et al., 2004; Makhija & Vavia, 2003; Liu et al., 2000a, 2000b). Among cellulose ethers, different grades of HPMC have been used more than others in core formulation. Microcrystalline cellulose (MCC) has also frequently used the core formulations as compressibility enhancer. MCC is one of the most compressibility enhancers that widely used in direct compression as well as wet granulation techniques for preparing various types of tablets, pellets and pills.

3.5. Application in enteric coated solid dosage form

Enteric coated solid dosage forms are the main groups of delayed release drug delivery systems which designed for releasing of their drug(s) content in the lower parts of gastrointestinal tract such as small intestine and colon. Enteric dosage forms can be considered as a type of oral site specific pharmaceuticals that initiate drug release after passing from stomach. Enteric oral dosage forms are suitable for formulation of acid-labile drugs or drugs with irritancy potential for inner protective layer of stomach such as non-steroidal anti inflammatory drugs (NSAIDs). The commonly used materials in enteric coated formulations are pH-dependent polymers containing carboxylic acid groups. These polymers remain un-ionized in low pH conditions like environment of stomach and become ionized with increasing of pH toward natural and light alkaline zone similar to the small intestine condition (Liu et al., 2011). These polymers also should have the good film forming properties to produce smooth coats with good integrity. Various polymers have been used for production of enteric coated dosage forms such as Eudragit® polymers and pH-dependent cellulose derivatives. Cellulose derivatives which commonly used as enteric coating polymers include cellulose acetate phthalate (CAP), cellulose acetate trimelitate (CAT), hydroxypropylmethyl cellulose phthalate (HPMCP), carboxymethylethyl cellulose (CMEC) and hydroxypropylmethyl cellulose acetate succinate (HPMCAP) (Williams III & Liu, 2000). Apart from the main enteric polymer, the type and amount o plasticizer(s) is very important for achieving uniform, smooth and resistant enteric films. Some of mostly used plasticizers in enteric coated formulations are diethyl phthalate, glyceryl triacetate, glyceryl monocaprylate and triethyl citrate (Williams III & Liu, 2000; Gosh et al., 2011). In some cases, hydrophilic cellulose ether derivatives especially HPMC are used with enteric polymer for improving the film forming and plasticity of main enteric polymer. HPMC is also used in enteric coating process as pre-coating or sub-coating polymer due to its very good film forming properties and suitable polymer-to-polymer adhesion with enteric coating polymers especially with cellulose ester derivatives such as CAP, HPMCP, HPMCAS, CMEC and CAT (Williams III & Liu, 2000). Three commercially available enteric coating preparations included solid forms of enteric polymers which should be dissolved in suitable organic solvent mixture before coating process, ready-to-use organic enteric coating solutions and aqueous polymeric dispersions. Aqueous nanodispersions of enteric coating polymers such as HPMCP have also been investigated for improving physicochemical and mechanical characteristics of coating (Kim et al., 2003).

3.6. Application as compressibility enhancers

More than 80 percent of all dosage forms available or administered to man are tablets. The main reason of this great popularity is the advantages of tablets over other forms of pharmaceuticals. Ease of manufacturing, convenience dosing and greater stability in compared with liquid or semisolid dosage forms are some of these advantages. Two common ways for tablet manufacturing are compression and molding. Except of a few cases, tablets are made by compression technique. The simplest and fastest kind of compression is named direct compression method in which the drug and all of other excipients are mixed and compressed in one-stage process with proper compression force to form tablet. This method commonly used for tabletting of medium to high potency drugs where the drug content in them is less than 30% of formulation (Jivari et al., 2000). In the other cases with higher amounts of low compactable drugs, dry or wet granulation techniques are used for preparing tablets. In dry granulation method, compression of ingredients are performed in two or multi-stage process to improve compressibility of the ingredients. Slugging and roller compaction techniques used for initial compression of powder mixtures before final tabletting process.

One of the common difficulties in direct compression and dry granulation is low compactability of the drug content especially when the drug amount is higher than 30% of formulation. In these cases, an efficient compressibility enhancer can help to achieving a good tablet with pharmaceutically accepted characteristics. Although, all of the cellulose based polymers are good compactable, however special grades of microcrystalline cellulose exhibit excellent compatibility. These grades can significantly improve compressibility of low compactable powder mixtures so they are widely used as compressibility enhancers in tablet manufacturing by direct compression and dry granulation methods. Various grades of MCC have different fundamental properties including their morphology, particle size, surface area, porosity and density (Rojas & Kumar, 2011). These physicochemical properties poses the different characteristics to them for example, smaller particle size MCC grades have good compressibility and poor flowability whereas the larger particle size grades have poor compressibility and excellent flowability. Particle size of MCC varies from 20 to 270 micrometer based on the manufacturer and type of application. MCC is available in three public brand names including Avicel® (FMC BioPolymer), VIVAPUR®/EMCOCEL® (JRS Pharma) and TABULOSE® (Blanver). Various grades of commercially available Avicel® and their particle size are shown in table 1 (Colorcon official website).

brand name	Application	MCC grade	Particle size
Avicel®	Roller compaction	DG	45
	Wet granulation	PH-101	50
	Direct compression	PH-102	100
		HFE-102	100
	Superior compactability	PH-105	20
	Superior Flowability	PH-102 SCG	150
		PH-200	180
	High Density	PH-301	50
		PH-302	100
	Low Humidity	PH-103	50
		PH-113	50
		PH-112	100
		PH-200LM	180
	Mouthfeel improvement	CE-15	75

Table 1. Various grenades of Avicel®

The effects of size, shape and porosity of MCC particles on flowability and compatibility have also been investigated by several researchers (Johansson et al., 2001). Various types of MCCs are extensively used in direct compression and dry granulation methods especially in roller compaction for preparing compressed tablets or pellets (Strydom et al., 2011; Bultmann, 2002). Microcrystalline cellulose type II (MCC-II) was recently introduced as new pharmaceutical excipients. MCC-II has a fibrous structure with lower compactability than MCC grades and suitable for using in rapid disintegrating dosage forms (Rojas et al., 2011; Reus-Medina & Kumar, 2006). In recent years, the new methods have been established for improving mechanical characteristics of MCCs. One of these innovative methods is lubricating or silisifying for improving compactability of low compressible grades of MCCs such as MCC-II or large particle size MCC grades. In this method, amorphous silicon dioxide ($SiO2$) is used as companion excipient for co-processing with low compressible MCC grades. Cellulose/SiO2 ratio is 98:2 and resulted product is called lubricated or silisified microcrystalline cellulose (SMCC). This method can be used for both MCC-I or MCC-II for production SMCC-I or SMCC-II (Rojas & Kumar, 2011; Van Veen et al., 2005). SMCC-I have excellent compaction properties and less stickiness to the lower punches over MCC-I or MCC-I/SiO2 physical mixtures (Rojas & Kumar, 2011). SMCC-II has also better mechanical properties especially higher compactability than MCC-II without detriment if it's self-disintegrating characteristics. SMCC-I grades are commercially available under trade name of ProSolv® (JRS Pharma) but SMCC-II is not commercialized yet. Apart from MCC, other forms of cellulose are existed such as powdered cellulose (PC) and low cristallinity powdered cellulose (LCPC). LCPC and MCC have agglomerated and PC has fibrous structure. PC applications in pharmaceutical industries is similar that MCC. It is widely used in direct compression formulation and in dry granulation by either slugging or roller compaction methods. LCPC is a new direct compression cellulose excipient which is prepared by controlled decrystallization and depolymerization of cellulose with phosphoric acid (Rojas & Kumar, 2011). LCPC was shown superior tabletting properties than direct compression grades of MCC like Avicel®PH-101 (Kothari et al., 2002).

3.7. Application as gelling agents

Gels are semisolid systems consisting of dispersions of very small particles or large molecules in an aqueous liquid vehicle rendered jellylike by the addition of a gelling agent. In recent decades, synthetic and semi-synthetic macromolecules are mostly used as gelling agents in pharmaceutical dosage forms. Some of these agents include: carbomers, cellulose derivatives and natural gums. Cellulose derivatives such as HPMC and CMC are the most popular gelling agents used in drug formulations. These polymers are less sensitive for microbial contamination than natural gelling agents such as tragacanth, acacia, sodium algininate, agar, pectin and gelatin. Cellulose derivatives generally dissolve better in hot water (except MC grades) and their mechanisms of jellification is thermal. For preparing gel, powder of these polymers with suitable amount initially dispersed in cold water by using mechanical mixture and then, the dispersion is heated to about 60-80°C and gradually cooled to normal room temperature to form a gel (except MC grades). The resulted gels from these polymers are single-phase gels. Adding of electrolytes in the low concentrations increase the viscosity of these gels by salting out mechanism and higher concentrations (above 3-4%) can precipitate the polymer and breakup the gel system (Allen, et al., 1995). Maximum stability and transparency of the gels prepared by these polymers is about neutral range (pH= 7-9) and acidic pHs can precipitate them from gel system. Minimum gel-forming concentrations of cellulose derivatives are different based on the type and the molecular weights of them but the medium range is about 4-6%w/v. The type of cellulose derivative in pharmaceutical gels can significantly affect drug release from gel formulations (Tas, et al., 2003). These gels also can be used as the base of novel drug delivery systems such as liposomal formulations (Gupta, et al., 2012).

3.8. Application as thickening and stabilizing agents

Cellulose derivatives are extensively used for thickening of pharmaceutical solutions and disperse systems such as emulsions and suspensions (Adibkia et al., 2007a, 2007b). Furthermore, these polymers can increase viscosity of non-aqueous pharmaceutical solution likes organic-based coating solutions. Viscosity enhancing of drug solutions poses many advantages such as improving consuming controllability and increasing residence time of drugs in topical and mucosal solutions which lead to improve bioavailability of topical, nasal or ocular preparations (Grove et al., 1990; Adibkia et al., 2007a, 2007b). It has been revealed that viscosity enhancement, in some cases, can increase absorption of some poorly-absorb drugs like insulin from oral dosage forms (Mesiha, M. & Sidhom, M.). Cellulose ethers in concentrations lower than minimum gel-forming amounts are used as thickening agents or viscosity builder. These polymers play an important role in stabilizing of pharmaceutical disperse systems especially in suspensions and coarse emulsions. By increasing the viscosity of suspension, based on the stock's equation, the sedimentation rate of dispersant decreased and thus, the uniformity of dispersion after shaking of product will improve. In the case of emulsions, these polymers can increase the shelf life and their resistance against mechanical and thermal shocks. Among cellulose derivatives, cellulose ethers especially their higher molecular weight grades are more suitable for using as viscosity enhancer and stabilizer for liquid pharmaceutical disperse systems such as suspensions and emulsions. There is a direct proportionality between viscosity of cellulose ether solutions and molecular weights of them.

3.9. Application as fillers in solid dosage forms

Cellulose and related polymers are commonly used in solid dosage forms like tablets and capsules as filler. Various forms of cellulose have been used in pharmaceutical preparations as multifunctional ingredients thus; they are concerned as precious excipients for formulation of solid dosage forms. Cellulose and its derivatives have many advantages in using as filler in solid pharmaceuticals such as their compatibility with the most of other excipients, pharmacologically inert nature and indigestibility by human gastrointestinal enzymes. These polymers do not cause any irritancy potential on stomach and esophagus protective mucosa. Various forms of pure cellulose and cellulose ether derivatives can be used as filler in these formulations.

3.10. Application as binders in granulation process

Binders are the essential components of solid drug formulations made by wet granulation process. In wet granulation process, drug substance is combined with other excipients and processed with the use of a solvent (aqueous or organic) with subsequent drying and milling to produce granules. Cellulose and some derivatives have excellent binding effects in wet granulation process. A number of MCC grades such as PH-101 are widely used as binder in wet granulation. Other cellulose derivatives such as MC, HPMC and HPC have good binding properties in wet granulation. Low substituted cellulose ethers such as low substituted HPC (L-HPC) also used as binder in wet granulation process (Desai et al., 2006; Wan & Prasad, 1988). Even though, low substituted cellulose ethers have lower water solubility compared with normal grades, however they have very good binding efficacy. Cross-linked cellulose (CLC) and cross-linked cellulose derivatives such as cross-linked NaCMC can be used as excellent binders in pharmaceuticals as well (Chebli & Cartilier, 1998).

3.11. Application as disintegrating agents

Solid oral dosage forms such as tablets undergo several steps before systemic absorption of the drug. Disintegration is the first step immediately after administration of oral dosage forms that breakup the dosage forms into the smaller fragments in an aqueous environment. Converting of solid dosage forms to smaller fragments, increase the available surface area and promote a more rapid release of the drug substances from dosage forms. The earliest known disintegrant is Starch. Corn Starch or Potato Starch was recognized as being the ingredient in tablet formulations responsible for disintegration as early as 1906. Due to low compressibility of starch, pre-gelatinized starch was invented for using as disintegrant. Pre-gelatinized starch and MCC are two main types of classic disintegrants. In recent years, the classic disintegrnts have been gradually replaced with the newer ones called super disintegrants. Super disintegrants can acts in lower concentrations than starch and have not detriment effect on compressibility and flowability of formulations. Three main groups of these excipients are: modified starches like sodium starch glycolate (Primogel®, Explotab®) with 4-6% effective concentration, cross-linked polyvinyl pyrrolidones like crospovidone (Polyplasdone XL, Kollidon CL) with 2-4% effective concentration and modifies cellulose like cross-linked sodium carboxymethyl cellulose or croscarmellose (Ac-Di-Sol™ and Nymcel) with 2-4% effective concentration in wet granulation process. Modified cellulose compounds are very efficient disintegrants and additionally, can accelerate the dissolution rate of drugs in aqueous environment (Chebli & Cartilier, 1998).

3.12. Application as taste masking agents

There are numerous drugs with unfavorable tastes. The most prevalent unpleasant taste of the drugs is bitter taste. Unpleasant-tasting dosage forms leads to lack of patient compliance of oral drug preparations. Various tastes are feeling by taste buds on the tongue. Taste buds are onion-shaped structures containing between 50 to 100 taste cells. Chemicals from food or oral ingested medicine are dissolved by the saliva and enter via the taste pore. They either interact with surface proteins known as taste receptors or with pore-like proteins called ion channels. These interactions cause electrical changes within the taste cells that trigger them to send chemical signals that translate into neurotransmission to the brain. Salt and sour responses are of the ion channel type of responses, while sweet and bitter are surface protein responses.

Taste masking is an important consideration in formulation of oral dosage forms especially in the case of high dose, poorly tasting drugs. Improving the taste of liquid dosage forms is more important because of better sensitivity and faster stimulation of taste receptors by liquids in compared than solids. Taste masking in solid dosage forms can be performed by coating (in the case of tablets, pellets, pills or coarse granules) or micro-coating (in the case of fine granules, powders or microcapsules) of them by a gastro-soluble polymeric coating. These coats can prevent from contacting of the drug with taste buds without detriment of release characteristics of the drug formulations in gastrointestinal tract. Soluble cellulose ether derivatives are suitable for this purpose. These polymers like HPMC, HEC, MC and

HPC are completely water soluble and they have very good film forming properties. Some grades of MCC also can improve tooth-feel such as Avicel® CE-15. These coats can produce additional benefits in drug formulations such as protection of the active ingredients against moisture, oxygen of the air and light due to their barrier effects. Masking of the taste in liquid dosage forms especially in drug solutions is more sophisticated. In these cases test receptor blockers, flavoring agents and viscosity enhancers are simultaneously needed.

Author details

Javad Shokri and Khosro Adibkia

Faculty of Pharmacy, Tabriz University of Medical Sciences, Tabriz, Iran

4. References

Adibkia, k.; Hamedeyazdan, S.; Javadzadeh, Y. (2011). Drug release kinetics and physicochemical characteristics of floating drug delivery systems. *Expert Opinion on Drug Delivery,* Volume 8, Issue 7 (July), Pages 891-903.

Adibkia, K.; Omidi, Y.; Siahi Shadbad, MR.; Nokhodchi, A.; Javadzedeh, A.; Barzegar-Jalali, M.; Barar, J.; Mohammadi, G. (2007b). Inhibition of endotoxin-induced uveitis by methylprednisolone acetate nanosuspension in rabbits. *Journal of Ocular Pharmacology and Therapeutics,* Volume 23, Issue 5 (May), Pages 421-432.

Adibkia, K.; Siahi Shadbad, MR.; Nokhodchi, A.; Javadzedeh, A.; Barzegar-Jalali, M.; Barar, J.; Mohammadi, G.; Omidi, Y. (2007a). Piroxicam nanoparticles for ocular delivery: Physicochemical characterization and implementation in endotoxin-induced uveitis. *Journal of Drug Targeting,* Volume 15, Issue 6 (June), Pages 407-416.

Allen, L.V.; Popovich, N.G.; Ansel, H.C. (1995). Ansels Pharmaceutical Dosage Forms and Drug Delivery ayatems (8th Edition), Lippincott Williams & Wilkins, ISBN: 0-7817-4612-4, United States of America

Arwidsson, H. & Nicklasson, M. (1990). Application of intrinsic viscosity and interaction constant as a formulation tool for film coating II. Studies on different grades of ethyl cellulose in organic solvent systems. *International Journal of Pharmaceutics,* Volume 58, Issue 1 (15 January), Pages 73-77.

Asnaashari, S.; Khoei, NS.; Zarrintan, MH.; Adibkia, K.; Javadzadeh, Y. (2011). Preparation and evaluation of novel metronidazole sustained release and floating matrix tablets. *Pharmaceutical Development and Technology,* Volume 16, Issue 4 (April), Pages 400-407.

Barzegar-Jalali, M.; Valizadeha, H.; Dastmalchi, S.; Siahi Shadbad, MR.; Barzegar-Jalal, A.; Adibkia, K.; Mohammadi, G. (2007). Enhancing dissolution rate of carbamazepine via cogrinding with crospovidone and hydroxy propyl methyl cellulose. *Iranian Journal of Pharmaceutical Research,* Volume 6, Issue 3 (March), Pages 159-165.

Barzegar-Jalali, M.; Valizadeha, H.; Siahi Shadbad, MR.; Adibkia, K.; Mohammadi, G.; Farahani, A.; Arash, Z.; Nokhodchi, A. (2010). Cogrinding as an approach to enhance dissolution rate of a poorly water-soluble drug (gliclazide). *Powder Technology,* Volume 197, Issue 3 (March), Pages 150-158.

Bultmann, J.M. (2002). Multiple compaction of microcrystalline cellulose in a roller compactor. *European Journal of Pharmaceutics and Biopharmaceutics*, Volume 54, pages 59–64.

Chebli, C. & Cartilier, R. (1998). Cross-linked cellulose as a tablet excipient: A binding/disintegrating agent. *International Journal of Pharmaceutics*, Volume 171, Pages 101–110.

Chopra, Sh.; Patil, G.V.; Motwani, S.K. (2007). Release modulating hydrophilic matrix systems of losartan potassium: Optimization of formulation using statistical experimental design. *European Journal of Pharmaceutics and Biopharmaceutics*, Volume 66, Issue 1 (April), Pages 73-82.

Colorcon   official   website, http//fmcbiopolymer.com/pharmaceutical/product/Avicelforsoliddosageforms/ASPX

Conti, S.; Maggi, L.; Segale, L.; Machiste, E.O.; Conte, U.; Grenier, P.; Vergnault, G. (2007). Matrices containing NaCMC and HPMC: 1. Dissolution performance characterization. *International Journal of Pharmaceutics*, Volume 333, Issues 1–2 (21 March), Pages 136-142.

Desai, D.; Rinaldi,F.; Kothari, S.; Paruchuri, S.; Li, M.; Lai, D.; Fung, S.; Both, D. (2006). Effect of hydroxypropyl cellulose (HPC) on dissolution rate of hydrochlorothiazide tablets *International Journal of Pharmaceutics*, Volume 308, Pages 40–45.

Deshpande, M.C.; Venkateswarlu, V.; Babu, R.K.; Trivedi, R.K. (2009). Design and evaluation of oral bioadhesive controlled release formulations of miglitol, intended for prolonged inhibition of intestinal α-glucosidases and enhancement of plasma glucagon like peptide-1 levels. *International Journal of Pharmaceutics*, Volume 380, Issues 1–2, (1 October), Pages 16-24.

Ferrero, C.; Massuelle, D.; Doelker, E. (2010). Towards elucidation of the drug release mechanism from compressed hydrophilic matrices made of cellulose ethers. II. Evaluation of a possible swelling-controlled drug release mechanism using dimensionless analysis. *Journal of Controlled Release*, Volume 141, Issue 2 (25 January), Pages 223-233.

Ferrero, C.; Massuelle, D.; Jeannerat, D.; Doelker, E. (2008). Towards elucidation of the drug release mechanism from compressed hydrophilic matrices made of cellulose ethers. I. Pulse-field-gradient spin-echo NMR study of sodium salicylate diffusivity in swollen hydrogels with respect to polymer matrix physical structure. *Journal of Controlled Release*, Volume 128, Issue 1 (22 May), Pages 71-79.

Frohoff-Hülsmann, M.A.; Lippold, B.C.; McGinity, J.W. (1999a). Aqueous ethyl cellulose dispersion containing plasticizers of different water solubility and hydroxypropyl methyl- cellulose as coating material for diffusion pellets II: properties of sprayed films. *European Journal of Pharmaceutics and Biopharmaceutics*, Volume 48, Issue 1 (1 July), Pages 67-75.

Frohoff-Hülsmann, M.A.; Schmitz, A.; Lippold, B.C. (1999b). Aqueous ethyl cellulose dispersions containing plasticizers of different water solubility and hydroxypropyl methylcellulose as coating material for diffusion pellets: I. Drug release rates from coated pellets. *International Journal of Pharmaceutics*, Volume 177, Issue 1 (15 January), Pages 69-82.

Ghosh, I.; Snyder, J.; Vippagunta, R.; Alvine, A.; Vakil, R.; Tong W-Q,; Vippagunta, S. (2011). Comparison of HPMC based polymers performance as carriers for manufacture of solid dispersions using the melt extruder. *International Journal of Pharmaceutics*, Volume 419, Issues 1–2, (31 October), Pages 12-19.

Grabovac, V.; Guggi, D.; Bernkop-Schnurch, A. (2005). Comparison of the mucoadhesive properties of various polymers. *Advanced Drug Delivery Reviews*. Volume 57, Pages 1713– 1723.

Grove, J.; Durr, M.; Quint, M-P.; Plazonnet. B. (1990). The effect of vehicle viscosity on the ocular bioavailability of L-653,328. *International Journal of Pharmaceutics*, Volume 66, Issues 1–3, (December 1990), Pages 23-28

Gupta, P.N.; Pattani, A.; Curran, R.M.; Kett, V.L.; Andrews, G.P.; Morrow, R.J.; Woolfson, A.D.; Malcolm, R.K. (2012). Development of liposome gel based formulations for ntravaginal delivery of the recombinant HIV-1 envelope protein CN54gp140, *European Journal of Pharmaceutical Sciences*, In Press, Corrected Proof, Available online 14 February 2012.

Heinämäki, J.T.; Iraizoz Colarte, A.; Nordström, A.J.; Yliruusi, J.K. (1994). Comparative evaluation of ammoniated aqueous and organic-solvent-based celluloseester enteric coating systems: a study on free films. *International Journal of Pharmaceutics*, Volume 109, Issue 1 (22 August), Pages 9-16.

Hjärtstam, J.; Borg, K.; Lindstedt, B. (1990). The effect of tensile stress on permeability of free films of ethyl cellulose containing hydroxypropyl methylcellulose. *International Journal of Pharmaceutics*, Volume 61, Issues 1–2 (11 June), Pages 101-107.

Javadzadeh, Y.; Hamedeyazdan, S.; Adibkia, K.; Kiafar, F.; Zarrintan, MH.; Barzegar-Jalali, M. (2010). Evaluation of drug release kinetics and physicochemical characteristics of metronidazole floating beads based on calcium silicate and gas forming agents. *Pharmaceutical Development and Technology*, Volume 15, Issue 4 (April), Pages 329–338.

Jivraj, M.; Martini, L.G.; Thomson, C.M. (2000). An overview of the different excipients useful for the direct compression of tablets. *Pharmaceutical Science & Technology Today*, Volume 3, Issue 2, (1 February), Pages 58-63.

Johansson, B. & Alderborn, G. (2001). The effect of shape and porosity on the compression behavior and tablet forming ability of granular materials formed from microcrystalline cellulose, *European Journal of Pharmaceutics and Biopharmaceutics*, Volume 52, Issue 3 (November), Pages 347-357.

Kangarlou, S.; Haririan, I.; Gholipour, Y. (2008). Physico-mechanical analysis of free ethyl cellulose films comprised with novelplasticizers of vitamin resources. *International Journal of Pharmaceutics*, Volume 356, Issues 1–2 (22 May), Pages 153-166.

Karasulu, Y.H.; Hilmioğlu, S.; Metin, D.Y.; Güneri T. (2004). Efficacy of a new ketoconazole bioadhesive vaginal tablet on Candida albicans.*Il Farmaco*, Volume 59, Issue 2 (February), Pages 163-167.

Karavas, E.; Georgarakis, E.; Bikiaris, D. (2006). Application of PVP/HPMC miscible blends with enhanced mucoadhesive properties for adjusting drug release in predictable pulsatile chronotherapeutics. *European Journal of Pharmaceutics and Biopharmaceutics*, Volume 64, Issue 1 (August), Pages 115-126.

Kim, H.; Park, J.H.; Cheong, I.W.; Kim, J.H. (2003). Swelling and drug release behavior of tablets coated with aqueous hydroxypropyl methylcellulose phthalate (HPMCP) nanoparticles. *Journal of Controlled Release*, Volume 89, Issue 2 (29 April), Pages 225-233.

Kothari, S.H.; Kumar, V.; Banker, G.S. (2002). Comparative evaluations of powder and mechanical properties of low crystallinity celluloses, microcrystalline celluloses, and powdered celluloses. *International Journal of Pharmaceutics*, Volume 232, Issues 1–2 (31 January), Pages 69-80.

Kumaravelrajan, R.; Narayanan, N.; Suba, V.; Bhaskar, K. (2010). Simultaneous delivery of Nifedipine and Metoprolol tartarate using sandwiched osmoticpump tablet system. *International Journal of Pharmaceutics*, Volume 399, Issues 1–2 (31 October), Pages 60-70.

Larsson, M.; Hjärtstam, J.; Berndtsson, J.; Stading, M.; Larsson, A. (2010). Effect of ethanol on the water permeability of controlled release films composed of ethyl cellulose and hydroxypropyl cellulose. *European Journal of Pharmaceutics and Biopharmaceutics*, Volume 76, Issue 3 (November), Pages 428-432.

Lecomte, F.; Siepmann, J.; Walther, M.; MacRae, R.J.; Bodmeier, R. (2003). Blends of enteric and GIT-insoluble polymers used for film coating: physicochemical characterization and drug release patterns. *Journal of Controlled Release*, Volume 89, Issue 3 (20 May), Pages 457-471.

Lehr, Claus-Michael. (2000). Lectin-mediated drug delivery: The second generation of bioadhesives *Journal of Controlled Release*. 65, Pages 19–29.

Liu, F.; Merchant, H.A.; Kulkarni, R.P.; Alkademi, M.; Basit, A.W. (2011). Evolution of a physiological pH 6.8 bicarbonate buffer system: Application to the dissolution testing of enteric coated products. *European Journal of Pharmaceutics and Biopharmaceutics*, Volume 78, Issue 1 (May), Pages 151-157.

Liu, J. & Williams, R.O. (2002). Long-term stability of heat–humidity cured cellulose acetate phthalate coatedbeads. *European Journal of Pharmaceut cs and Biopharmaceutics*, Volume 53, Issue 2 (March), Pages 167-173.

Liu, L.; Khang, G.; Rhee, J.M.; Lee, H.B. (2000a). Monolithic osmotic tablet system for nifedipine delivery. *Journal of Controlled Release, Volume 67, Issues 2–3 (3 July), Pages 309-322.*

Liu, L.; Ku, J.; Khang, G.; Lee, B.; Rhee, J.M.; Lee, H.B. (2000b). Nifedipine controlled delivery by sandwiched osmotic tabletsystem. *Journal of Controlled Release*, Volume 68, Issue 2 (10 August), Pages 145-156.

Ludwig, Annick.; (2005). The use of mucoadhesive polymers in ocular drug delivery. *Advanced Drug Delivery Reviews*. Volume 57, Pages 1595– 1639.

Makhija S.N. & Vavia, P.R. (2003). Controlled porosity osmotic pump-based controlled release systems ofpseudoephedrine: I. Cellulose acetate as a semipermeable membrane. *Journal of Controlled Release*, Volume 89, Issue 1 (14 Apri), Pages 5-18.

Marucci, M.; Ragnarsson, G.; Nilsson, B.; Axelsson, A. (2010). Osmotic pumping release from ethyl–hydroxypropyl– cellulose -coated pellets: A new mechanistic model. *Journal of Controlled Release*, Volume 142, Issue 1 (25 February), Pages 53-60.

Mesiha, M.; Sidhom, M. (1995). Increased oral absorption enhancement of insulin by medium viscosity hydroxypropyl cellulose. *International Journal of Pharmaceutics*, Volume 114, Issue 2, (14 February), Pages 137-140.

Movassaghian, S.; Barzegar-Jalali, M.; Alaeddini, M.; Hamedyazdan, S.; Afzalifar, R.; Zakeri-Milani, P.; Mohammadi, G.; Adibkia K. (2011). Development of amitriptyline buccoadhesive tablets in management of pain in dental procedures. *Drug Development and Industrial Pharmacy*, Volume 37, Issue 7 (June), Pages 1-12.

Nerurkar, J.; Jun, H.W.; Price, J.C.; Park, M.O. (2005). Controlled-release matrix tablets of ibuprofen using cellulose ethers and carrageenans: effect of formulation factors on dissolution rates. *European Journal of Pharmaceutics and Biopharmaceutics*, Volume 61, Issues 1–2 (September), Pages 56-68.

Nokhodchi, A.; Momin, M.N.; Shokri, J.; Shahsavari, M.; Rashidi, P.A. (2008). Factors Affecting the Release of Nifedipine from a Swellable Elementary Osmotic Pump. *Drug Delivery*, volume 15, Pages 43–48.

Okimoto, K.; Ohike, A.; Ibuki, R.; Aoki, O.; Ohnishi, N.; Rajewski, R.A.; Stella, V.J.; Irie, T.; Uekama, K. (1999). Factors affecting membrane-controlled drug release for an osmotic pump tablet (OPT) utilizing (SBE)7m-β-CD as both a solubilizer and osmotic agent. *Journal of Controlled Release*, Volume 60, Issues 2–3 (5 August), Pages 311-319.

Perioli, L.; Ambrogi, V.; Rubini, D.; Giovagnoli, S.; Ricci, M.; Blasi, P.; Rossi, C. (2004). Novel mucoadhesive buccal formulation containing metronidazole for the treatment of periodontal disease. *Journal of Controlled Release*, Volume 95, Issue 3, 24 (March), Pages 521-533.

Prabakaran, D.; Singh, P.; Kanaujia, P.; Jaganathan, K.S.; Rawat, A.; Vyas, S.P. (2004). Modified push–pull osmotic system for simultaneous delivery of theophylline andsalbutamol: development and in vitro characterization. *International Journal of Pharmaceutics*, Volume 284, Issues 1–2, (13 October), Pages 95-108.

Pygall, S.R.; Kujawinski, S.; Timmins, P.; Melia, C.D. (2009). Mechanisms of drug release in citrate buffered HPMC matrices. *International Journal of Pharmaceutics*, Volume 370, Issues 1–2 (31 March), Pages 110-120.

Reus Medina, M.L. & Kumar, V. (2006). Evaluation of cellulose II powders as a potential multifunctional excipient in tablet formulations. *International Journal of Pharmaceutics*, Volume 322, Pages 31–35.

Rojas, J. & Kumar, V. (2011). Comparative evaluation of silicified microcrystalline cellulose II as a direct compression vehicle. *International Journal of Pharmaceutics*, Volume 416, Issue 1 (15 September), Pages 120-128.

Rowe, R.C. (1986). The effect of the molecular weight of ethyl cellulose on the drug release properties ofmixed films of ethyl cellulose andhydroxypropylmethylcellulose. *International Journal of Pharmaceutics*, Volume 29, Issue 1 (March), Pages 37-41.

Roy, D.S. & Rohera, B.D. (2002). Comparative evaluation of rate of hydration and matrix erosion of HEC and HPC and study of drug release from their matrices. *European Journal of Pharmaceutical Sciences*, Volume 16, Issue 3, August 2002, Pages 193-199.

Saša, B.; Odon, P.; Stane, S.; Julijana, K. (2006). Analysis of surface properties of cellulose ethers and drug release from their matrix tablets. *European Journal of Pharmaceutical Sciences*, Volume 27, Issue 4 (March), Pages 375-383.

Sensoy, D.; Cevher, H.; Sarıcı, A.; Yılmaz, M.; Özdamar, A.; Bergişadi, N. (2009). Bioadhesive sulfacetamide sodium microspheres: Evaluation of their effectiveness inthe treatment of bacterial keratitis caused by Staphylococcus aureus and *Pseudomonasaeruginosa* in a rabbit model. *European Journal of Pharmaceutics and Biopharmaceutics*, Volume 72, Issue 3 (August), Pages 487-495.

Shokri, J.; Ahmadi, P.; Rashidi, P.; Shahsavari, M.; Rajabi-Siahboomi, A.; Nokhodchi A. (2008a). Swellable elementary osmotic pump (SEOP): An effective device for delivery of poorly water-soluble drugs. *European Journal of Pharmaceutics and Biopharmaceutics*, volume 68, Pages 289–297.

Shokri, J.; Alizadeh, M.; Hassanzadeh, D.; Motavalli, F. (2008b). Evaluation of various parameters on release of indomethacin from two-layered core osmotic pump. *Pharmaceutical Sciences (Tabriz Faculty of Pharmacy Journal)*, Volume 3 (winter), Pages 13-22.

Shokri, M.H.; Arami, Z.; Shokri, J. (2011). Evaluation of formulation related parameters on the release of gliclazide from controlled porosity osmotic pump system. *Pharmaceutical Sciences (Tabriz Faculty of Pharmacy Journal)*, 2011, Volume 16 (4), Pages 249-260.

Strydom, S.J.; Otto, D.P.; Liebenberg, W.; Lvov, Y.M.; Villiers, M.M. (2011). Preparation and characterization of directly compactible layer-by-layer nanocoated cellulose. *International Journal of Pharmaceutics*, Volume 404, Issues 1–2 (14 February), Pages 57-65.

Tas, Ç.; Özkan, Y.; Savaser, A.; Baykara, T. (2003). In vitro release studies of chlorpheniramine maleate from gels prepared by different cellulose derivatives. *Il Farmaco*, Volume 58, Issue 8 (August), Pages 605-611.

Van Veen, B.; Bolhuis, G.K.; Wu, Y.S.; Zuurman, K.; Frijlink, H.W. (2005). Compaction mechanism and tablet strength of unlubricated and lubricated (silicified) microcrystalline cellulose, *European Journal of Pharmaceutics and Biopharmaceutics*, Volume 59, Issue 1 (January), Pages 133-138.

Venkatesan, N.; Yoshimitsu, J.; Ohashi, Y.; Ito, Y.; Sugioka, N.; Shibata, N.; Takada, K. (2006) Pharmacokinetic and pharmacodynamic studies following oral administration of erythropoietin mucoadhesive tablets to beagle dogs. *International Journal of Pharmaceutics*, Volume 310, Issues 1–2, 9 (March), Pages 46-52.

Wan, L.S.C. & Prasad, K.P.P. (1988). Effect of microcrystalline cellulose and cross-linked sodium carboxymethylcellulose on the properties of tablets with methylcellulose as a binder, *Internafronal Journal of Pharmaceutics*,Volume 41, Pages 159-167.

Wang, C.Y.; Ho, H-O.; Lin, L-H.; Lin, Y-K.; Sheu, M-T. (2005). Asymmetric membrane capsules for delivery of poorly water-soluble drugs byosmotic effects. *International Journal of Pharmaceutics*, Volume 297, Issues 1–2 (13 June), Pages 89-97.

Williams III, R.O & Liu, J. (2000). Influence of processing and curing conditions on beads coated with an aqueous dispersion of cellulose acetate phthalate. *European Journal of Pharmaceutics and Biopharmaceutics*, Volume 49, Pages 243-252.

Chapter 6

Magnetic Responsive Cellulose Nanocomposites and Their Applications

Shilin Liu, Xiaogang Luo and Jinping Zhou

Additional information is available at the end of the chapter

1. Introduction

Magnetically responsive cellulose materials are specific subset of smart materials, in which magnetic nanoparticles are embedded in the polymer matrix, which can adaptively change their physical properties due to an external magnetic field. These kind of materials are expected to exhibit interesting magnetic field-dependent mechanical behavior with a wide range of potential applications, such as fibers and fabrics for protective clothing for military use (Raymond et al., 1994), magnetic filters (Pinchuk et al., 1995), sensors (Epstein & Miller, 1996), information storage, static and low frequency magnetic shielding (Dikeakos et al., 2003) and health care or biomedical products (Wang et al., 2004). In general, magnetic cellulose materials can be prepared with different morphologies, such as films, fibers, microspheres, hydrogels and aerogels, and they respond differently to externally applied magnetic field because of the different natures and structures. The main purpose of the present review is to overview on recent advances in the development of magnetic field-responsive cellulose composites with emphasis on the fabrication, properties and possible applications.

2. Magnetic cellulose fibers

Natural cellulose fibers are composed of microfibrils of 10-30nm width and three-dimensionally connected with each other (Mark & Kroschwitz, 1985). The surface of the fiber is rough and consists of pores with diameter of 30-70 nm, with specific surface area of 30-55 $m^2 \cdot g^{-1}$ (Kaewprasit et al., 1998). These nanopores may allow guest molecules to penetrate into their inner spaces. The preparation of magnetically responsive fibers based on cellulose and magnetic nanoparticles has been investigated by several approaches. In the past 2 to 3 decades, magnetically responsive cellulose fibers have been prepared by vigorously agitation of cellulose pulp in a concentrated suspension of iron oxide particles such as magnetite and maghemite particles, followed by a mild washing step to remove all the unbound-magnetic particles. This preparation process is called lumen-loading (Chia et al., 2009; Rioux et al., 1992). The process proceeds in three stages. Firstly, a short initial stage with an advancing diffusion front penetrating the lumen with negligible deposition of filler particles on the internal surface. Secondly, a quasi-steady state regime (the main stage) in which the number of fillers particles entering the lumen equals the number being deposited on the lumen wall, and for which the concentration of filler particles suspended in the lumen is approximately constant, and finally the rate determining step switches from being the rate at which particles can enter into the lumen, and particle deposition occurs on the remaining empty spots of the lumen. The lumen-loading is a physical approach, and the diffusion kinetics of the method is mainly limited by the transport of filler particles through the pit apertures in the fiber walls (Zakaria et al., 2004 a, 2004b). It often results in a heterogeneous composite with deleterious particle dispersion, aggregation and therefore inferior performance. In contrast, the latter co-deposits both matrices and particles simultaneously from a premixed precursor offer more homogeneous and uniform composites. One common procedure is the integrating of pre-synthesized Fe_3O_4 particles into the lumens of disintegrated cellulose fibers with the aid of certain retention agent (Chia et al., 2006; Zakaria et al., 2005). After impregnation with an agitation and washing step to remove the unwanted particles, the filler particles are introduced exclusively into the lumen of the fibers while leaving the external surfaces free of filler. The filler is protected by the cell wall from dislodgement and the particles do not interfere with interfiber bonding. In addition, the resultant material shows relatively higher saturation magnetization and coercivity. However, the particles are spatially aggregated presumably due to the effect of the magnetic dipole within the short range. This represents a phenomenon commonly encountered in particle nanocomposite. Moreover, this morphology also lowers the mechanical properties, such as tensile strength and results in brittle material as compared to the host matrix (Zakaria et al., 2004; Middleton & Scallan, 1989).

To circumvent these problems, a modified pathway has been performed by using surface coating method. In this process, a colloidal suspension of magnetic nanoparticles is prepared firstly, and then cellulose fibers are dispersed in it and stirred vigorously. After successive washing and sonication, the particles remain bonded to the surface of the fibers. One significant finding is that, a new bonding phase of α-FeOOH is formed at the interface between the Fe_3O_4 particles and cellulose fibers. The formation of such a bridge is crucial to the integrity of heterogonous hybrid materials in processing and practical applications. In the meanwhile, it allows the inherent properties of the fiber, e.g. tensile strength and flexibility, to be retained while enabling the magnetic properties to the matrix. In addition, the surface of the fibers is completely and uniformly encapsulated by the nanoparticles (Small & Johaston, 2009).

Another approach to prepare magnetically responsive cellulose fibers involves synthesis of iron oxide particles within the cellulosic matrix itself by vigorously agitation of cellulose pulp in iron ion solution, and then iron ions are converted to iron oxide particles within cellulosic matrix by the addition of an excess NaOH solution. This preparation method is called in situ co-precipitation method. This method offers better control of both the magnetic properties and the variety of magnetic particles that are incorporated into the final product than the lumen-loaded method (Marchessault et al., 1992a, 1992b). This method is widely used for the preparation of magnetic cellulose materials. Recently, a modification of the in situ co-precipitation method has been carried out. In this method, bacterial cellulose pellicles are firstly dipped in a solution of $FeCl_2 \bullet 4H_2O$, followed by dipping in a fresh solution of NaOH. The suspension is then heated in a water bath at 65 °C, followed by adding H_2O_2 solutions. Finally, samples are washed with distilled water. By using this method, the individual reaction is exclusively occurred step by step inside the bacterial cellulose. However, this stepwise dipping process still has some drawbacks. The obtained samples show the non-uniform dispersion of the precipitated nanoparticles across the cross-sectional area of bacterial cellulose. Formation of the darker skin at the surface results from the predominant forming of ferrites at the surface of the processed bacterial cellulose. Moreover, the dipping process is done under ambient condition. As a result, the presence of oxygen gas in the atmospheric air promotes the formation of maghemite (γ-Fe_2O_3) and hematite (α-Fe_2O_3). In order to make homogeneous dispersion and control the crystalline phase of magnetic nanoparticles in cellulose matrix, ammonia gas-enhancing in co-precipitation method operated in a closed system without oxygen has been used (Katepetch & Rujiravanit, 2011). The use of ammonia gas, instead of conventional aqueous basic solutions, can prevent the magnetic particles from accumulation at the surface of cellulose fibers, resulting in the homogeneous dispersion of the magnetic nanoparticles throughout the cellulose matrix. Accordingly, the as-prepared magnetic nanoparticles-incorporated cellulose sheet exhibits the uniform magnetic properties throughout the cellulose matrix. Moreover, the homogeneous dispersion of the magnetic nanoparticles throughout the cellulose matrix can enhance the percent incorporation of magnetic nanoparticles into cellulose samples, leading to high and uniform magnetic properties throughout the matrix of cellulose. Regarding to the use of cellulose pellicle and ammonia gas-enhancing in situ co-precipitation method, magnetic particles in the crystal form of magnetite (Fe_3O_4) are obtained and the diameter of the as-synthesized magnetic particles are ranged in nanoscale. The average particle size of the magnetic nanoparticles is in the range of 20–39 nm. The particle size and particle size distribution of magnetic nanoparticles are controllable by adjusting the concentration of aqueous iron ion solution. The saturation magnetization of the magnetic nanoparticle-incorporated cellulose sheet ranges from 1.92 to 26.20 emu•g^{-1} with very low remnant magnetization (0.15–2.67 emu•g^{-1}) and coercive field (40–65 G) at room temperature. Moreover, the responsiveness to an externally applied magnetic field of the magnetic nanoparticle-incorporated bacterial cellulose sheet is exhibited by its deflection in the direction of increasing magnetic field.

Magnetic cellulose fibers can also be obtained by adding ferro-magnetic powders into the cellulose solution, and then spun into fibers. This technique is one of the most effective methods of imparting new features to fibers as it guarantees stability of their properties, due to the fact that the stabile magnetic modifier is firmly integrated in the polymer matrix, and its percentage content does not change while using the fibers. N-methylmorpholine-N-oxide hydrate (NMMO) is a direct solvent for cellulose, a ferromagnetic compound can be added into the solution, and magnetic cellulose fibers can be spun directly from the mixed solution

(Rubacha et al., 2007). The obtained composite fibers can be used to build textile magnetic coils with a textile core. The magnetic properties of the composite fibers depend on the kind of implemented magnetic filler and the percentage content by volume in the fiber matter, and the composite fibers have an increase in the efficiency of shielding the magnetic field. However, mechanical mixing magnetic fills into dissolved cellulose solution often results in an inhomogeneous dispersion of particles in the cellulose matrix, thus considerable attention has been paid to the in situ chemical synthesis of metal nanoparticles in polymer matrices. In our previous works, 7wt% NaOH/12wt% urea aqueous solvent at low temperature is used for cellulose dissolving, and regenerated cellulose fibers can be spun from this solution (Chen et al., 2006; Mao et al., 2008). The cellulose fiber at swollen state exhibits an interpenetrating macroporous structure with a mean pore diameter of about 150 nm. This unique structure makes the porous structured cellulose fibers can be used as reacting sites for the in situ synthesis of inorganic nanoparticles. Magnetic Fe_2O_3 nanoparticles can be synthesized in situ in the cellulose fibers for the preparation of magnetic cellulose fibers (Liu et al., 2008a, 2008b), as it is shown in Fig. 1. The synthesized Fe_2O_3 nanoparticles with a mean diameter of 18 nm are uniformly dispersed in the cellulose matrix. There has strong interaction between Fe_2O_3 nanoparticles and cellulose matrix, the composite fibers are kept in water for a long time, and the Fe_2O_3 nanoparticles can hardly move out from the composite fibers. The composite fibers exhibit improved mechanical strength and a strong capability to absorb UV rays, superparamagnetic properties, as well as a relatively high dielectric constant; it indicated that the composite fibers can be used as protective materials for low frequency magnetic shielding.

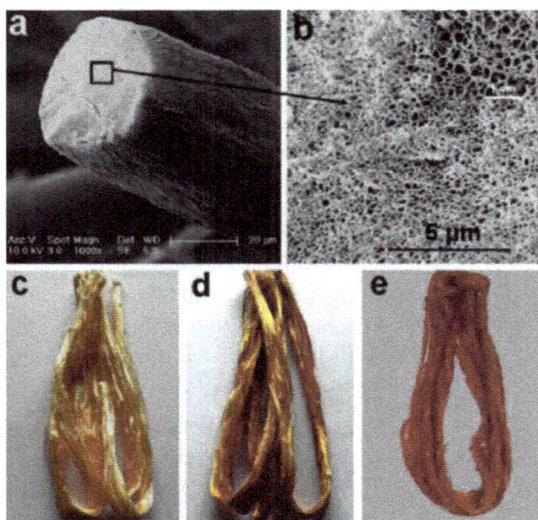

Figure 1. SEM images of surface (a) and cross section (b) for a single swollen RC (regenerated cellulose) fiber (insert is its enlarged image and the scale bar is 1 μm), as well as photographs of the composite fibers F001(FeCl₃, 0.01M) (c), F01 (FeCl₃, 0.1M) (d), and F05(FeCl₃, 0.5M) (e), respectively.

3. Magnetic cellulose films

Cellulose products are used traditionally in paper, packagings and artificial fibers, but technologies such as magnetic nanopapers open up opportunities for entirely new product areas. By using the lumen-loading technology, commercially available magnetic pigments can be introduced into the lumens of softwood fibers from which magnetic paper may be prepared. Lumen-loaded fibers act as magnetic dipoles allowing manipulation of fiber orientation in papermaking (Marchessault et al., 1992b). Another classic mixing of magnetic nanoparticles in the cellulose solution often results in the aggregation of magnetic nanoparticles in the composite films, because of the interparticle dipolar forces worsens their dispersion, which often decreases the properties of the composites and the single function of the magnetic nanomaterials. Precipitation of nano-sized ferrite (Fe_3O_4 and $CoFe_2O_4$) particles with the presence of cellulose fibers has also been used to produce films with good magnetic properties (Chia et al., 2008). The magnetic properties of the films increased with the loading of the magnetic particles. The coercivity of the magnetic films prepared with $CoFe_2O_4$ is higher than that with Fe_3O_4, and the thermal stability of the magnetic film depends on the degree of crystallinity of the precipitated particles. The magnetic particles deposited on the surface of the fibers have detrimental effects on the film strength. However, it is difficult to control the dispersion or particles size of the loaded magnetic particles for the above mentioned methods.

An interesting advance in the development of nanofibril cellulose (NFC) template materials may further enable nanocomposites to have tunable properties and open up many new multifunctional utilities. In-situ precipitation of the magnetic nanoparticles onto the individual cellulose nanofibrils has been used for the preparation of magnetic cellulose films (Galland, 2012). In this process, aqueous nanofibril suspension is used for magnetic functionalization. This method is based on aqueous co-precipitation of cobalt and iron species by forced hydrolysis to form cobalt-ferrite ($CoFe_2O_4$) magnetic nanoparticles. The stable suspension of NFC is favorable for the black suspension of magnetic NFC, and it is then used for membrane formation by a suitable vacuum filtration procedure. In the magnetically functionalized cellulose nanofibril networks, the processing conditions have a major effect on size distribution of magnetic nanoparticles, with the interesting observation that presence of NFC during precipitation results in smaller particle formation. In turn, this directly has an influence on magnetic properties of the material, with e.g. reduced coercivity for materials with smaller nanoparticles. The introduction of nanoparticles results in increased porosity and reduced interaction between fibers, which acts negatively on stiffness and strength of the magnetic nanopaper membranes. But ductility is preserved leading to remarkable tough nanocomposite materials.

Another alternative method is the in situ co-precipitation method by using the porous structured cellulose films as reacting sites. The cellulose hydrogel films prepared from LiOH/urea and NaOH/urea aqueous solution have unique fibrous network structure. The fibrous network structure, apart from providing high mechanical strength, offers macro/mesoporous spaces which can be used as reaction chambers for precipitation of nanoparticles with the fibers providing a support structure to hold the particles. (Liu et al., 2006, 2011a, 2012a). In this method, cellulose films are immersed into $FeCl_2$ or $FeCl_3$ solution firstly and metal ions can be readily impregnated into the cellulose films through the pores. The incorporated Fe^{2+} ions can be bound to cellulose macromolecules via electrostatic interaction, because the electron-rich oxygen atoms of polar hydroxyl of cellulose are expected to interact with electropositive transition metal cations. When the films are treated with aqueous NaOH solution, Fe_2O_3 nanoparticles can be synthesized in the cellulose scaffolds in situ. The obtained Fe_2O_3 nanoparticles in the composite films prepared from $FeCl_2$ or $FeCl_3$ solution are γ-Fe_2O_3. The Fe_2O_3 nanoparticles are plate-like, and distribute randomly in the cellulose matrix before drying. It is different from those reported works about in situ synthesis of inorganic nanoparticles in a polymer matrix. In order to clarify the mechanism for the formation of the plate-like inorganic nanoparticles, Fe_3O_4, CdS, $Co(OH)_2$ nanoparticles have been synthesized from different precursors through the same pathway (Liu et al., 2011b, 2011c; Zhou et al., 2009). It indicates that the prepared nanoparticles are irregular particles and are homogeneously dispersed in the cellulose matrix. The possible mechanism for the formation of plate-like magnetic nanoparticles is ascribed to the non-negligible magnetic dipole-dipole interactions between the magnetic nanoparticles, transforming from $Fe(OH)_3$ or $Fe(OH)_2$ to Fe_2O_3. The cellulose films that immersed into $FeCl_2$ or $FeCl_3$ solution and Fe^{2+}-cellulose or Fe^{3+}-cellulose are formed, when treated with NaOH. $Fe(OH)_3$ or $Fe(OH)_2$ are obtained in the cellulose matrix firstly. After drying, they transform into Fe_2O_3 nanoparticles. During this process, there is an anisotropic growth happens to the nanoparticles. While for the preparation of Fe_3O_4/cellulose films, Fe_3O_4 nanoparticles are directly formed in the cellulose matrix when treated with NaOH solution, therefore, the morphology of the magnetic nanoparticles is irregular particles, which agrees well with the reported works about the preparation of inorganic nanoparticles in polymer matrix from one-step method. The concentration of $FeCl_2$ or $FeCl_3$ solution has little influence on the crystal structure and morphology of the Fe_2O_3 nanoparticles, but had an obvious influence on the content of the Fe_2O_3 nanoparticles in cellulose films. The Fe_2O_3 nanoparticles in composite films that dried at ambient conditions distribute in a regular way, and the composite films have an obvious magnetic anisotropy property, while for the freeze-dried composite films, the Fe_2O_3 nanoparticles distribute randomly, and the resulting composite films displayed superparamagnetic properties without magnetic anisotropy. This interesting phenomenon may be ascribed to the shrinkage of the composite films during drying process and the magnetic dipolar-dipolar interactions between the magnetic nanoparticles. The size of the Fe_2O_3 nanoparticles is far smaller than that of the macropores of the wet cellulose film. Magnetic nanoparticles can rotate freely, and randomly align within the pores at wet state. When it is fixed and dried in air, the composite films only shrink in the longitudinal direction from 300 μm in wet state to about 30 μm in dry state. Furthermore, there is non-negligible magnetic dipole-dipole interaction between the magnetic nanoparticles, leading to the regulative distribution of the nanoparticles. In order to clarify the regular distribution behavior of the Fe_2O_3 nanoparticles in the cellulose matrix, an exo-magnetic field (static and dynamic magnetic field) is applied during the drying process of the composite films at ambient conditions. The regular distribution of Fe_2O_3

nanoparticles in cellulose matrix has been destroyed in the exo-magnetic field, and the dynamic exo-magnetic field has a more obvious effect on the distribution of Fe_2O_3 nanoparticles than that of static exo-magnetic field. The composite films prepared from static-magnetic field have weak ferromagnetic properties, while the composite films prepared from dynamic exo-magnetic field display superparamagnetic properties without magnetic anisotropy, which indicates that magnetic filed has an influence on the distribution of the Fe_2O_3 nanoparticles in cellulose matrix. The effects of different forces on the distribution of the magnetic nanoparticles in cellulose matrix are characterized by using TEM, as it is shown in Fig. 2. Moreover, the influence of uniaxial drawing on the distribution of Fe_2O_3 nanoparticles has been investigated. Interestingly, there is no rearrangement of the Fe_2O_3 nanoparticles in cellulose matrix happened after being drawing, and the distribution of Fe_2O_3 nanoparticles are destroyed, as it is shown in Fig. 3. With an increase of the draw ratio, the irregularity of the Fe_2O_3 nanoparticles in cellulose matrix is increased, and the magnetic anisotropy of the resulting composite films is decreased. The porous structure of the regenerated cellulose filmsare destroyed by uniaxial drawing, therefore, the rotation of the Fe_2O_3 nanoparticles in the pores of cellulose films is hindered, leading to the irregular distribution of the Fe_2O_3 nanoparticles in cellulose matrix. These results support that a transformation process took place in the synthesized Fe_2O_3 from $Fe(OH)_3$ or $Fe(OH)_2$. The magnetic dipole-dipole interaction between the Fe_2O_3 nanoparticles is the important factor of the regular distribution of the Fe_2O_3 nanoparticles in cellulose matrix.

Bacterial cellulose also can be used for the preparation of magnetic cellulose films. Bacterial cellulose is synthesized in the form of fibrous structure which constitutes a three-dimensional non-woven network of nanofibers with diameters less than 100 nm, and it is much smaller than the diameter of typical plant cellulose bundles (ca. 100nm). Bacterial cellulose fiber has the same chemical structure as plant cellulose, but has higher specific surface area than the cellulose nanofibers, indicating that bacterial cellulose has much more surface hydroxyl and ether groups than plant cellulose. These hydroxyl groups make up of active sites for metal ion adsorption (Li et al., 2009). Moreover, the porous structure of nanofibrous bacterial cellulose provides large amount of sub-micron pores. The precipitated metal nanoparticles are stabilized by the sub-micron pores of the bacterial cellulose, leading to good dispersion of the as-synthesized nanoparticles (Zhang et al., 2011), the detail preparation process is shown in Fig. 4. Magnetic ferrite nanoparticles loaded bacterial cellulose (BC) membranes have been synthesized earlier. The size of the ferrite particles varies from ~2 to 20 nm with the particles existing both individually and as aggregates in the matrix. These magnetic BC membranes are superparamagnetic at room temperature with no coercivity or remanence (Raymond et al., 1994, 1995; Sourty et al., 1998; Small & Johnston, 2009).

The precipitation reaction method of loading magnetic particles in the BC matrix does not allow complete control of particles formation-nucleation, growth, aggregation, and density, if surfactant molecules which cap the metal particles are not used. As a result, the synthesized particles may have a broad distribution of sizes and different interparticle

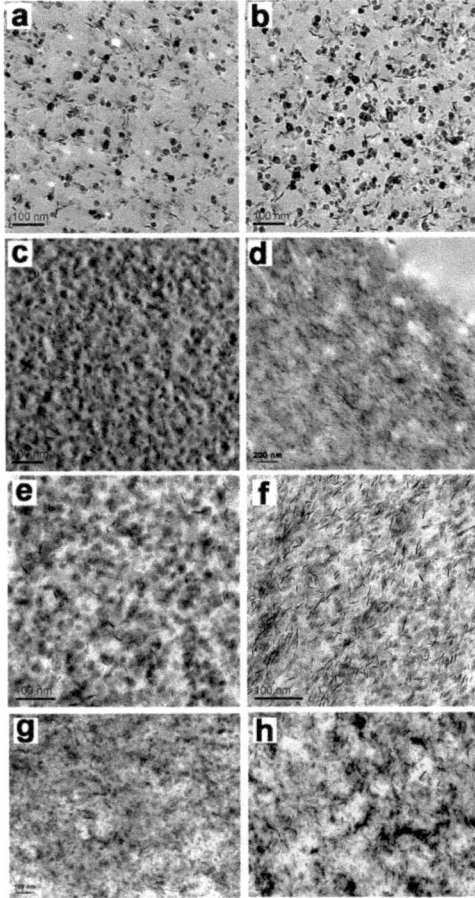

Figure 2. TEM images of the composite film prepared by freeze-dried (a, b) and being dried at ambient conditions (c, d); (a, c), the slice was parallel to the film plane; (b, d), the slice was perpendicular to the film plane, and the composite film fixed on a PMMA plate with a static magnetic field and air-dried. (e), the slice was parallel to the film plane; (f), the slice was perpendicular to the film plane, and the composite film dried on a rotating magnetic field at room temperature. (g), the slice was parallel to the film plane; (h), the slice was perpendicular to the film plane.

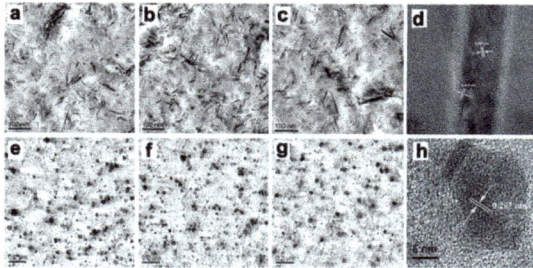

Figure 3. TEM images of the composite films with draw ratios different from 1.04 to 1.22: a, b, c are the slices that perpendicular to the surface of the films, e, f, g are the slices that parallel to the surface of the films, d and e were HRTEM images of c and g, respectively.

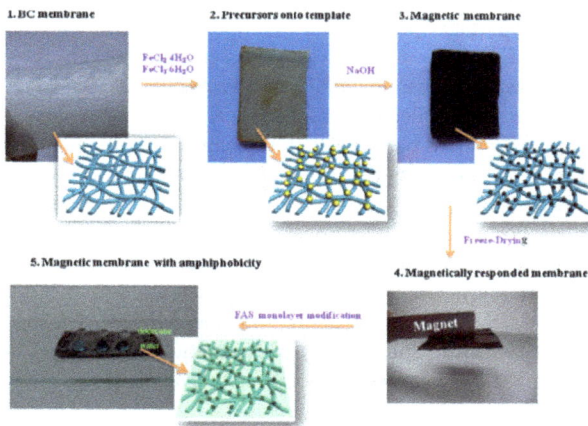

Figure 4. Schematic illustration for the flexible magnetic nanohybrid membrane from bacterial cellulose nanofibers.

distances. Recently, metallic magnetic particles have also been loaded into BC by precipitating from the respective salt solutions using $NaBH_4$, KH_2PO_2, and NaH_2PO_2 as reducing agents. The reaction conditions, type of reducing agent and the reduction medium are found to influence the chemical composition, crystallinity, and size of the metal particles. The saturation magnetization of Ni impregnated cellulose at room temperature is found to be very small compared with that of bulk due to P and B contamination in Ni as well as lack of crystallinity, amorphous structure. The amorphous Ni transforms into crystalline Ni on heat treatment. In the case of Co, NaH_2PO_2 reduction is found to result in the formation of ordered crystals with size of 5-6 nm and aggregates with micron size. The Co loaded cellulose is found to be ferromagnetic at room temperature with obvious coercivity (Pirkkalainen et al., 2007, 2008). The room temperature reduction in an aqueous environment is also an effective method for the controlling synthesis of magnetic nanoparticle with small particle size (Vitta et al., 2010). The formation of crystalline Ni nanoparticles with controlled particle size inside the bacterial cellulose along with $Ni(OH)_2$ has been performed. The nanocrystals have an equiaxed shape and are found both as individual particles as well as small aggregates depending on the porous network structure of cellulose matrix. The bacterial cellulose does not undergo any change and retains its crystal structure even after chemical reduction reaction. The Ni loaded bacterial cellulose is found to be ferromagnetic at room temperature with a saturation magnetization of 2.81 $emu \bullet g^{-1}$ which increases by an order of magnitude to 21.8 $emu \bullet g^{-1}$ at 1.8 K. The coercive field also increases by two orders of magnitude from 28 G at 300 K to 2900 G at 1.8 K. The zero field cooled magnetization however exhibits a superparamagnetic behavior with a peak at 20 K, the blocking temperature and this behavior is observed even in ac magnetization. The magnetization decreases with the increase of temperature up to 400 K, when extrapolated to high temperature using a power law indicates a Curie transition at 500 K, which is much lower than the Curie temperature of bulk Ni. The fraction of isolated superparamagnetic nanoparticles present in the composite is estimated from the saturation magnetization and is found to be ~88%. These results clearly highlight the presence of two separate magnetic phases, superparamagnetic, and ferromagnetic, and the role of various magnetic interactions in the collective magnetic behavior of Ni nanoparticles in the composite structure. Freeze-dried bacterial cellulose nanofibers can form porous structured scaffolds, which can also be used as a template for in-situ chemical reactions to form Co nanoparticles for the generation of a tunable multifunctional nanocomposite film. In this process, an extremely porous so-called aerogel which consists of only 2% cellulose fibrils and 98% pores is obtained, and then dip the porous cellulose aerogel into a saline solution and create the magnetic particles. The magnetic particles are 40 nm large and consist of cobalt ferrite. The particles bind very strongly with the cellulose. During the production process, it is possible to control the amount of magnetic particles that are formed. The porous nanopaper can also be compressed to different levels of porosity, due to obtain the required strength and the flexibility.

4. Magnetic cellulose microspheres

The use of superparamagnetic microspheres (SM) has been widely reported in various applications, such as biomedical research and technology (Šafařík & Šafaříková, 1999) and environmental protection (Yang et al., 2008; Atia et al., 2009), due to their remote responses to external magnetic fields. Ideal magnetic polymer microspheres must have high specific saturation magnetization, small size, narrow particle size distribution, biocompatibility, biodegradable, good chemical stability, rich functional groups on surface and simple process in preparation, cheap as shell. Cellulose, with high biocompatibility, good hydrophilic properties, and biodegradable natures, is a good candidate for magnetic nanocomposite.

Encapsulation of magnetic particles with cellulose polymers is the simple and classical method to prepare magnetic cellulose particles (Luo & Zhang, 2009). In our previous works, maghemite (γ-Fe_2O_3) nanoparticles are prepared. Subsequently, by blending it with cellulose solution and then millimeter-scale magnetic cellulose beads can be prepared via an optimal dropping technology. In this method, other fillers, such as activated carbon (AC) can be added into the mixed solution for the preparation of composite cellulose beads with improved properties. The cellulose beads containing Fe_2O_3 nanoparticles exhibit sensitive magnetic response, and their recovery can be facilitated by applying a magnetic field. Dyes are adsorbed effectively by the AC/magnetic cellulose beads. The Fe_2O_3 nanoparticles and AC in the AC/magnetic cellulose beads can play important roles in both the formation of spherical shape beads and the improvement of the adsorption capacity. Furthermore, the sorbent can be regenerated and used repeatedly. The magnetic properties of the beads allow the separation from the effluent by applying a magnetic field, leading to the development of a clean and safe process for water pollution remediation.

As for the magnetic cellulose microspheres (MCMS), most of the attentions are focused on the absorption performance. The absorption capacity of MCMS is important for the applications point of view. According to Langmuir theory, the smaller the MCMS, the larger the surface area, thereby the stronger the absorption capacity. However, the crystalline structure of cellulose, due to the hydrogen bonds between and within its chains and the high molecular weight, make the treatment of cellulose very difficult. On one hand cellulose exhibits zero solubility in water and common organic solvents, except for some solutions (Fischer et al., 2003; Ass et al., 2006; Yoshida et al., 2005). On the other hand an ionic solution of cellulose shows considerable viscosity, which causes great difficulty in the dispersal of material into small drops. Under conditions without protection from chemical cross-linking or contact, magnetic fluids dispersed in cellulose solution are impacted in their stability as liquid drops and will easily collapse or coalesce, if dispersing strength solutions are used. As a consequence, the average size distribution of MCMS is most often reported from dozens to hundreds of micrometers (Wolf, 1997; Guo & Chen, 2005). If a critical condition could be found in which the drops could maintain both the smallest diameter and stability, MCMS can be fabricated with the maximum absorption capability. A theory of "the smallest critical size"(SCS) is proposed and tested for the first time as a guideline for forming the desired MCMS (Tang et al., 2010). It indicates that the diameter of the SCS of MCMS is 5.82 µm, while the IgG absorption capability of the MCMS with SCS is 186.8 mg/mL. An innovative approach developed by us has shown that a more controlled nanostructured morphology and geometrical shape of the magnetic cellulose microspheres can be obtained (Luo et al., 2009a), and the incorporated magnetic nanoparticles has an obvious influence on the microstructure of the obtained magnetic microspheres, as it is shown in Fig. 5. In this study, cellulose drops in a precooled aqueous solution of sodium hydroxide and urea are utilized to form regenerated cellulose microspheres (RCS) using sol-gel process. These porous beads are then used as microreactors, which permits in situ co-precipitation of Fe_3O_4 nanoparticles into the cellulose pores of RCS in a solution mixture of $FeCl_3$ and $FeCl_2$ to finally form magnetic regenerated cellulose microspheres (MRCS). This process is able to create MRCS about 6 um in diameter with embedded nanoparticles with particle size of 20 nm. Transmission electron microscopy (TEM) clearly shows that embedded Fe_3O_4 nanoparticles are dispersed uniformly inside MRCS matrix. These ideal nanostructures ensure that the micron-sized cellulose beads maintain the superparamagnetic property. Such beads can be magnetized and attracted to that field in the presence of an external

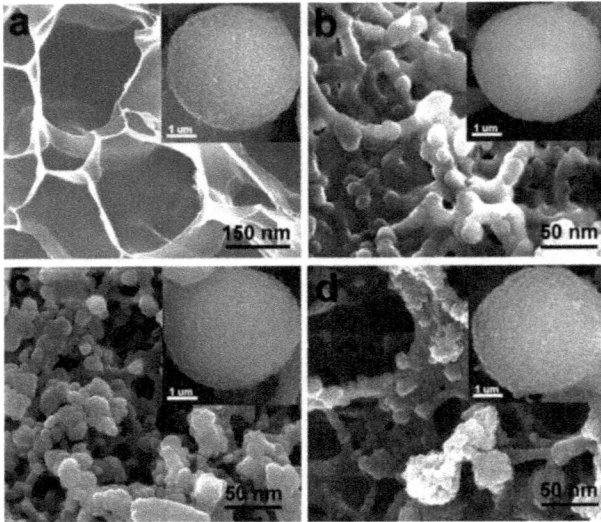

Figure 5. SEM images of the surface of the regenerated cellulose microspheres (RCS) (a), M10 (mixtured solutions of FeCl$_3$/FeCl$_2$ with concentration of 10mmol/5.2mmol) (b), M20(mixtured solutions of FeCl$_3$/FeCl$_2$ with concentration of 20mmol/10.6mmol) (c) and M30(mixtured solutions of FeCl$_3$/FeCl$_2$ with concentration of 30mmol/15.9mmol) (d). The insert illustrates the morphology of RCS and magnetic regenerated cellulose microspheres (MRCS).

magnetic field; however, they will not retain any magnetization when the external field is removed and can flow with the carrier medium like non-magnetic beads. As an embodiment, it demonstrates the magnetic-induced transference for targeting protein delivery and release using these nanocomposite beads.

In recent research aiming at a biomedical application, magnetic carriers based on proteins immobilized onto magnetic cellulose microspheres (MCMS) have also been widely used. The biospecific connection using a specific binding between a protein and the biomatrices displays an excellent repertoire of advantages, including convenient and simple preparation, elimination of toxic compounds, and highly efficient antibody utilization (Luo & Zhang, 2010; Hornes & Korsnes, 1990). The resulting extraction of a pure target molecule by this technique is both convenient and efficient. A novel method has developed for immobilizing antibodies onto MCMS using a cellulose binding domain–protein A (CBD–ProA) linkage, which allows for a one-step isolation of mRNA from eukaryotic cells and tissues (Gao et al., 2009). The produced CBD–SA fusion proteins display binding activities for both cellulose and biotin, and are endowed with superior attributes in the linkage between the MCMS and biotinylated oligo(dT). Using SA–CBD–MCMS for this application allows efficient and rapid isolation of mRNA from eukaryotic cells and tissues and represents an improvement over conventional, available techniques. The stability of the particles, especially the bound CBD to MCMS has been evaluated (Cao et al., 2007).

5. Magnetic cellulose hydrogels and aerogels

Magnetic responsive gels have become an interesting subject of study for several research groups. Cellulose hydrogels can be made stimuli-responsive which makes their study more interesting. Preparation of magnetic cellulose gels is similar to that of other filler-loaded networks. One way is to prepare and characterize magnetic particles separately and then to mix them with polymers, and the cross-linking takes place after mixing the polymer solution and the magnetic sol (Haas et al., 1993), one can precipitate well-dispersed colloidalsized particles in the polymeric material. The in situ precipitation can be made before, during, and after the cross-linking reaction (Mark, 1985). As these gels respond to magnetic stimuli, it can be readily applied in the areas of biotechnology/biomedicine, health care, catalysis, magnetic resonance imaging, and so on. Chatterjee and coworkers have developed a two-step synthesis of magnetic gel (Chatterjee et al., 2004). In the first step, hydroxypropyl cellulose particles are formed with surfactant-modified maghemite by the emulsion method. In the second step, these particles are cross-linked by a commercial cross-linking agent Zirmel M to give a network structure. By this mechanism of network formation, it is possible to introduce a homogeneous distribution of maghemite into the polymer matrix. These magnetic gels have a network of nanoparticles of hydroxypropyl cellulose (30–100 nm) and a homogeneous distribution of nanosized maghemite (~7 nm). The magnetic gel has magnetic moment in an applied field, due to the size distribution in maghemite, single blocking temperature can not be obtained at temperatures below room temperature. Superparamagnetic behavior is observed for the gel. The magnetic gel does not show any unique behavior in terms of the magnetic property though there is a possibility for obtaining different magnetic properties due to the restricted motion (Chatterjee et al., 2003). In order to improve the magnetic properties of the magnetic gels, a modified method has been performed. In this process, HPC is mixed with hexadecyl trimethyl ammonium bromide (CTAB) modified γ-Fe_2O_3 powder in sodium hydroxide solution (pH=12). The solution is then sonicated for 30s. When the solution become homogeneous, solid crushed pellets of NaOH are added to this solution and mixed thoroughly. A reddish brown gel is formed at pH 13 and can separate from the solution. With the decrease in pH (at pH 9), the gel break down and form a homogeneous dispersion of the HPC–γ-Fe_2O_3 complex. This brown dispersion again transformed into a gel with an increase in the pH to 13. The HPC can be loaded as much as 100% of its weight of iron oxide to form a complex structure. Therefore, a large value of magnetic moment is obtained from these gels. By this process of gelation with a magnetic material along with the cellulose polymer, heavy metals/metal oxides can be captured and separated with the help of an external magnet.

Cellulose aerogels, consisting of three-dimensional networks, are typically obtained by removing the liquid in cellulose gels under freeze drying or supercritical conditions (Li et al., 2011; Liebner et al., 2010; Kettunen et al., 2011). The unique properties like high internal surface, high porosity, low density, and with the additional advantages and characteristics of the renewable biopolymer cellulose makes the cellulose aerogel an interesting candidate for various applications. The combination of cellulose nanofibers and magnetic nanoparticles allows for the preparation of ultra-flexible porous magnetic aerogels. Strong cellulose nanofibrils derived from bacteria or wood can form ductile or tough networks that are suitable as functional materials. A bacterial cellulose hydrogel with a large measured surface area is first freeze-dried into a porous cellulose nanofibril aerogel. The dried aerogel template is then immersed in an aqueous $FeSO_4/CoCl_2$ solution at room temperature before heating the system to 90 °C to thermally precipitate the non-magnetic metal hydroxides/oxides on the template. Heating changes the color from transparent to translucent orange. The precipitated precursors are converted into ferrite crystal nanoparticles on immersion in $NaOH/KNO_3$ solution at 90 °C, resulting in highly flexible magnetic aerogels that can sustain large deformations, as it is shown in Fig. 6. Micrographs of freeze-dried samples show that the nanoparticles are located on the bacterial cellulose nanofibril surfaces. Unlike solvent-swollen gels and ferrogels, the magnetic aerogel is dry, lightweight, porous (98%), flexible, and can be actuated by a small household magnet. Moreover, it can absorb water and release it upon compression. Owing to the flexibility, high porosity and surface area, these aerogels are expected to be useful in microfluidics devices and as electronic actuators (Olsson et al., 2010).

In our previous works, regenerated cellulose films that prepared from LiOH/urea or NaOH/urea aqueous solution also can be used as templates for the preparation of magnetic cellulose aerogels (Liu et al., 2012b). In this process, cellulose hydrogel films are immersed in freshly prepared aqueous solutions of $FeCl_3$ and $CoCl_2$ with a molar ratio of

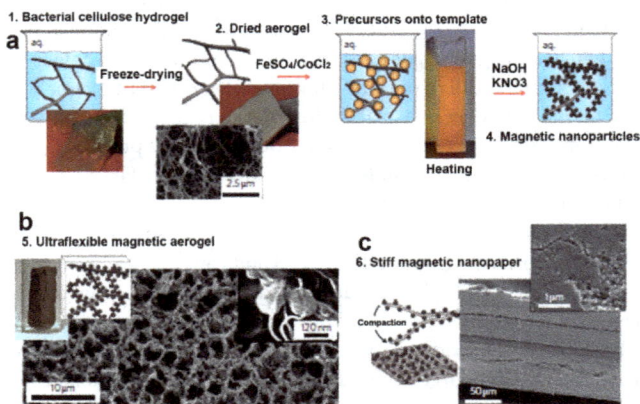

Figure 6. Synthesis of elastic aerogel magnets and stiff magnetic nanopaper. a, Schematic showing the synthetic steps. 1, Bacterial cellulose hydrogel (1 vol%) is produced by Acetobacter xylinum FF-88. 2, Photograph, scanning electron microscopy (SEM) image and schematic of a cellulose aerogel after freeze-drying. 3, Immersion of the dry aerogel in aqueous $FeSO_4/CoCl_2$ solution for 15 min followed by heating to 90 °C for 3 h transforms soluble Fe/Co hydroxides into insoluble complexes. 4, Cellulose networks subjected to $NaOH/KNO_3$ solutions at 90 °C immediately change in colour from red to orange to black as nanoparticles precipitate on the cellulose nanofibrils. b, Representative SEM image of a 98% porous magnetic aerogel containing cobalt ferrite nanoparticles after freeze-drying. Right inset: nanoparticles surrounding the nanofibrils. Left insets: photograph and schematic of the aerogel. c, SEM image of a stiff magnetic nanopaper obtained after drying and compression. Inset: higher magnification image.

[Fe]/[Co] = 2 for 24 h, it allows a homogeneous distribution of the precursor solution obtained inside the cellulose networks. The hydrogel films contain precursor solution is subsequently treated with NaOH solution (2 mol•L^{-1}). The color of the samples changes from red/orange to black immediately, and inorganic nanoparticles can be formed in the cellulose matrix. The composite cellulose films are washed with water to remove counter-ions, and then freeze-dried, and magnetic composite aerogels are obtained. The magnetic aerogels are light-weight, flexible. The porosities of the composite aerogels are ranged from 78 to 52%. The internal specific surface areas and densities of the aerogels are around 300–320 m^2•g^{-1} and 0.25–0.39 g•cm^{-3}, respectively. The content of the incorporated CoFe$_2$O$_4$ nanoparticles increases with the increase of the CoFe$_2$O$_4$ precursor concentration, but the particle size change hardly. The incorporated CoFe$_2$O$_4$ nanoparticles changed the microstructure of the cellulose aerogels obviously, making them different to those of the composite aerogels. The hybrid aerogels show superparamagnetic behaviors, improved mechanical properties with respect to the corresponding inorganic aerogel. Because the concepts of the process are simple and cellulose is sustainable and readily available in large quantities from plants (wood). Thus the suggested route is suitable for industrial-scale production and may be used with many types of nanoparticles, which will open up the new application fields of cellulose based functional materials.

6. Conclusion

The magnetic cellulose composites hold great promise in offering both multiple functionalities and economic functions. Cellulose is not only abundant, renewable, biodegradable, but also the production process is simple. Thus it is very likely to become an economically viable technology, in addition to having profound benefits to sustainable technology and also to our environment. With the versatile properties and a large variety of potential applications revealed in early developments, these new materials have the potential to impact many advanced multifunctional areas such as electromagnetically driven printing, "smart" magnetic biochips, novel localized drug delivery and other applications yet to be envisioned. Furthermore, this research area is expected to grow rapidly, and dramatic improvements in materials' functions will be achieved in the years to come, especially in furthering advanced applications as well as the pursuit of environmental friendly green technologies worldwide.

Author details

Shilin Liu*

College of Chemical and Material Engineering, Jiangnan University, Wuxi, Jiangsu, China

Xiaogang Luo

Key Laboratory of Green Chemical Process of Ministry of Education, Hubei Key Laboratory of Novel Chemical Reactor and Green Chemical Technology, Wuhan Institute of Technology, Wuhan, China

---

* Corresponding Author

Jinping Zhou
Department of Chemistry, Wuhan University, Wuhan, China

Acknowledgement

This work was supported by National Natural Science Foundation of China (51003043).

7. References

Ass BAP, Ciacco G T, & Frollini E. (2006). Cellulose acetates from linters and sisal: correlation between synthesis conditions in DMAc/LiCl and product properties. Bioresour. Technol, Vol. 97, No. 14, PP.1696–1702, ISSN. 0960-8524.

Atia A A, Donia A M, & Al-Amrani W A. Adsorption/desorption behavior of acid orange 10 on magnetic silica modified with amine groups. Chem. Eng. J 2009; 150(1) 55–62, ISSN. 1385-8947.

Cao Y, Zhang Q, Wang C, Zhu YY, & Bai G. (2007). Preparation of novel immunomagnetic cellulose microspheres via cellulose binding domain-protein A linkage and its use for the isolation of interferon α-2b. J. Chromatogr. A, Vol. 1149, No. 2, PP. 228-235, ISSN. 0021-9673.

Chia C H, Zakaria S, Ahamd S, Abdullah M, & Jani S M. (2006). Preparation of Magnetic Paper from Kenaf: Lumen Loading and in situ Synthesis Method. Am. J. Appl. Sci, Vol. 3, No. 3, PP. 1750-1754, ISSN. 1546-9239.

Chia C H, Zakaria S, & Bguyen K L, Abdullah M. (2008). Utilisation of unbleached kenaf fibers for the preparation of magnetic paper. Industrial crops and products, vol. 28, No. 3, PP. 333-339, ISSN. 1011-1344.

Chia C H, Zakaria S, Nguyen K L, Dang V Q, & Duong T D. (2009). Characterization of magnetic paper using fourier transform infrared spectroscopy. Mater. Chem. Phys, Vol. 113, No. 2-3, PP.768-772, ISSN. 0254-0584.

Chen X, Burger C, Fang D, Ruan D, Zhang L, Hsiao B S, & Chu B. (2006). X-ray studies of regenerated cellulose fibers wet spun from cotton linter pulp in NaOH/thiourea aqueous solutions. Polymer, Vol. 47, No. 8, PP.2839-2848, ISSN. 0032-3861.

Chatterjee J, Haik Y, & Chen C J. (2003). Biodegradable magnetic gel: synthesis and characterization. Colloid. Polym. Sci, Vol.281, No. 9, PP. 892-896, ISSN. 0303-402X.

Chatterjee J, Haik Y, & Chen C. (2004). pH-reversible magnetic gel with a biodegradable polymer. J. Appl. Polym. Sci, Vol. 91, No. 5, PP. 3337–3341, ISSN. 0021-8995.

Dikeakos M, Tung L D, Veres T, Stancu A, Spinu L, & Normandin F. (2003). Fabrication and characterization of tunable magnetic nanocomposite materials. Materials Research Society Symposium Proceedings, Vol. 734, No. 134, PP. 315–320, ISSN. 0272-9172.

Epstein A J, & Miller J S. (1996). Molecule- and polymer-based magnets, a new frontier. Synth. Met, Vol. 80, No. 2, PP. 231–237, ISSN: 0379-6779.

Fischer S, Leipner H, Thümmler K, Brendler E, & Peters J. (2003). Inorganic molten salts as solvents for cellulose. Cellulose, Vol. 10, No. 3, PP. 227–236, ISSN. 09690239.

Galland S. (2012). Cellulose network materials-compression molding and magnetic functionalization. Licentiate thesis, Stockholm, Sweden, ISSN. 1654-1081.

Guo X, & Chen F. (2005). Removal of Arsenic by Bead Cellulose Loaded with Iron Oxyhydroxide from Groundwater. Environ. Sci. Technol, vol. 39, No. 17, pp. 6808-6818, ISSN. 0013-936X.

Gao Z, Zhang Q, Cao Y, Pan P, Bai F, & Bai G. (2009). Preparation of novel magnetic cellulose microspheres via cellulose binding domain–streptavidin linkage and use for mRNA isolation from eukaryotic cells and tissues. J. Chromatogr. A, Vol. 1216, No. 45, PP. 7670–7676, ISSN. 0021-9673.

Haas W, Zirnyi M, Kilian HG, & Heise B. (1993). Structural analysis of anisometric colloidal iron(III)-hydroxide particles and particle-aggregates incorporated in poly(vinyl-acetate) networks. Colloid. Polym. Sci, Vol. 271, No. 12, PP.1024-1034, ISSN. 0303-402X.

Hornes E, & Korsnes L. (1990). Magnetic DNA hybridization properties of oligonucleotide probes attached to superparamagnetic beads and their use in the isolation of poly(A) mRNA from eukaryotic cells. Genet. Anal. Tech. Appl , vol. 7, No. 6, pp. 145-150, ISSN. 1050-3862.

Kaewprasit C, Hequet E, Abidi N, & Gourlot J P. (1998). Application of Methylene Blue Adsorption to Cotton Fiber Specific Surface Area Measurement: Part I. Methodology. J. Cotton Sci. Vol. 2, No. 4, PP.164-173, ISSN. 1523-6919.

Katepetch C, & Rujiravanit R. (2011). Synthesis of magnetic nanoparticle into bacterial cellulose matrix by ammonia gas-enhancing in situ co-precipitation method. Carbohydrate Polymers, Vol. 86, No. 1, PP. 162-170, ISSN. 0144-8617.

Kettunen M, Silvennoinen R J, Houbenov N, Nykänen A, Ruokolainen J, Sainio J, PoV, Kemell M, Ankerfors M, Lindström T, Ritala M, Ras RHA, & Ikkala O. (2011). Photoswitchable superabsorbency based on nanocellulose aerogels. Adv. Funct. Mater, Vol.21, No3, PP. 510-517, ISSN, 1616-301X.

Liu S, Zhou J, Zhang L, Guan J, & Wang J. (2006). Synthesis and alignment of iron oxide nanoparticles in a regenerated cellulose film, Macromol. Rapid Commun, vol. 27, No. 6, PP. 2084-2089, ISSN. 1022-1336.

Liu S, Zhang L, Zhou J, & Wu R. (2008a). Structure and properties of cellulose/Fe$_2$O$_3$ nanocomposite fibers spun via an effective pathway. J. Phys. Chem. C, Vol. 112, No. 12, PP. 4538-4544, ISSN. 1932-7447.

Liu S, Zhang L, Zhou J, Xiang J, Sun J, & Guan J. (2008b). Fiber like Fe$_2$O$_3$ Macroporous nanomaterials fabricated by calcinating regenerate cellulose composite fibers, Chem. Mater, Vol. 20, No. 11, PP. 3623-3628, ISSN. 0897-4756.

Liu S, Zhou J, & Zhang L. (2011a). In situ synthesis of plate-like Fe$_2$O$_3$ nanoparticles in porous cellulose films with obvious magnetic anisotropy. Cellulose, Vol. 18, No. 3, pp. 663-673, ISSN. 1572-882X.

Liu S, Ke D, Zeng J, Zhou J, Peng T, & Zhang L. (2011b). Construction of inorganic nanoparticles by micro-nano-porous structure of cellulose matrix. Cellulose, Vol. 18, No. 4, PP. 945-956, ISSN. 0969-0239.

Liu S, Hu H, Zhou J, & Zhang L. (2011c).Cellulose scaffolds modulated synthesis of Co_3O_4 nanocrystals: preparation, characterization and properties. Cellulose, Vol. 18, No. 5, PP. 1273-1283, ISSN. 1572-882X.

Liu S, Li R, Zhou J, & Zhang L. (2012a). Effects of external factors on the arrangement of plate-liked Fe_2O_3 nanoparticles in cellulose scaffolds. Carbohydr. Polym, Vol. 87, No. 1, pp. 830-838, ISSN. 0144-8617.

Liu S, Yan Q, Tao D, Yu T, & Liu X. (2012b). Highly Flexible Magnetic Composite Aerogels Prepared by Using Cellulose nanofibril networks as Templates. Carbohydr. Polym, doi:10.1016/j.carbpol. 2012.03.046.

Li X, Chen S, Hu W, Shi S, Shen W, Zhang X, & Wang H. (2009).In situ synthesis of CdS nanoparticles on bacterial cellulose nanofibers. Carbohydr. Polym, Vol. 76, No. 4, PP. 509–512, ISSN. 0144-8617.

Li J, Lu Y, Yang D, Sun Q, Liu Y, & Zhao H. (2011). Lignocellulose Aerogel from Wood-Ionic Liquid Solution (1-Allyl-3-methylimidazolium Chloride) under Freezing and Thawing Conditions. Biomacromolecules, Vol. 12, No. 5, PP. 1860-1867, ISSN. 1525-7797

Luo X, & Zhang L. (2009a). High effective adsorption of organic dyes on magnetic cellulose beads entrapping activated carbon. J. Hazard. Mater, Vol. 171, No. 1-3, PP. 340–347, ISSN. 0304-3894.

Luo X, Liu S, Zhou J, & Zhang L. (2009b). In situ synthesis of Fe_3O_4/cellulose microspheres with magnetic-induced protein delivery. J. Mater. Chem, Vol. 19, No. 21, PP. 3538–3545, ISSN. 0959-9428.

Luo X, & Zhang L. (2010). Immobilization of Penicillin G Acylase in Epoxy-Activated Magnetic Cellulose Microspheres for Improvement of Biocatalytic Stability and Activities. Biomacromolecules, Vol. 11, No. 11, pp. 2896–2903, ISSN. 1525-7797.

Liebner F, Haimer E, Wendland M, Neouze M-A, Schlufter K, Miethe P, Heinze T, Potthast A, & Rosenau T. (2010). Aerogels from unaltered bacterial cellulose: application of $scCO_2$ drying for the preparation of shaped, ultra-lightweight cellulosic aerogels. Macromol. Biosci, Vol. 10, No. 4, PP. 349-352, ISSN, 1616-5187.

Mark H F, & Kroschwitz J I. (Eds.) (1985). Encyclopedia of Polymer Science and Technology; John Wiley & Sons: New York; p 60, ISSN. 9780470569696.

Mark J E. (1985).Bimodal networks and networks reinforced by the in situ precipitation of silica. British. Poly. J, Vol. 17, No.2, PP. 144-148, ISSN. 0007-1641.

Middleton S R, & Scallan A M. The effect of cationic starch on the tensile strength of paper. J. Pulp Pap. Sci 1989; 15(5) J229–J234, ISSN. 0826-6220.

Marchessault R H, Richard S, & Rioux P. (1992a). In situ synthesis of ferrites in lignocellulosics. Carbohydr. Res, Vol. 224, PP.133–139, ISSN. 00323861.

Marchessault R H, Rioux P, & Raymond L. (1992b). Magnetic cellulose fibers and paper: Preparation, processing and properties. Polymer, Vol. 33, No. 19, PP. 4024–4028, ISSN. 00323861.

Mao Y, Zhang L, Cai J, Zhou J, & Kondo T. (2008). Effects of Coagulation Conditions on Properties of Multifilament Fibers Based on Dissolution of Cellulose in NaOH/Urea Aqueous Solution. Ind. Eng. Chem. Res, Vol. 47, No. 22, PP. 8676-8683, ISSN. 0888-5885.

Olsson RT, Azizi Samir M A S, Salazar-Alvarez G, Belova L, Ström V, Berglund L A, Ikkala O, Nogués J, & Gedde U W. (2010). Making flexible magnetic aerogels and stiff magnetic nanopaper using cellulose nanofibrils as templates. Nature Nanotechnology, vol.5, No.8, PP. 584-588, ISSN. 1748-3387.

Pinchuk L S, Markova L V, Gromyko Y V, Markov E M, & Choi U S. (1995). Polymeric magnetic fibrous filters. J. Mater. Process. Technol, Vol. 55, No. 3-4, PP. 345–350, ISSN. 0924-0136.

Pirkkalainen K, Vainio U, Kisko K, Elbra T, Kohout T, Kotelnikova N E, & Serimaa R J. (2007). Structure of nickel nanoparticles in a microcrystalline cellulose matrix studied using anomalous small-angle X-ray scattering. J. Appl. Crystallogr, Vol. 40, PP. s489-s494, ISSN. 0021-8898

Pirkkalainen K, Leppanen K, Vainio U, Webb M A, Elbra T, Kohout T, Nykanen A, Ruokolainen J, Kotelnikova N, & Serimaa R. (2008). Nanocomposites of magnetic cobalt nanoparticles and cellulose. Eur. Phys. J. D, Vol. 49, No. 3, PP. 333-342, ISSN. 1434-6060.

Raymond L, Revol J F, Ryan D H, & Marchessault R H. (1994). In situ synthesis of ferrites in cellulosics. Chem. Mater, Vol. 6, No. 2, PP. 249–255, ISSN. 0897-4756.

Rioux P, Ricard S, & Marchessault R H. (1992). The preparation of magnetic papermaking fibers. J. Pulp. Pap. Sci, Vol. 18, No. 1, PP. 39–43, ISSN. 0826-6220.

Rubacha M, & Zięba J. (2007). Magnetic cellulose fibers and their application in textronics. Fibers & Textiles in Eastern Europe, Vol. 5-6, No. 64-65, PP.101-104, ISSN. 1230-3666.

Raymond L, Revol J-F, Ryan D H, & Marchessault R H. (1994). In situ synthesis of ferrites in cellulosics. Chem. Mater, Vol. 6, No. 2, PP. 249-255, ISSN. ISSN: 08974756.

Raymond L, Revol J-F, Marchessault R H, & Ryan D H. (1995). In situ synthesis of ferrites in ionic and neutral cellulose gels. Polymer, Vol. 36, No. 26, PP. 5035-5043, ISSN. 0032-3861.

Small A C, & Johaston J H. (2009). Novel hybrid materials of magnetic nanoparticles and cellulose fibers. J. Colloid. Interface. Sci, Vol. 331, No. 1, PP. 122–126, ISSN. 0021-9797.

Sourty E, Ryan D H, & Marchessault R H. (1998). Characterization of magnetic membranes based on bacterial and man-made cellulose. Cellulose, Vol. 5, No. 1, PP. 5-17, ISSN. 0969-0239.

Šafařík I, & Šafaříková M. (1999). Use of magnetic techniques for the isolation of cells. J. Chromatogr. B, Vol. 722, No. 1-2, PP. 33–53, ISSN. 1570-0232.

Tang Y, Zhang Q, Wang L, Pan P-W, & Bai G. (2010). Preparation of cellulose magnetic microspheres with "the smallest critical size" and their application for microbial immunocapture. Langmuir, Vol. 26, No. 13, PP. 11266-11271, ISSN. 0743-7463.

Vitta S, Drillon M, & Derory A. (2010). Magnetically responsive bacterial cellulose: Synthesis and magnetic studies. J. Appl. Phys, Vol. 108, No. 5, PP. 053905-053911, ISSN. 0021-8979

Wang M, Singh H, Hatton T A, & Rutledge G C. (2004). Field-responsive superparamagnetic composite nanofibers by electrospinning. Polymer, Vol. 45, No. 16, PP. 5505–5514, ISSN. 0032-3861.

Wolf B. (1997). Bead cellulose products with film formers and solubilizers for controlled drug release. Intl. J. Pharm, Vol. 156, No. 1, PP. 97–107, ISSN. 0378-5173.

Yang N, Zhu S, Zhang D, & Xu S. (2008). Synthesis and properties of magnetic Fe_3O_4-activated carbon nanocomposite particles for dye removal. Mater. Lett, Vol. 62, No.4-5, PP. 645–647, ISSN. 0167-577X.

Yoshida Y, Yanagisawa M, Isogai A, Suguri N, & Sumikawa N. (2005). Preparation of Polymer Brush-Type Cellulose b-Ketoesters using LiCl/1,3-Dimethyl-2-imidazolidinone as a Solvent. Polymer, Vol. 46, No. 8, PP. 2548–2557, ISSN. 0032-3861.

Zakaria S, Ong B H, & Van de Ven T G M. (2004a). Lumen loading magnetic paper II: mechanism and kinetics. Colloids. Surf. A, Vol. 251, No. 1-3, PP.31–36, ISSN. 09277757.

Zakaria S, Ong B H, & Van de Ven T G M. (2004b). Lumen loading magnetic paper I: flocculation. Colloids. Surf. A, Vol. 251, No. 1-3, PP. 1–4, ISSN. 09277757.

Zakaria S, Ong B H, Ahmad S H, Abdullah M, & Yamauchi T. (2005). Preparation of lumen-loaded kenaf pulp with magnetite (Fe_3O_4). Mater. Chem. Phys, Vol. 89, No.2-3, PP. 216 - 220, ISSN. 0254-0584.

Zhou J, Li R, Liu S, Li Q, & Zhang L. (2009). Structure and Magnetic Properties of Regenerated Cellulose/Fe_3O_4 Nanocomposite Films. J. Appl. Polym. Sci, Vol. 111, No. 5, PP. 2477-2484, ISSN. 0021-8995.

Zhang W, Chen S, Hu W, Zhou B, Yang Z, Yin N, & Wang H. (2011). Facile fabrication of flexible magnetic nanohybrid membrane with amphiphobic surface based on bacterial cellulose. Carbohydr. Polym, Vol. 86, No.4, PP. 1760-1767, ISSN. 0144-8617.

Chapter 7

Probing the Interaction Between Cellulose and Cellulase with a Nanomechanical Sensor

Jun Xi, Wenjian Du, and Linghao Zhong

Additional information is available at the end of the chapter

1. Introduction

1.1. Cellulose and cellulose biomass

A cellulose molecule is a linear polymer of D-anhydroglucopyranose units linked by β-1, 4-glucosidic bonds (Figure 1). On its reducing end, a cellulose molecule has an unsubstituted hemiacetal. On its non-reducing end, it has a hydroxyl group.

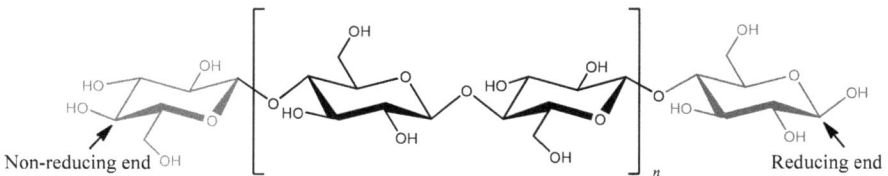

Figure 1. Molecular structure of a cellulose molecule.

Cellulose is the skeleton structure of almost all green plants. It is particularly abundant in non-food plants like trees and grasses, which typically have 40-60% cellulose, 20-40% hemicellulose, and 10-25% lignin (Lynd et al., 2002; Yang et al., 2007). There are four major polymorphs of cellulose: I, II, III, and IV. Cellulose I, often found in native cellulose, contains allomorphs I_α (bacteria and algae) and I_β (higher plants) (Kontturi et al., 2006; Pérez & Samain, 2010). Cellulose I, when treated with a concentrated alkaline solution, turns into cellulose II, a thermodynamically more stable crystalline form than cellulose I. Cellulose III_I can be obtained when cellulose microcrystal is subjected to supercritical ammonia. The structure of another allomorph of cellulose III, III_{II} is still being debated. Cellulose IV_I and IV_{II} are formed when cellulose III is heated in glycerol at 260°C (Peter, 2001).

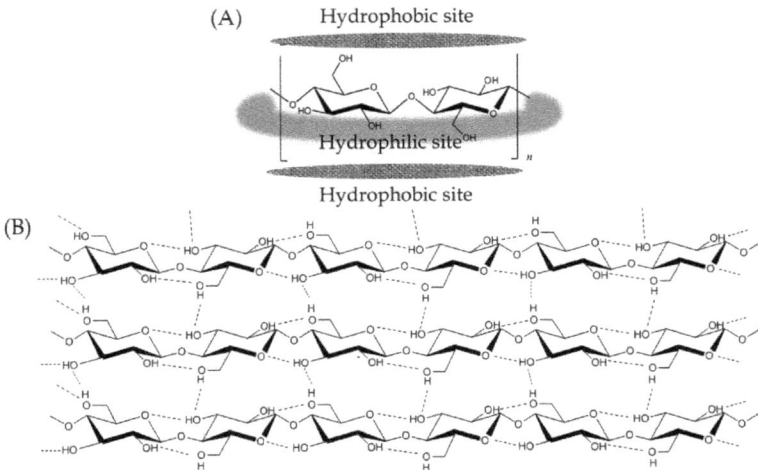

Figure 2. (A) Hydrophilic and hydrophobic sites of cellulose. (B) Schematic drawing of the intrasheet hydrogen-bonding network in cellulose Iα.

Various noncovalent interactions such as hydrogen bonding and van der Waals interactions are present in the ultrastructure of cellulose. While the OH-O hydrogen bonding is mostly responsible for cellulose intrasheet interactions, both the weaker CH-O hydrogen bonding and van der Waals interactions contribute to cellulose intersheet interactions (Li Q. & Renneckar, 2011). Figure 2 shows the arrangement of the intrasheet hydrogen bonding network in cellulose Iα and the resulting hydrophilic and hydrophobic sites of the ring plane (Brown & Saxena, 2007). Overall, because of these noncovalent interactions, cellulose chains aggregate into various forms of ultrastructure, which do not melt or dissolve in any common solvents. Such aggregation prevents the potential cleavage sites (i.e., glycosidic bonds) of a cellulose chain from being accessed by cellulase.

1.2. Degradation of cellulose in biomass conversion

The biomass conversion is the key step to produce biofuel from cellulosic biomass. Such conversion is often accomplished either through biochemical methods or thermochemical methods, where the polysaccharides in cellulosic biomass are hydrolyzed by biochemical agents such as cellulase enzyme, or by thermal treatment such as gasification to produce simple sugars that are fermentable to produce biofuel products (Dwivedi et al., 2009). For biochemical methods (Gray et al., 2006), cellulases are usually employed to convert the solid cellulosic biomass into glucose or small sugar polymers that can be readily fermented with microorganism to produce ethanol. Compared to thermochemical methods which often require a large amount of acid and energy, biochemical methods are more environmentally friendly and economically feasible because of their better conversion efficiencies and milder operating conditions. By far the enzyme-based biochemical methods are considered as the most promising technologies for biomass conversion. However, because of biomass recalcitrance and high cost of cellulase in biomass conversion, the current process for biofuel production is not yet a viable option for the large-scale production (Dwivedi et al., 2009). Much research and developmental efforts have been dedicated to the improvement of the efficiency of cellulase in biomass conversion. One feasible approach is through the incorporation of new features (mutations) into cellulase that accelerate key steps (e.g., rate limiting step) of the enzymatic process. This approach requires a comprehensive mechanistic understanding of cellulose hydrolysis by cellulase.

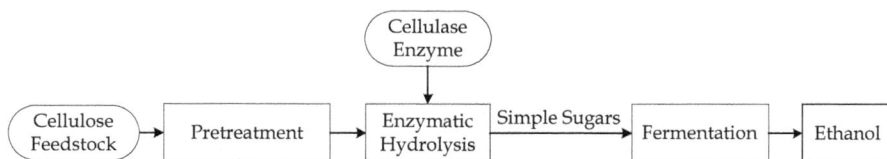

Figure 3. The process of biomass conversion through the enzyme-based biochemical method.

1.3. Cellulase and enzymatic hydrolysis of cellulose

Cellulase (Lynd et al., 2002; Mosier et al., 1999; Wilson & Irwin, 1999), like all glycosyl hydrolase enzymes found in bacteria, fungi, plants and some invertebrate animals, breaks down β-1, 4-glycosidic bonds of cellulose through general acid/base catalysis. There are mainly three kinds of cellulases: exo-β-1, 4-D-glucanase, endo-β-1, 4-D-glucanase and β-D-glucosidase. Each enzyme alone cannot hydrolyze the complex crystalline cellulose efficiently but working synergistically with other types of cellulases can increase the rate of hydrolysis significantly (Dwivedi et al., 2009; Lynd et al., 2002).

- Exo-β-1, 4-D-glucanase can access individual cellulose chains from the exposed reducing end or non-reducing end and cleave two to four glucose units at a time to produce tetrasaccharides or disaccharides (i.e., cellobioses) (Figure 4).
- Endo-β-1, 4-D-glucanase breaks internal glycosidic bonds of individual cellulose chains to disrupt the network structure of cellulose and expose individual polysaccharide chains (Figure 4).
- β-D-Glucosidase or cellobiase hydrolyzes cellobiose to release D-glucose units.

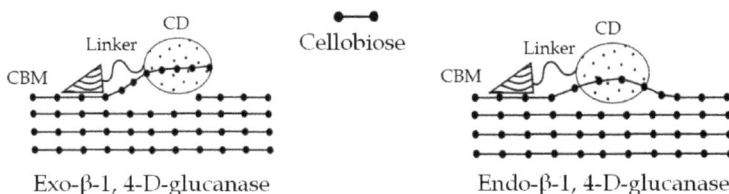

Figure 4. Schematic illustration of cellulose hydrolysis by exo-β-1, 4-D-glucanase and endo-β-1, 4-D-glucanase.

Most of fungal cellulases have a domain like structure that contains a catalytic domain (CD) and a carbohydrate binding module (CBM) (Martin, 2000). These two domains are connected by a peptide linker, which is known to maintain the separation between the CD and the CBM (Figure 5). The CD contains the enzyme active site that is responsible for cellulose hydrolysis. The CBM is a contiguous amino acid sequence that anchors the CD onto the surface of cellulose through hydrogen bonding and van der Waals interactions (Boraston et al., 2004; Guillén et al., 2010). Cellulases can be grouped into families according to sequence similarities of their amino acid residues within their CDs and CBMs.

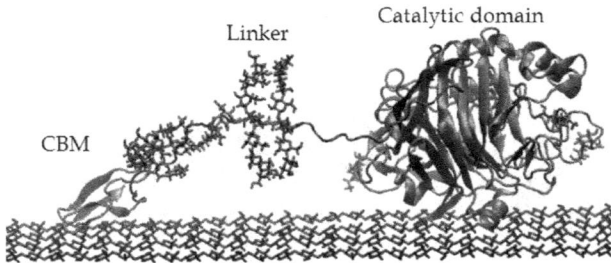

Figure 5. The domain like structure of the cellulase cellobiohydrolase I (CBH I) that is bound to cellulose I_β microfibril. (Zhong et al., 2008)

Cellulose hydrolysis by cellulase is a multi-step process (Chundawat et al., 2011) that is initiated with the binding of cellulase (E) onto the surface of cellulose (S), shown in Figure 6. After this "initial binding" step, a single cellulose chain will be separated from the cellulose aggregate by cellulase and pulled into the active site of cellulase. This is the "decrystallization" step which forms a pseudo-Michaelis complex (E*S). E*S will then undergo "hydrolytic cleavage" to produce cellobiose as the product (P).

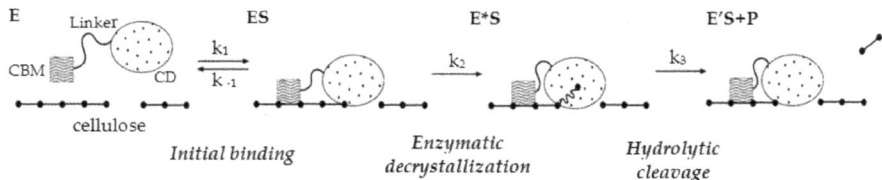

Figure 6. Mechanism of cellulose hydrolysis by cellulase.

For cellulose hydrolysis, the substrate cellulose is water insoluble and resistant to the attack by biological agents. This makes the formation of E*S much more difficult compared to those formed with soluble substrates.

1.4. Knowledge gap

The initial binding (step 1 in Figure 6) has been extensively studied over the years. A Langmuir equation is widely used as a simplified mechanistic model to describe the formation of ES resulting from the initial interaction between the cellulase and the cellulose, which often reaches steady-state within half an hour (Lynd et al., 2002; Zhang Y. H. & Lynd, 2004). Other equilibrium binding models and dynamic binding models have also been proposed to account for the complexities of the binding process including the partially irreversible adsorption of cellulase, multiple types of adsorption site, and so on (Lynd et al., 2002).

For cellulose hydrolysis, most recent studies have been focused on elucidation of the mechanism of the hydrolytic cleavage reaction (step 3 in Figure 6) (Divne et al., 1998; Li Y. et al., 2007; Parsiegla et al., 2008). Some of the key amino acid residues involved in cleavage reactions have been identified. The distinctions between endo- and exo-glucanases, and between retention and inversion for the stereo configuration of the products have been made after years of biochemical and biophysical studies. The majority of studies on the hydrolytic cleavage step were done by measuring the concentration of the sugars (P) released during cleavage of soluble cellodextrins or insoluble cellulose.

Meanwhile, very little success has been achieved in obtaining fundamental knowledge of the enzymatic decrystallization reaction in step 2 (DOE/SC-0095, 2006). In particular, the importance of enzymatic decrystallization was largely unnoticed until very recently (Chundawat et al., 2011; DOE/SC-0095, 2006; Wilson, 2009). Several pieces of biochemical and physical evidence have indicated the presence of such an activity. Back to 1997, Wilson and his coworker showed that cleavage of the β-1, 4-glycosidic linkage in crystalline cellulose is not the rate-limiting step for $T.$ $fusca$ endoglucanase E2 (Zhang S. & Wilson, 1997). They speculated that the binding of a cellulose chain from a microfibril into the active site of a cellulase is the rate-limiting step for degradation of crystalline cellulose (Wilson, 2009). Lee and coworkers found indentations and paths on the surface of cotton fibers that had been treated with a cellulase that was incapable of hydrolytic cleavage (Lee et al., 2000). The evidence suggests that such surface modifications on cellulose are likely caused by the decrystallization activity of cellulase.

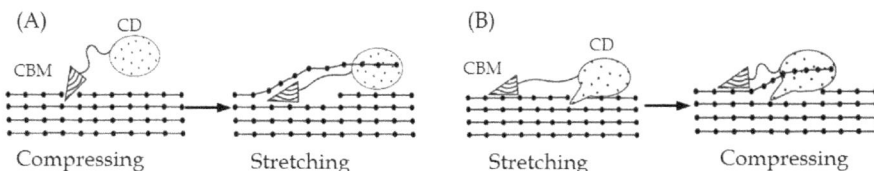

Figure 7. Mechanistic models of enzymatic decrystallization. (A) The CBM model: the CBM serves as a wedge to assist the release of a single cellulose chain. (B) The CD model: the protrusion of the CD domain serves as a wedge to assist the release of a single cellulose chain.

The mechanistic model for enzymatic decrystallization proposed by Reilly and coworkers is shown in Figure 7A (Mulakala & Reilly, 2005). In this model, the CBM is inserted under a cellulose chain like a wedge to separate the chain from the cellulose network. Then the released cellulose chain is pulled into the active site of the CD along the top face of the CBM to achieve the decrystallization. In this model, the CBM is essential to enzymatic decrystallization of cellulose. Numerous biochemical studies, however, have shown that in the absence of the CBM, the CD domain alone still retains 20~50% of the hydrolytic activity on crystalline cellulose (Reinikainen et al., 1992; Srisodsuk et al., 1993; Van Tilbeurgh et al.,

1986). To resolve this discrepancy, we propose an alternative mechanistic model, where a wedge-like structure at the bottom of the CD can be inserted under the cellulose chain to lift it into the active site of the CD (Figure 7B). Since decrystallization by cellulase has been speculated to be the rate limiting step for the degradation of crystalline cellulose (DOE/SC-0095, 2006; Wilson, 2009), understanding the mechanism of this enzymatic activity becomes essential to a comprehensive understanding of cellulose hydrolysis by cellulase.

2. Innovative approach and microcantilever

2.1. Existing technologies and their technical limitations

The conventional approaches to study the cellulose hydrolysis by cellulase are based on the measurement of the concentration of glucose or other simple sugars that are produced in the hydrolytic process. These approaches are not suitable for studying the decrystallization process because no new product is formed and released from this process. Spectroscopic techniques such as Fourier transform infrared spectroscopy (Fengel et al., 1995), Raman spectroscopy (Schenzel et al., 2005), and x-ray photoelectron spectroscopy (Ahola et al., 2008; Fardim et al., 2005) have been used to study the structural change of cellulose fibers. All these techniques focus on the global variations of cellulose. Since cellulose decrystallization only occurs on the outer layer within a relative small region of the surface of cellulose, these techniques are not sensitive enough for such study. Quartz crystal microbalance has been used to study the enzymatic hydrolysis of cellulose (Ahola et al., 2008; Rojas Orlando et al., 2007; Turon et al., 2008), however, its suitability for studying decrystallization has not yet been demonstrated.

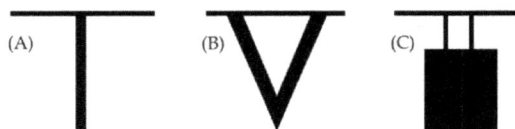

Figure 8. Various shapes of AFM probes.

Atomic force microscopy (AFM) has been used to examine changes of surface morphology of cellulose. Typical size of AFM probe is 200 μm x 40 μm x 1 μm (length x width x thickness) with various shapes (Figure 8). Lee and coworkers used the tapping mode AFM to exam the effects of three different cellulases on the surface of cotton fibers (Lee et al., 2000). Li and coworkers used the AFM to detect structural changes of cellulose microfibril fragments caused by sonication (Li Q. & Renneckar, 2011). With this powerful surface imaging tool, we can investigate the surface change caused by enzymatic activities. However, AFM imaging is mostly limited to the surface analysis at discrete time points. Additionally a high quality image requires that the surface area of cellulose be prepared prior to imaging. Any extensive handling may potentially alter the surface properties and delay the timely analysis under the assay conditions. Therefore, AFM imaging is not an ideal approach for analyzing the dynamic impact of cellulase on cellulose. Both scanning electron microscopy and transmission electron microscopy have similar limitations. This has led to the development of the microcantilever technique for cellulase study.

2.2. Microcantilever and its applications

The microcantilever is a highly sensitive nanomechanical sensor that originates from a micro-fabricated AFM probe. Typical dimensions of a microcantilever, which are similar to AFM probes, are 200 μm long, 1 μm thick, and 20 μm wide (Goeders et al., 2008; Lavrik et al., 2004). The microcantilever is capable of detecting minute changes in interaction energy between individual molecules in the thin film of a polymer coating (cellulose, protein, DNA, or polymer brush) in the form of a measurable bending (10^{-6} to 10^{-12} m) of the microcantilever (Moulin et al., 1999; Mukhopadhyay et al., 2005; Shu et al., 2005; Yan et al., 2006; Zhao et al., 2010; Zhou et al., 2006). The microcantilever bending can be measured based on the deflection of a laser beam reflecting from the tip of the microcantilever in the AFM (Figure 9). Overall, the microcantilever detection has high specificity, high sensitivity, and quick response.

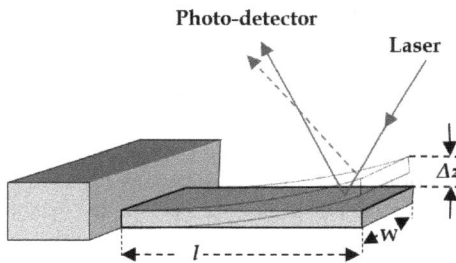

Figure 9. Working scheme of the microcantilever.

Dynamic mode (Vashist, 2007): When an analyte is adsorbed on an oscillating microcantilever, the microcantilever will oscillate at a lower frequency. The difference in frequency can be used to measure properties of adsorbates, such as viscosity and density, etc.

Static mode (Vashist, 2007): Adsorbates tend to induce a significant change in surface stress of a microcantilever, which results in a deflection of the microcantilever (Lavrik et al., 2004; Shuttleworth, 1950). The surface stress and the deflection of the microcantilever are quantitatively related (Yan et al., 2006):

$$\Delta z = \frac{3(1-v)l^2}{Et^2}\Delta\sigma \qquad (1)$$

Where Δz is the deflection of the microcantilever at the end of the microcantilever, v and E are Poisson's ratio and Young's modulus of the microcantilever, t and l are the thickness and length of the microcantilever, and $\Delta\sigma$ is the differential stress on the microcantilever. Using the static mode microcantilever sensor, Ji and coworkers investigated the conformational change of calmodulin (Yan et al., 2006). Sauers and coworkers successfully detected 2-mercaptoethanol using a gold-coated microcantilever (Datskos & Sauers, 1999). Zhou and coworkers successfully demonstrated the use of the microcantilever bending as a means to probe changes in internal structure of polymer brushes in response to changes in pH and electrolyte concentration (Zhou et al., 2006).

2.3. Microcantilever in cellulose study

In cellulose, the interaction energy arises from noncovalent interactions (hydrogen bonding and van der Waals interactions) between tightly packed cellulose chains. To release cellulose chains, surface-adsorbed cellulase must break up noncovalent interactions between cellulose chains through decrystallization, which results in a change in overall interaction energy in the cellulose. If cellulose is deposited onto a microcantilever, such dynamic change in interaction energy will be transduced from the cellulose coating into the microcantilever, and will result in a measurable bending of the microcantilever. Thus, the unique link between the enzymatic decrystallization and the resulting energy alteration in cellulose measured by the microcantilever bending has provided a novel strategy to experimentally examine this unusual enzymatic activity.

Such adsorbate-induced bending often occurs in the presence of a specific interaction (e.g., decrystallization) between an adsorbate (e.g., cellulase) and a substrate (e.g., cellulose) that can alter the internal energy of the substrate. Since a mere adsorption such as the initial binding of cellulase on cellulose has a very minimum impact on the internal energy of cellulose, it will not induce a measurable bending even though the surface-bound cellulase changes the overall gravity (mass) on the microcantilever. Therefore the microcantilever bending can be attributed primarily to enzymatic actions (e.g., enzymatic decrystallization and hydrolytic cleavage) after the initial binding. Because the extent of bending is linearly proportional to the effective concentration of adsorbed species on the microcantilever (Berger et al., 1997; Desikan et al., 2006; Velanki & Ji, 2006), the real-time measurement of the microcantilever bending will reveal the kinetics for enzymatic actions including enzymatic decrystallization by cellulase. To our knowledge, this is the first use of a nanomechanical sensor to study mechanistic enzymology and heterogeneous enzymatic catalysis that involves a solid substrate (e.g., cellulose).

3. Investigation of the interaction between cellulase and cellulose

3.1. Materials and methods

Materials: Microcrystalline cellulose, dimethyl sulfoxide (DMSO), and N-methylmorpholine -N-oxide (NMMO) were purchased from Sigma-Aldrich (St. Louis, MO). Polyvinylamine (PVAM) was purchased from BASF (Florham Park, NJ). The microcantilevers (200 μm × 25 μm × 2 μm, 0.1 N/m) were home-made.

Preparation of the cellulose-coated microcantilever (Zhao et al., 2010): The cellulose II model surface was prepared on the front side of a microcantilever that was made of SiO₂. The surface of the microcantilever was first treated with UV ozone for 20 min. It was then immersed in 0.22% PVAM for 60 min followed by rinsing with water. Both the front side and the back side of the microcantilever were coated with PVAM. A suspension of 0.5 mg of microcrystalline cellulose powder in 25 mL of 50% NMMO was heated while stirring until a transparent brown solution was obtained. While still warm, DMSO was added to afford a cellulose solution with a final concentration of 1%. This solution was first evenly applied onto the surface of the front side of the PVAM-coated microcantilever and then allowed to sit for about 1 h. Afterwards a drop of water was added to form the cellulose film. The resulting cellulose-coated microcantilever was soaked in water for additional 4 h, during which the water was replaced every 30 min. Finally the cellulose-coated microcantilever was incubated in an oven at 80°C for 1 h to complete the surface coating.

AFM imaging (Zhao et al., 2010): To characterize the coverage, morphology, roughness, and thickness of surface coatings, AFM imaging was performed using a Nanoscope IIIa multimode scanning probe microscope from Digital Instruments, Inc. The samples were scanned in contact mode in air using silicon nitride cantilevers (MLCT) manufactured by Veeco (Camarillo, CA), with a nominal spring constant of 0.05 N/m. Images were obtained from at least three different surface areas of the sample with a typical size of 5 μm × 5 μm. The images of both height and deflection modes were captured and the surface morphology was analyzed using the image-processing software.

Microcantilever measurement: All the experiments were performed using a modified commercial Nanoscope III scanning probe microscope (Digital Instruments/Veeco, Santa Barbara, CA). The cellulose-coated microcantilever was mounted with the coating facing down in a liquid cell of the AFM with a volume of 50 μL. The solutions were introduced through injection. The cellulose-coated microcantilever was usually allowed to equilibrate in the assay buffer (25 mM sodium acetate, pH 5.5) for at least 2 h prior to any addition. The enzyme solution was prepared in the same buffer 30 min prior to use. Each assay was run against a reference to allow the subtraction of the background signal and to control for the bulk solvent effect. A desktop PC, running programs written in LabView (National Instruments, Austin, TX), was used to record the microcantilever deflection signal from the AFM via a data acquisition board with a maximum data acquisition rate of 300 kHz. The deflection measurement was monitored using a 5 mW laser diode with a wavelength of 680 nm, and a split position sensitive detector. During a measurement, 100,000 data points were taken every 30 seconds at a rate of 100 kHz. The bending of the microcantilever was obtained by simply averaging the data points.

3.2. Preparation of a cellulose-coated microcantilever

The surface of a microcantilever was coated with a model film of cellulose II having a thickness between 10 to 20 nm. It is generally agreed that the natural substrate is often too complicated to be useful for detailed characterizations of cellulases (Kontturi et al., 2006). Various cellulose model surfaces have therefore been developed in the past decade (Kontturi et al., 2006) and utilized extensively in a variety of studies including the investigation of the interaction between cellulose and cellulase. Among all the model surfaces, cellulose II has been used most often and there is much technical information available about it. In addition, the surface of cellulose II is easy to prepare and characterize. All of these make the cellulose II film an ideal model surface for the current study.

For attaching cellulose to the surface of a microcantilever, an anchoring layer of PVAM was first prepared (Figure 10A). The cellulose dissolved in a hot mixture of NMMO and DMSO was then deposited onto the top of the anchoring layer (Falt et al., 2003; Zhao et al., 2010). The cellulose coating on the microcantilever was characterized with AFM imaging and the results of the modification of the surface topography after each coating step were shown in the Figures 10B to D. A typical PVAM-coated surface is shown as a layer of small oval particles in Figure 10C. The smoothness and thickness of the coating can be adjusted by changing the coating time. The cellulose II model film on the microcantilever exhibits a cement-like characteristic with a cover of thick, short fibers (Figure 10D).

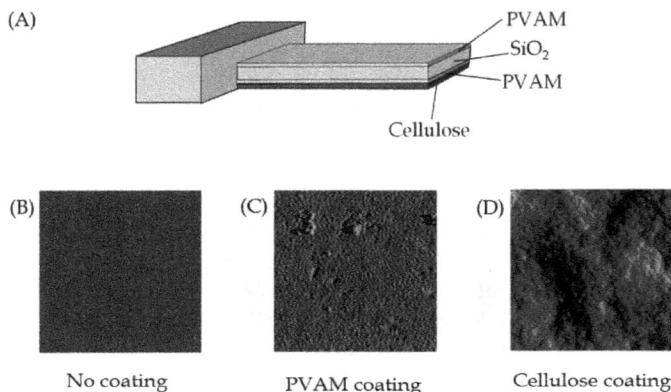

Figure 10. Surface coating on the microcantilever (Zhao et al., 2010). (A) The coating scheme. (B) to (D) AFM images of surface coatings (5 μm × 5 μm).

3.3. Examination of morphological and structural changes of cellulose surface

The microcantilever coated with cellulose II was treated with a successive exposure of different water/salt solutions. The result of bending of the microcantilever is presented in Figure 11 (Zhao et al., 2010). In the experiment, the cellulose-coated microcantilever was allowed to equilibrate in water for 2 h to achieve a stable baseline. The bending of the microcantilever was measured based on the deflection of the microcantilever at its apex. The measurement was initiated with the injection of water and a constant bending was observed during the first 45 min. The level of bending remained virtually constant even after the injection of a solution of 0.1 M NaCl. Upon the subsequent treatment of higher concentrations of NaCl (0.5 and 1 M), a continuous rise in bending was detected at a pace of roughly 1 nm/min and a cumulative bending of more than 100 nm was observed. To compensate for bulk effects of the buffer and the salt, we measured the differential bending, termed simply "bending" hereafter, which is defined as the difference in deflection of the microcantilever with the cellulose coating and without the cellulose coating.

The observed bending of the microcantilever can be attributed to the change in interaction energy within the outlayer of the cellulose surface. It has been shown previously that a cellulose model film undergoes a change in internal charge density when exposed to an electrolyte solution. And the magnitude of the change in charge density depends on the concentration of the electrolyte (Ahola et al., 2008; Freudenberg et al., 2007; Tammelin et al., 2006). Such change likely alters the intermolecular repulsion among cellulose molecules, which leads to the change in interaction energy. When the cellulose is deposited onto one side of the microcantilever (Figure 10A), the change in the interaction energy in the cellulose coating can exert a differential mechanical stress between the opposite surfaces of the microcantilever, leading to the continuous bending of the microcantilever (Figure 11A). Meanwhile, the change in interaction energy also causes the change in surface morphology of the cellulose coating as indicated by the image shown in Figure 11D.

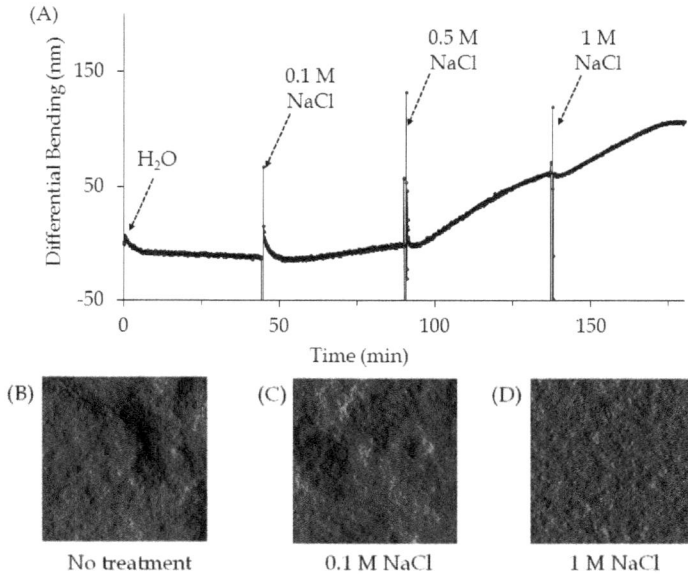

Figure 11. The morphological changes of the cellulose coating monitored by means of the microcantilever technique (Zhao et al., 2010). (A) The bending of the microcantilever increased with the increasing NaCl concentration. (B) AFM image of the surface of the cellulose coating on the microcantilever. (C) AFM image of the cellulose surface (as shown in (B)) after being treated with 0.1 M NaCl. (D) The cellulose surface (as shown in (C)) after being treated with 1.0 M NaCl. Mean roughness: 2.59 nm with a Z range of 25 nm.

Overall, this study validates that the microcantilever technique is highly sensitive and specific in detecting real-time changes in interaction energy in the surface layer of the cellulose coating on the microcantilever. So, it is feasible of using the microcantilever technique to monitor the dynamic change in interaction energy in the surface layer of the cellulose caused by enzymatic decrystallization. The microcantilever bending also correlates well with the change in molecular structure of the surface region of the cellulose film.

3.4. Detection of the enzymatic decrystallization by cellulase on cellulose

The cellulose coating on a microcantilever was treated with the cellulase CBH I (cellobiohydrolase I, $k_d = 1$ μM, mw = 66 kD), an exoglucanase from *Trichoderma reesei* in 25 mM sodium acetate buffer, pH 5.5 at 25°C. Immediately after the addition of 0.15 μM of CBH I, a bell-shaped bending curve was obtained (Figure 12A), which implies that the cellulase is capable of inducing bending of the cellulose-coated microcantilever.

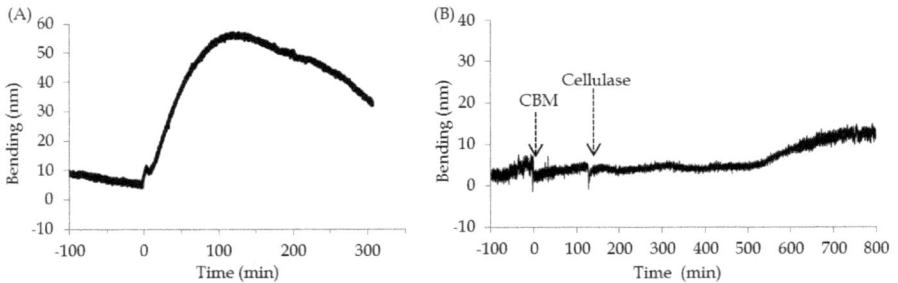

Figure 12. (A) The progress curves of cellulase actions on cellulose measured by the microcantilever sensor. (B) The progress curves of CBM and cellulase actions on cellulose measured by the microcantilever sensor.

Next, the cellulose coating was treated with 0.9 μM of the carbohydrate binding module (CBM, k_d = 0.6 μM, mw = 17 kD) from *Clostridium cellulovorans*, the domain that anchors cellulase to cellulose at 25°C. Figure 12B clearly shows that exposing the cellulose to the CBM does not generate any measurable bending in microcantilever over the course of 120 min. Over the same time frame of the previous experiment, CBH I that contains both CD and CBM domains did induce bending, as shown in Figure 12A. Notably, more weight was probably adsorbed onto the surface of the cellulose coating in the presence of the CBM than in the presence of CBH I due to the difference in protein concentration used in each experiment. This result confirmed that the change in mass (gravity) due to the initial binding of the CBM on the surface of the cellulose does not generate any bending in microcantilever, which was fully expected for a mere protein binding (physisorption) in liquid media. Thus, the cellulase-induced bending shown in Figure 12A was not caused by the initial binding between the cellulase (via the CBM of the cellulase) and the cellulose coating. This bending can therefore be attributed to the result of the cellulase actions that occurred after its initial binding on cellulose. Such actions may include enzymatic decrystallization and/or subsequent hydrolytic cleavage (Figure 6). After the addition of the CBM, the subsequent addition of CBH I (cellulase) did not generate any bending until approximately 6 h later (Figure 12B). This result verified that the CBM was indeed bound to the surface of the cellulose and the cellulose-bound CBM prevented the subsequent binding of CBH I to the cellulose. We believe that the bending in microcantilever beginning after 500 min was due to a slow displacement of the cellulose-bound CBM by CBH I.

4. Conclusion

These studies have demonstrated that a nanomechanical sensor in microcantilever is capable of detecting the interaction between cellulase and cellulose in real time. More specifically, this technique can be used to probe the dynamic process of the enzymatic decrystallization of cellulose by cellulase. The bending of the microcantilever is likely a result of the change in interaction energy within the cellulose caused by the interaction between cellulase and cellulose (e.g., enzymatic decrystallization), not by the adsorption of cellulase onto cellulose. The innovative microcantilever sensor approach will be used to determine the kinetics of the enzymatic decrystallization by cellulase.

Author details

Jun Xi and Wenjian Du
Drexel University, Department of Chemistry, Philadelphia, USA

Linghao Zhong
Penn State University, Mont Alto, USA

Acknowledgement

This work was supported in part by a grant from the National Science Foundation (NSF) CBET-0843921, Drexel University Career Development award, and subcontract XCO-4-33099-01 from the National Renewable Energy Laboratory funded by the U.S. DOE Office of the Biomass Program.

5. References

Ahola, S., Salmi, J., Johansson, L. S., Laine, J. & Österberg, M. (2008). Model films from native cellulose nanofibrils. Preparation, swelling, and surface interactions. *Biomacromolecules*, Vol.9, No.4, pp. 1273-1282, ISSN 1525-7797

Ahola, S., Turon, X., Österberg, M., Laine, J. & Rojas, O. J. (2008). Enzymatic hydrolysis of native cellulose nanofibrils and other cellulose model films: Effect of surface structure. *Langmuir*, Vol.24, No.20, pp. 11592-11599, ISSN 0743-7463

Berger, R., Delamarche, E., Lang, H. P., Gerber, C., Gimzewski, J. K., Meyer, E. & Guntherodt, H.-J. (1997). Surface stress in the self-assembly of alkanethiols on gold. *Science (Washington, D. C.)*, Vol.276, No.5321, pp. 2021-2024, ISSN 0036-8075

Boraston, A. B., Bolam, D. N., Gilbert, H. J. & Davies, G. J. (2004). Carbohydrate-binding modules: Fine-tuning polysaccharide recognition. *Biochem J*, Vol.382, No.Pt 3, pp. 769-781, ISSN 1470-8728 (Electronic)

Brown, R. M. J. & Saxena, I. M. (2007). *Cellulose: Molecular and structural biology*, Springer, ISBN 978-1-4020-5332-0, New York, NY

Chundawat, S. P. S., Beckham, G. T., Himmel, M. E. & Dale, B. E. (2011). Deconstruction of lignocellulosic biomass to fuels and chemicals. *Annual Review of Chemical and Biomolecular Engineering*, Vol.2, No.1, pp. 121-145, ISSN

Datskos, P. G. & Sauers, I. (1999). Detection of 2-mercaptoethanol using gold-coated micromachined cantilevers. *Sensors and Actuators B: Chemical*, Vol.61, No.1–3, pp. 75-82, ISSN 0925-4005

Desikan, R., Lee, I. & Thundat, T. (2006). Effect of nanometer surface morphology on surface stress and adsorption kinetics of alkanethiol self-assembled monolayers. *Ultramicroscopy*, Vol.106, No.8-9, pp. 795-799, ISSN 0304-3991

Divne, C., Stahlberg, J., Teeri, T. T. & Jones, T. A. (1998). High-resolution crystal structures reveal how a cellulose chain is bound in the 50 .Ang. Long tunnel of cellobiohydrolase i from trichoderma reesei. *Journal of Molecular Biology*, Vol.275, No.2, pp. 309-325, ISSN 0022-2836

DOE/SC-0095. (2006). Breaking the biological barriers to cellulosic ethanol: A joint research agenda. *DOE/SC-0095*, U.S. Departemnt of Energy Office of Science and Office of Energy Efficiency and Renewable Energy (www.doegenomestolife.org/biofuels/), ISSN

DOE/SC-0095. (2006). Breaking the biological barriers to cellulosic ethanol: A joint research agenda. *DOE/SC-0095*, Available from: <http://genomicscience.energy.gov/biofuels/b2bworkshop.shtml>

Dwivedi, P., Alavalapati, J. R. R. & Lal, P. (2009). Cellulosic ethanol production in the united states: Conversion technologies, current production status, economics, and emerging developments. *Energy for Sustainable Development*, Vol.13, No.3, pp. 174-182, ISSN 0973-0826

Falt, S., Waagberg, L. & Vesterlind, E.-L. (2003). Swelling of model films of cellulose having different charge densities and comparison to the swelling behavior of corresponding fibers. *Langmuir*, Vol.19, No.19, pp. 7895-7903, ISSN 0743-7463

Fardim, P., Gustafsson, J., von Schoultz, S., Peltonen, J. & Holmbom, B. (2005). Extractives on fiber surfaces investigated by xps, tof-sims and afm. *Colloids and Surfaces A: Physicochemical and Engineering Aspects*, Vol.255, No.1–3, pp. 91-103, ISSN 0927-7757

Fengel, D., Jakob, H. & Strobel, C. (1995). Influence of the alkali concentration on the formation of cellulose-ii - study by x-ray-diffraction and ftir spectroscopy. *Holzforschung*, Vol.49, No.6, pp. 505-511, ISSN 0018-3830

Freudenberg, U., Zimmermann, R., Schmidt, K., Behrens, S. H. & Werner, C. (2007). Charging and swelling of cellulose films. *Journal of Colloid and Interface Science*, Vol.309, No.2, pp. 360-365, ISSN 0021-9797

Goeders, K. M., Colton, J. S. & Bottomley, L. A. (2008). Microcantilevers: Sensing chemical interactions via mechanical motion. *Chemical Reviews*, Vol.108, No.2, pp. 522-542, ISSN 0009-2665

Gray, K. A., Zhao, L. & Emptage, M. (2006). Bioethanol. *Current Opinion in Chemical Biology*, Vol.10, No.2, pp. 141-146, ISSN 1367-5931

Guillén, D., Sánchez, S. & Rodríguez-Sanoja, R. (2010). Carbohydrate-binding domains: Multiplicity of biological roles. *Applied Microbiology and Biotechnology*, Vol.85, No.5, pp. 1241-1249, ISSN 0175-7598

Kontturi, E., Tammelin, T. & Osterberg, M. (2006). Cellulose-model films and the fundamental approach. *Chemical Society Reviews*, Vol.35, No.12, pp. 1287-1304, ISSN 0306-0012

Lavrik, N. V., Sepaniak, M. J. & Datskos, P. G. (2004). Cantilever transducers as a platform for chemical and biological sensors. *Review of Scientific Instruments*, Vol.75, No.7, pp. 2229-2253, ISSN 0034-6748

Lee, I., Evans, B. R. & Woodward, J. (2000). The mechanism of cellulase action on cotton fibers: Evidence from atomic force microscopy. *Ultramicroscopy*, Vol.82, No.1–4, pp. 213-221, ISSN 0304-3991

Li, Q. & Renneckar, S. (2011). Supramolecular structure characterization of molecularly thin cellulose i nanoparticles. *Biomacromolecules*, Vol.12, No.3, pp. 650-659, ISSN 1525-7797

Li, Y., Irwin, D. C. & Wilson, D. B. (2007). Processivity, substrate binding, and mechanism of cellulose hydrolysis by thermobifida fusca cel9a. *Appl. Environ. Microbiol.*, Vol.73, No.10, pp. 3165-3172, ISSN 0099-2240

Lynd, L. R., Weimer, P. J., van Zyl, W. H. & Pretorius, I. S. (2002). Microbial cellulose utilization: Fundamentals and biotechnology. *Microbiol Mol Biol Rev*, Vol.66, No.3, pp. 506-577, table of contents, ISSN 1092-2172 (Print)

Martin, S. (2000). Protein engineering of cellulases. *Biochimica et Biophysica Acta (BBA) - Protein Structure and Molecular Enzymology*, Vol.1543, No.2, pp. 239-252, ISSN 0167-4838

Mosier, N. S., Hall, P., Ladisch, C. M. & Ladisch, M. R. (1999). Reaction kinetics, molecular action, and mechanisms of cellulolytic proteins. *Advances in Biochemical Engineering/Biotechnology*, Vol.65, No.Recent Progress in Bioconversion of Lignocellulosics, pp. 23-40, ISSN 0724-6145

Moulin, A. M., O'Shea, S. J., Badley, R. A., Doyle, P. & Welland, M. E. (1999). Measuring surface-induced conformational changes in proteins. *Langmuir*, Vol.15, No.26, pp. 8776-8779, ISSN 0743-7463

Mukhopadhyay, R., Sumbayev, V. V., Lorentzen, M., Kjems, J., Andreasen, P. A. & Besenbacher, F. (2005). Cantilever sensor for nanomechanical detection of specific protein conformations. *Nano Lett.*, Vol.5, No.12, pp. 2385-2388, ISSN 1530-6984

Mulakala, C. & Reilly, P. J. (2005). Hypocrea jecorina (trichoderma reesei) cel7a as a molecular machine: A docking study. *Proteins: Structure, Function, and Bioinformatics*, Vol.60, No.4, pp. 598-605, ISSN 1097-0134

Parsiegla, G., Reverbel, C., Tardif, C., Driguez, H. & Haser, R. (2008). Structures of mutants of cellulase cel48f of clostridium cellulolyticum in complex with long hemithiocellooligosaccharides give rise to a new view of the substrate pathway during processive action. *Journal of Molecular Biology*, Vol.375, No.2, pp. 499-510, ISSN 0022-2836

Pérez, S. & Samain, D. (2010). Structure and engineering of celluloses. *Advances in Carbohydrate Chemistry and Biochemistry*, Vol.Volume 64, 25-116, ISSN 0065-2318

Peter, Z. (2001). Conformation and packing of various crystalline cellulose fibers. *Progress in Polymer Science*, Vol.26, No.9, pp. 1341-1417, ISSN 0079-6700

Reinikainen, T., Ruohonen, L., Nevanen, T., Laaksonen, L., Kraulis, P., Jones, T. A., Knowles, J. K. C. & Teeri, T. T. (1992). Investigation of the function of mutated cellulose-binding domains of trichoderma reesei cellobiohydrolase i. *Proteins: Structure, Function, and Bioinformatics*, Vol.14, No.4, pp. 475-482, ISSN 1097-0134

Rojas Orlando, J., Jeong, C., Turon, X. & Argyropoulos Dimitris, S. (2007). Measurement of cellulase activity with piezoelectric resonators, In: *Materials, chemicals, and energy from forest biomass*, pp. 478-494, American Chemical Society, ISBN 0-8412-3981-9,

Schenzel, K., Fischer, S. & Brendler, E. (2005). New method for determining the degree of cellulose i crystallinity by means of ft raman spectroscopy. *Cellulose*, Vol.12, No.3, pp. 223-231, ISSN 0969-0239

Shu, W., Liu, D., Watari, M., Riener, C. K., Strunz, T., Welland, M. E., Balasubramanian, S. & McKendry, R. A. (2005). DNA molecular motor driven micromechanical cantilever arrays. *J. Am. Chem. Soc.*, Vol.127, No.48, pp. 17054-17060, ISSN 0002-7863

Shuttleworth, R. (1950). The surface tension of solids. *Proceedings of the Physical Society. Section A*, Vol.63, No.5, pp. 444, ISSN 0370-1298

Srisodsuk, M., Reinikainen, T., Penttilä, M. & Teeri, T. T. (1993). Role of the interdomain linker peptide of trichoderma reesei cellobiohydrolase i in its interaction with crystalline cellulose. *Journal of Biological Chemistry*, Vol.268, No.28, pp. 20756-20761, ISSN 0021-9258

Tammelin, T., Saarinen, T., Österberg, M. & Laine, J. (2006). Preparation of langmuir/blodgett-cellulose surfaces by using horizontal dipping procedure. Application for polyelectrolyte adsorption studies performed with qcm-d. *Cellulose*, Vol.13, No.5, pp. 519-535, ISSN 0969-0239

Turon, X., Rojas, O. J. & Deinhammer, R. S. (2008). Enzymatic kinetics of cellulose hydrolysis: A qcm-d study. *Langmuir*, Vol.24, No.8, pp. 3880-3887, ISSN 0743-7463

Van Tilbeurgh, H., Tomme, P., Claeyssens, M., Bhikhabhai, R. & Pettersson, G. (1986). Limited proteolysis of the cellobiohydrolase i from trichoderma reesei: Separation of functional domains. *FEBS Letters*, Vol.204, No.2, pp. 223-227, ISSN 0014-5793

Vashist, S. K. (2007). A review of microcantilevers for sensing applications. Available from: <http://dx.doi.org/10.2240/azojono0115>

Velanki, S. & Ji, H.-F. (2006). Detection of feline coronavirus using microcantilever sensors. *Meas. Sci. Technol.*, Vol.17, No.11, pp. 2964-2968, ISSN 0957-0233

Wilson, D. B. & Irwin, D. C. (1999). Genetics and properties of cellulases. *Advances in Biochemical Engineering/Biotechnology*, Vol.65, No.Recent Progress in Bioconversion of Lignocellulosics, pp. 1-21, ISSN 0724-6145

Wilson, D. B. (2009). Cellulases and biofuels. *Current Opinion in Biotechnology*, Vol.20, No.3, pp. 295-299, ISSN 0958-1669

Yan, X., Hill, K., Gao, H. & Ji, H.-F. (2006). Surface stress changes induced by the conformational change of proteins. *Langmuir*, Vol.22, No.26, pp. 11241-11244, ISSN 0743-7463

Yang, H., Yan, R., Chen, H., Lee, D. H. & Zheng, C. (2007). Characteristics of hemicellulose, cellulose and lignin pyrolysis. *Fuel*, Vol.86, No.12–13, pp. 1781-1788, ISSN 0016-2361

Zhang, S. & Wilson, D. B. (1997). Surface residue mutations which change the substrate specificity of thermomonospora fusca endoglucanase e2. *Journal of Biotechnology*, Vol.57, No.1–3, pp. 101-113, ISSN 0168-1656

Zhang, Y. H. & Lynd, L. R. (2004). Toward an aggregated understanding of enzymatic hydrolysis of cellulose: Noncomplexed cellulase systems. *Biotechnol Bioeng*, Vol.88, No.7, pp. 797-824, ISSN 0006-3592 (Print)

Zhao, L., Bulhassan, A., Yang, G., Ji, H.-F. & Xi, J. (2010). Real-time detection of the morphological change in cellulose by a nanomechanical sensor. *Biotechnology and Bioengineering*, Vol.107, No.1, pp. 190-194, ISSN 1097-0290

Zhong, L., et al. (2008). Interactions of the complete cellobiohydrolase i from trichodera reesei with microcrystalline cellulose iβ. *Cellulose*, Vol.15, No.2, pp. 261-273, ISSN 0969-0239

Zhou, F., Shu, W., Welland, M. E. & Huck, W. T. S. (2006). Highly reversible and multi-stage cantilever actuation driven by polyelectrolyte brushes. *Journal of the American Chemical Society*, Vol.128, No.16, pp. 5326-5327, ISSN 0002-7863

Chapter 8

Cellulose and Its Derivatives Use in the Pharmaceutical Compounding Practice

Flávia Dias Marques-Marinho and Cristina Duarte Vianna-Soares

Additional information is available at the end of the chapter

1. Introduction

For centuries, the pharmaceutical profession has provided services of fundamental value to society, such as the procurement, storage, compounding and dispensing of drugs. In recent decades, the focus of the pharmacist's role has shifted from compounding medicines to ensuring their safe and effective use by providing information and advice [1,2]. Although compounding activity has decreased over time, it is incontestable that this service is essential in certain patient specific situations, where industrially produced medicine is not available or is inappropriate for a particular reason [3,4]. Thus, compounded medicines are mainly important for paediatric and geriatric patients, and patients with special needs such as those with dermatological diseases [3,5,6,7,8]. In many countries, nowadays, the activity of compounding is a complementary practice to the production of medicines in alternative amounts and diversified dosage forms (liquid, semi-solid, solid) in community pharmacies (United States of America, The Netherlands), as well as in hospital pharmacies (Canada, France, Belgium, Croatia, Denmark, England, Finland, Germany, Ireland, Italy, Norway, Scotland, Slovenia, Spain, Sweden, Switzerland) [3,4,8-10].

Interestingly, compounded medicines were estimated to make up 10-15% of all dispensed drugs in the Netherlands in the early 90's, 5.5% in 1994 and 6.6% in 1995. By the year 2000 this estimate was 3.7-5.5% [3]. On the other hand, the compounding field appears to have been a considerable and growing business since the 1990's in the United States. These products represented around 1% of all prescriptions dispensed yearly and according to this estimate, 30 million medications would have been compounded in 2003 [4]. This shows that the population has recognized that compounding pharmacies can provide individualized drug therapy benefits [11].

The practice of compounding requires not only the drug(s) (active pharmaceutical ingredient, API), but also, the excipient(s) (pharmacological inert component) in order to obtain the final medicine. The excipients are chosen according to the characteristics of the required dosage form [5]. Each excipient exerts specific functions in the formulation, as, for instance, a diluent for hard capsules or powders, a coating agent for solid oral dosage forms, a suspending, thickening or stabilizing agent for oral liquids, etc. The excipient function depends on the concentration in a particular pharmaceutical formulation [12,13].

Cellulose (Figure 1) and its derivatives (ether and ester) are among the excipients frequently used in pharmaceutical compounded and industrialized products with various purposes. Among their uses, the most frequently reported are as suspending agents in oral liquid extemporaneous preparation and as viscosity increasing agents in topical formulations, exemplified in Tables 1, 2 [5,7,8,14]. Particularly, in oral solid dosage forms, cellulose and its derivatives (also known as cellulosics) can render distinct drug delivery property patterns: immediate, controlled/sustained or delayed release [15,16]. In addition, cellulosics show several interesting characteristics such as low cost, reproducibility, biocompatibility, and recyclability [16]. The latter is currently an important aspect considering the need for green technology.

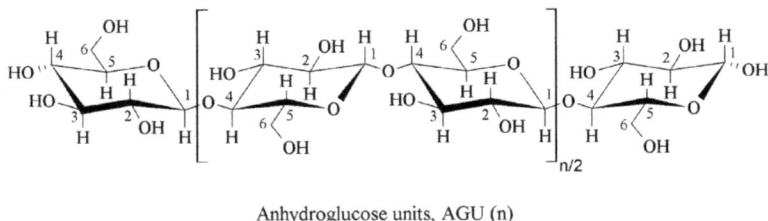

Anhydroglucose units, AGU (n)

Figure 1. The chemical structure of cellulose with two β-1,4 linked anhydroglucose units.

Cellulose is the most abundant biopolymer. It is present in the cell walls of a great diversity of organisms, from bacteria (Cyanobacteria), prokaryotes (*Acetobacter, Rhizobium, Agrobacterium*) to eukaryotes (fungis, amoebae, green algae, freshwater and marine algae, mosses, ferns, angiosperms, gymnosperms). It is also produced by some animals, the tunicates (urochordates), members of the subphylum Tunicata in the Chordata phylum [17-19]. Native cellulose made by biosynthesis in living organisms is composed only of glucose monomers, as anhydroglucose (AGU) or glucan units ($C_6H_{10}O_5$, n) with β-1,4 linkages (Figure 1). It usually exists as cellulose I (in most plants) and rarely as cellulose II (in several algae and some bacteria) allomorphs, in which the glucan chains are oriented parallel and antiparallel respectively [20, 21]. Cellulose allomorphs (I, II, III_I, III_{II}, IV_I, IV_{II}) have structural variations regarding unit cell dimensions, degree of intra/interchain hydrogen bonding per unit cell and polarity of adjacent cellulose sheets [22]. Cellulose I allomorphs consist of distinct numbers of parallel glucan chains arranged to form nanofibrils. Native crystalline cellulose I has two suballomorphs, α and β, which exist as a single chain triclinic unit cell and a two chain monoclinic unit cell, respectively. Cellulose I β is rarely synthesized in nature as a pure form (except by tunicates) and is more thermodynamically stable [20]. Cellulose I can be altered by a strong alkali treatment in order to produce other crystalline forms, II, III and IV. Cellulose II is the allomorph that is thermodynamically most stable [16,23-24]. Cellulose III can be prepared by liquid ammonia or (mono, di, tri) amine treatment of cellulose I and II [25]. The cellulose IV crystalline form is obtained by immersion in glycerol and heating of cellulose III [26]. Cellulose is an excipient widely employed by both pharmaceutical companies (tablet processing) and compounding pharmacies; it is available in powdered (n≈500) and microcrystalline (n≈220) forms, the latter being obtained by acid hydrolysis of the amorphous regions of the cellulose nanofibrils.

Cellulosics, such as methyl, ethyl, hydroxyethyl, hydroxyethylmethyl, hydroxypropyl (HP), hydroxypropyl methyl (HPM, also denominated hypromellose) and carboxymethyl ethers cellulose (Figure 2) are formed by hydroxyl etherification with the appropriate alkyl halide (R-Cl, see Figures 2, 3) from previously alkalinized cellulose usually obtained from wood pulp [16]. The degree of substitution (DS) in these ether derivatives indicates the average number of R groups present in each glucan unit along the chain. The maximum DS is three, since it is the number of hydroxyl groups that can be substituted on each glucan unit. DS affects cellulose derivatives' physical properties such as solubility [12].

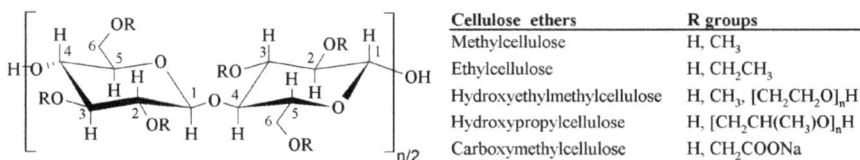

Cellulose ethers	R groups
Methylcellulose	H, CH_3
Ethylcellulose	H, CH_2CH_3
Hydroxyethylmethylcellulose	H, CH_3, $[CH_2CH_2O]_nH$
Hydroxypropylcellulose	H, $[CH_2CH(CH_3)O]_nH$
Carboxymethylcellulose	H, CH_2COONa

Figure 2. Chemical structure of cellulose ether derivatives.

Cellulose derivatives are employed as excipients in pharmaceutical industrial products for oral, topical or parenteral administration [12,16,27]. Their most relevant application, as observed in pharmaceutical industrial products, is to create matrix systems for solid oral dosage forms. Due to their aqueous swelling, the drug release is controlled by its diffusion through the hydrogel layers that are formed. For instance, the use of carboxymethyl cellulose (CMC) sodium salt as an excipient sustains the release in solid oral dosage forms.

Cellulosics, such as the cellulose esters acetate, acetate trimellitate, acetate phthalate (CAP), HPM phthalate, HPM acetate succinate are formed by hydroxyl esterification with either acetic, trimellitic, dicarboxylic phthalic or succinic acids, or a combination of them, as represented in Figure 3. The reaction usually occurs in the presence of a strong acid that promotes the acid catalysis.

Among these cellulosics, CAP was one of the earliest and most effective solutions to pH-controlled release, and its use still continues today [15]. These cellulosics are usually resistant to acid environments such as that of the stomach and are thus very useful as enteric coatings for capsules or tablets [12,16]. Cellulose esters require plasticizers (acetylated monoglyceride, butyl phtalylbutyl glycolate, dibutyl tartrate, diethyl phthalate, dimethyl phthalate, ethyl phthalylethyl glycolate, glycerin, propylene glycol, triacetin, triacetin citrate, triethylcitrate, tripropionin) soluble in organic solvents (ketones, esters, ether alcohols, cyclic ethers) or in their mixtures, such as methanol/chloroform and ethyl acetate/isopropanol in order to produce more effective coating films [12,28,29]. Some of the cellulose esters are employed either in industrial or compounded pharmaceutical preparations.

Some cellulosics, if they are to be applied in distinct drug delivery formulations, may require special large scale processing and equipment normally only installed in pharmaceutical industry plants. This is one of the reasons why not all commercially available cellulosics are employed in compounding pharmacies. A description of some cellulosics and their applications in compounded medicines is presented in the following sections.

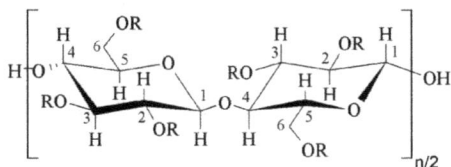

(Cellulose) esters	R groups
Acetate	H, I
Acetate trimellitate	H, I, II
Acetate phthalate	I, III
Hydroxypropylmethylphthalate	H, CH_3, $CH_2CH(OH)CH_3$, III, IV
Hydroxypropylmethylphthalate acetate succinate	H, CH_3, $CH_2CH(OH)CH_3$, I, V

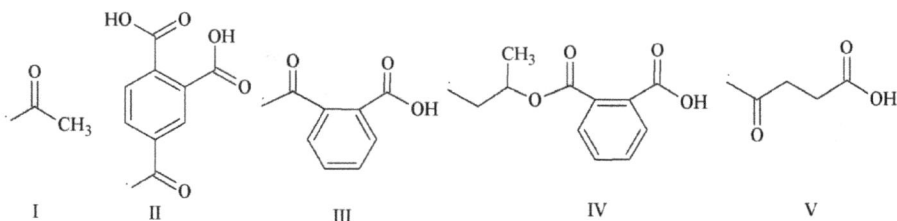

Figure 3. Chemical structures of cellulose ester derivatives.

2. Cellulose and its derivatives in compounded medicines

2.1. Cellulose

Powdered cellulose and microcrystalline cellulose come from α-cellulose (cellulose free of hemi-celluloses and lignin) pulp from fibrous plant materials; they differ in regard to their manufacturing processes. Powdered cellulose is obtained by α-cellulose purification and mechanical size reduction. Crystalline cellulose is obtained by controlled hydrolysis of α-cellulose with mineral acid solutions (2 to 2.5 N), followed by hydrocellulose purification by filtration and spray-drying of the aqueous portion [12].

In compounded medicines, powdered cellulose and microcrystalline cellulose are used as an adsorbent, a suspending agent, a capsule diluent (5-30% and 20-90%, respectively). Powdered cellulose is also used as a thickening agent.

The applications of the powdered cellulose and the microcrystalline cellulose in compounding pharmacies include the oral solid dosage form (capsules) as a bulking agent to increase the mass in formulations containing small amounts of the active ingredient. The powdered cellulose is a base material for powder dosage forms, a suspending agent for aqueous peroral delivery and an adsorbent and thickening agent for topic preparations [12]. Moreover, the microcrystalline cellulose is a constituent of the vehicle used for oral suspension [27].

2.2. Cellulose ether derivatives

2.2.1. Methylcellulose (MC)

In this cellulose ether derivative approximately 27–32% of hydroxyl groups are changed to the methyl ether (CH_3O) form. MC is practically insoluble in most organic solvents. Various grades of MC can be found with degrees of polymerization in the range of 50 to 1000 and molecular weights (number average) in the range 10 000 to 220 000 Da [12].

In compounded medicines, MCs function as emulsifying agents (1-5%), suspending agents (1-2%), capsule disintegrants and viscosity increasing agents.

In compounding pharmacies, MCs of different viscosity grades, low and high, have been applied in oral liquid (oil emulsions, suspensions, solutions) and topical (creams, gels) formulations respectively. MC is often used instead of sugar-based syrups and other suspension bases. MC delays the settling of suspensions and increases the contact time of drugs in the stomach [12].

2.2.2. Ethylcellulose (EC)

This cellulose derivative is partially or completely ethoxylated, yielding 44-51% of ethoxyl groups (OCH_2CH_3). Full substitution (DS=3) of cellulose units produce $C_{12}H_{23}O_6(C_{12}H_{22}O_5)_nC_{12}H_{23}O_5$, where n can vary, thus providing a wide variety of molecular weights. EC is a long-chain polymer of ethyl-substituted β-glucan units joined together by glycoside linkages [12].

In compounded medicines, EC functions as a flavouring and as a viscosity increasing agent.

In compounding pharmacies, EC finds applications in oral and topical (creams, lotions, gels) formulations. For oral use, it works as an active delivering agent and for topical dosage forms as a thickening agent. It has been evaluated as a stabilizer for emulsions [12].

2.2.3. Hydroxyethylcellulose (HEC)

This cellulose derivative is a partially substituted hydroxyethyl (CH_2CH_2OH) ether of cellulose. It is found in various viscosity grades, with respect to the DS and molecular weight. Some grades are modified so as to improve aqueous dispersion. HEC is insoluble in most organic solvents.

In compounded medicines, HEC has the following functions: a suspending, a thickening and a viscosity-increasing agent.

It is widely employed in topical formulations (gel) and cosmetics due to its nonionic and water-soluble polymer characteristics. The main use is as a thickening agent [12].

2.2.4. Hydroxypropylcellulose (HPC)

This cellulose derivative is partially hydroxypropylated, yielding 53.4–80.5% of hydroxypropyl groups [$OCH_2CH(OH)CH_3$]. Because the added hydroxypropyl contains a hydroxyl group which can also be etherified during the preparation, the degree of substitution of hydroxypropyl groups can be higher than three. HPC is found in different grades that provide solutions with various viscosities. Its molecular weight has a range of 50 000 to 1 250 000. HPC with an value of moles of substitution of approximately four is necessary in order to have good water solubility.

In compounded medicines, HPC is used as an emulsifying, a stabilizing, a suspending, a thickening or a viscosity-increasing agent.

In compounding pharmacies, HPC is also employed in topical formulations (gel) and especially in cosmetics, as an emulsifier and a stabilizer [12].

2.2.5. Hydroxypropylmethylcellulose (HPMC)

This cellulose derivative, also called hypromellose, is a partly O-methylated and O-(2-hydroxypropylated) cellulose. HPMC is found in various grades with different viscosities and extents of substitution. The content of methoxyl (OCH_3) and hydroxypropyl groups [$OCH_2CH(OH)CH_3$] affects the HPMC molecular weight, which ranges from 10 000 to 1 500 000.

HPMC has many different functions in compounded medicines as a dispersing, an emulsifying, a foaming, a solubilizing, a stabilizing, a suspending (0.25-5%) and a thickening (0.25-5%) agent. In addition, HPMC can be applied as a controlled-release and sustained-release agent.

In compounding pharmacies, HPMC has found application for nasal (liquid) and topical (gel, ointment) formulations as a thickening, a suspending, an emulsifying and a stabilizing agent. The aqueous solution produced with HPMC presents greater clarity and fewer undissolved fibres compared with MC. HPMC can prevent droplets and particles from coalescing or agglomerating, thus inhibiting the formation of sediment. In addition, it is also widely used in cosmetics [12].

2.2.6. *Carboxymethyl cellulose (CMC)*

It is available as calcium and sodium salt forms of a polycarboxymethyl (CH_2COOX, X=Ca or Na) ether of cellulose. Only sodium CMC is commonly used in compounded preparations. The degree of substitution can be estimated by a sodium assay, which must be between 6.5-9.5%.

CMC-Na acts as a capsule disintegrant and a stabilizing, a suspending, an emulsifying (0.25-1%), a gel-forming (3-6%) and a viscosity-increasing (0.1-1%) agent in compounded medicines.

In compounding pharmacies, CMC-Na has applications in oral (liquid, solid) and topical (liquid, gel, emulsion) formulations, primarily for its viscosity-increasing properties. Viscous aqueous solutions are used to suspend powders intended for either topical or oral use. In emulsions, CMC may be used as stabilizer. At higher concentrations, a CMC of intermediate-viscosity grade forms gels that are employed as a base for cosmetics or other drug formulations [12]. Similarly to microcrystalline cellulose, CMC-Na is also described as a constituent of vehicles used for oral suspension [27].

2.3. Cellulose ester derivatives

2.3.1. *Cellulose acetate*

This cellulose derivative has partially or completely acetylated ($COCH_3$) hydroxyl groups. Cellulose acetate is available in a wide range of acetyl levels (29-44.8%) and chain lengths, with molecular weights ranging from 30 000 to 60 000.

Cellulose acetate is used as a capsule diluent, a filler and as a taste-masking agent in compounded medicines [12].

2.3.2. *Cellulose acetate phthalate (CAP)*

CAP acid form is a cellulose derivative obtained by the reaction of phthalic anhydride and a partial acetate ester of cellulose. It contains 21.5–26% of acetyl ($COCH_3$) and 30-36% of phthalyl (o-carboxybenzoyl, COC_6H_4COOH) groups.

In compounded medicines, CAP confers gastro-resistance, and is thus used as an enteric coating agent (0.5-9%) [12].

In compounding pharmacies, CAP has applications in oral solid dosage forms either by film coating from organic (ketones, esters, ether alcohols, cyclic ethers) or aqueous solvent systems. Such coatings resist prolonged contact with the strongly acidic gastric fluid, but dissolve in the mildly acidic or neutral intestinal environment. The addition of plasticizers improves the water resistance of such coating materials, making formulations with this derivative more effective [12,29].

As mentioned, cellulose and its derivatives, primarily intended for use in the pharmaceutical industry, are also relevant in compounding practice. The common uses of cellulosics in compounding practices as a diluent in solid dosage forms, a thickening and a suspending agent in liquid dosage forms, an emulsifying agent in semi-solid preparations and others require well known manufacturing techniques, which do not need sophisticated apparatus [5]. Nevertheless, specific equipment may be necessary when cellulosics are employed to impart special dosage form properties, such as in modified release systems. Among the modified drug delivery systems, mostly delayed and controlled (extended or slow) release have been described in compounding practice. In the former, the systems are frequently employed to prevent drug degradation in acid environments after oral administration, to protect the stomach mucosa from drug irritation and to release the drug in the intestine. In controlled release, systems are used to prevent side effects and to reduce the number of daily administrations [29,30].

CAP is the cellulosic most frequently mentioned in compounded delayed-release dosage forms [31-33]. The use of sodium carboxymethylcellulose or hydroxypropyl methylcellulose have also been reported in extended or slow-release systems [34-40]. These systems are most commonly obtained by the simple mixing of the drug with an appropriate inert matrix [39]. The Food and Drug Administration (FDA) warns that in some instances, compounders may lack sufficient control techniques and resources (equipment, training, testing or facilities) to assure product quality or to compound more elaborated products such as modified release drugs [4]. Since obtaining high quality, safe and effective products is fundamental, compounding techniques must be developed and standardized. Thus, coating techniques used to obtain delayed release compounded capsules by the beaker flask method, by dipping or by spraying have been proposed [41]. In beaker flask coating, a small amount of coating material is added to the beaker and heated until melted. Subsequently, a few capsules are added away from the heat and the beaker is manually rotated to coat them. Small quantities of coating material are continuously added in order to prevent the capsules from sticking. The immersion or dipping method consists of heating the coating material in a recipient that permits the dipping of the capsules with the aid of tweezers in the coating solution and subsequent hardening. This process is repeated until all the capsules have a homogenous film. The vaporization or atomization method, also called spraying, consists in preparing a solution of the coating material in alcohol, ether, or keto-alcoholic solvents and transferring it to a spray bottle. The capsules are held over a screen, under ventilation. The coating solution is applied in multiple thin layers that are allowed to dry between applications. A small scale piece of machinery exists for this coating process [33].

3. Review of the use of cellulosics in compounding

The use of cellulose and/or its derivatives as part of compounded formulations has been cited in a great number of studies in the literature. A review of their properties and usage in pharmaceutical preparations is presented in the following section. Focus is given to the products that require more elaborate techniques in compounding pharmacies.

3.1. Cellulose derivatives in oral liquid (suspensions) extemporaneous preparation

It is well known that children and the elderly have difficulties in swallowing solid dosage forms, due to their size or texture. Such population groups benefit from oral liquid administration; hence, this is a preferred means of administration. Frequently, drugs in a concentration appropriate for paediatric use are unavailable or extemporaneous preparations from commercial products become a necessity. Thus, patients with special needs can be provided with drugs easily administered in the hospitals if extemporaneous preparations are compounded from drugs that are industrially produced.

Methylcellulose (1%) and simple syrup NF (as described in the United States Pharmacopeia National Formulary monograph) mixtures have been used as a vehicle for many extemporaneous oral drug suspensions prepared from commercial products (tablets or capsules) [27,41]. Compounded oral preparations can be obtained by finely grinding tablets or the content of capsules in a mortar and pestle, with the gradual addition of small volumes of the vehicle being mixed. The final volume can be adjusted in a graduated glass cylinder. Afterwards, the suspension is transferred to an appropriate plastic or glass bottle protected (amber) or not (transparent) from light.

Stability studies conducted at 4 °C and 25 °C for distinct drugs over at least 8 weeks are shown in Table 1 [42-52].

In such listed studies, suspensions containing MC did not show substantial changes in pH, odour or physical appearance in the period during which the drug content was assayed and found not to be less than 90% of the original concentration. This is a demonstration that MC formulations provide satisfactorily safe and stable products.

Oral extemporaneous suspensions of other drugs (amiodarone, granisetron, trimethoprim, and verapamil salt) prepared in methylcellulose and simple syrup (MC:SS) were also reported to have adequate stability [14].

Drug [Ref.]	Drug dosage form (mg)	Suspension (mg/mL)	MC:SS ratio	Stability (days) in different temperatures	
				4 °C	25 °C
Clonazepam [42]	Tablets (0.5, 1, 2)	0.1	NI	60c	NE
Gabapentin [43]	Capsules (100, 300, 400)	100	1:1	91c	56c
Nifedipine [44]	Capsules (10, 20)	4	1:13	91b	91b
Procainamide hydrochloride [45]	Capsules, tablets (250, 375, 500)	5, 50	70:30 Cherry syrup	180b	180b
Propylthiouracil [46]	Tablets (50)	5	1:1	91c	70c
Pyrazinamide [47]	Tablets (500)	100	1:1	60b	60b
Pyrimethamine [48]	Tablets (25)	2	1:1	91a,c	91a,c
Sildenafil citrate [49]	Tablets (25, 50, 100)	2.5	1:1	91c	91c
Sotalol hydrochloride [50,51]	Tablets (80, 120, 160, 240)	5	1:9	91b	91b
			2.4:1	84a	84a
Tiagabine hydrochloride [52]	Tablets (2, 4, 6, 8, 10, 12, 16)	1	1:6	91c	42c

a: amber glass, b: plastic; c: amber plastic bottles; NI, not informed; NE, not evaluated.

Table 1. Stability of different drugs in oral suspension extemporaneously prepared in a mixture of 1% methylcellulose:simple syrup (MC:SS) from commercial available products.

Another cellulose derivative frequently used as part of a vehicle in oral extemporaneous suspensions is sodium carboxymethylcellulose. It is a constituent of a commercial product, Ora-Plus®, which is employed in a 1:1 mixture with Ora-Sweet® (with sugar) or Ora-Sweet SF® (sugar free) [7,8]. Ora-Sweet® is a sugar-based citrus-berry flavoured syrup (see Table 2 footnote). Researchers have performed stability studies for various oral suspensions of drugs with this vehicle. The results of drug stability tests for periods of at least 8 weeks at refrigerated (3-5°C) and room (22-25°C) temperatures, are presented in Table 2 [43,46,49-77].

Drug [Ref.]	Dosage form (mg)	Suspension (mg/mL)	CMC or Ora-Plus added (1:1) of	Stability (days) in different temperatures	
				3-5°C	22-25°C
Acetazolamide [53,54]	Tablets (125, 250)	25[a]	Ora-Sweet® Ora-Sweet SF®	60[g]	60[g]
Allopurinol [54]	Tablets (100, 300)	20[a]			
Alprazolam [55]	Tablets (0.25, 0.5, 1, 2)	1[a]		60[h]	60[h]
Azathioprine [54]	Tablets (50)	50[a]		60[g]	60[g]
Baclofen [56]	Tablets (10, 20)	10[a]		60[h]	60[h]
Bethanechol chloride [57]	Tablets (5, 10, 25, 50)	5[a]	Ora-Sweet® Ora-Sweet SF®		
Captopril [56]	Tablets (12.5, 25, 50, 100)	0.75[a]		<10[h]	<10[h]
Chloroquine phosphate [55]	Tablets (250, 500)	15[a]		60[h]	60[h]
Cisapride [55]	Tablets (10, 20)	1[a]			
Clonazepam [54]	Tablets (0.5, 1, 2)	0.1[a]		60[g]	60[g]
Dapsone [58]	Tablets (25, 100)	2	Ora-Sweet®	91[e]	91[e]
Diltiazem HCl [56]	Tablets (30, 60, 90, 120)	12[a]	Ora-Sweet® Ora-Sweet SF®	60[h]	60[h]
Dipyridamole [56]	Tablets (25, 50, 75)	10[a]			
Dolasetron mesylate [59]	Tablets (50, 100)	10	Strawberry syrup Ora-Sweet SF®	90[e]	90[e]
Enalapril maleate [55]	Tablets (2.5, 5, 10, 20)	1[a]	Ora-Sweet® Ora-Sweet SF®	60[h]	60[h]
Famotidine [60]	Tablets (10, 20, 40)	8	Ora-Sweet®	NE	95[h]
Flecainide acetate [56]	Tablets (50, 100, 150)	20[a]	Ora-Sweet® Ora-Sweet SF®	60[h]	60[h]
Flucytosine [54]	Tablets (250, 500)	10[a]		60[g]	60[g]
Gabapentin [43]	Capsules (100, 300, 400)	100	Ora-Sweet®	91[e]	56[e]

Drug [Ref.]	Dosage form (mg)	Suspension (mg/mL)	CMC or Ora-Plus added (1:1) of	Stability (days) in different temperatures	
				3-5°C	22-25°C
Hydralazine HCl [55]	Tablets (10, 25, 50, 100)	4[a]		1[h]	unstable [h]
				2[h]	
Ketoconazole [61]	Tablets (200)	20[a]	Ora-Sweet® Ora-Sweet SF®	60[h]	60[h]
Labetalol HCl [62]	Tablets (100, 200, 300)	40[a]			
Lamotrigine [63]	Tablets (25, 100, 150, 200)	1		91[h]	91[h]
Levodopa + carbidopa [64]	Tablets (100+25)	5 2.5	Ora-Sweet®	42[e]	28[e]
Levofloxacin [65]	Tablets (200, 500, 750)	50	Strawberry syrup	57[e]	57[e]
Metolazone [61]	Tablets (2.5, 5, 10)	1[a]	Ora-Sweet® Ora-Sweet SF®	60[h]	60[h]
Metoprolol tartrate [62]	Tablets (50, 100)	10[a]		90[e]	90[e]
Moxifloxacin [66]	Tablets (400)	20			
Mycophenolate mofetil [67]	Capsules (250) Tablets (500)	50, 100	Ora-Sweet®	91[d]	91[d]
Naratriptan HCl [68]	Tablets (1, 2.5)	0.5	Ora-Sweet® Ora-Sweet SF®	90[e]	7[e]
Nifedipine [44]	Capsules (10, 20)	4	Ora-Sweet®	91[d]	91[d]
Norfloxacin [69]	Tablets (400)	20[b]	Strawberry syrup	56[d]	56[d]
Procainamide HCl [45,61]	Capsule/tablets (250, 375, 500)	50[a]	Ora-Sweet® Ora-Sweet SF®	60[h]	60[h]
Propylthiouracil [46]	Tablets (50)	5		91[e]	70[e]
Pyrazinamide [57]	Tablets (500)	10[a]			
Quinidine sulphate [57]	Tablets (200, 300)	10[a]		60[h]	60[h]
Rifabutin [70]	Capsules (150)	20	Ora-Sweet®	84[g]	84[g]
Rifampin [57]	Capsules (150, 300)	25[a]	Ora-Sweet® Ora-Sweet SF®	28[h]	28[h]
Sildenafil citrate [49]	Tablets (25, 50, 100)	2.5	Ora-Sweet®	91[e]	91[e]
Sotalol HCl [50,51]	Tablets (80, 120, 160, 240)	5	Ora-Sweet® Ora-Sweet SF®	91[d], 84[c] 84[c]	
Spironolactone [61,71]	Tablets (25, 50, 100)	1	Simple syrup	91[c]	91[c]
		2.5[a]	Ora-Sweet® Ora-Sweet SF®	60[h]	60[h]
Spironolactone + hydrochlorothiazide [62]	Tablets (25+25)	(5+5)[a]	Ora-Sweet® Ora-Sweet SF®	60[h]	60[h]

Drug [Ref.]	Dosage form (mg)	Suspension (mg/mL)	CMC or Ora-Plus added (1:1) of	Stability (days) in different temperatures	
				3-5°C	22-25°C
Sunitinib malate [72]	Capsules (50)	10	Ora-Sweet®	60[c]	60[c]
Tacrolimus [73-75]	Capsules (0.5, 1, 5)	0.5	Simple syrup NF	NE	56[c,e]
Terbinafine HCl [76]	Tablets (250)	25	Ora-Sweet®	42[h]	42[i]
Tetracycline HCl [57]	Capsules (250, 500)	25[a]	Ora-Sweet®	28[g]	28[h]
			Ora-Sweet SF®	10[g]	7[h]
Theophylline [77]	Capsules (125, 200, 300)	5	Ora-Sweet® Ora-Sweet SF®	NE	90[e]
Tiagabine HCl [52]	Tables (2, 4, 6, 8, 10, 12, 16)	1	Ora-Sweet®	91[d]	70[e]

a: storage in the dark, b: at fluorescent lighting, c: amber glass, d: plastic, e: amber plastic, f: polyvinyl chloride, g: polyethylene terephthalate (PET), h: amber PET; i: amber high density polyethylene bottles; NE, not evaluated; HCl, hydrochloride.

Ora-Plus® constituents: CMC-Na, citric acid, flavouring, methylparaben, microcrystalline glucose, potassium sorbate, purified water, simethicone, sodium phosphate, xanthan gum, pH 4.2. Ora-Sweet® constituents: citric acid, flavouring, glycerin, methylparaben, purified water, sodium phosphate, sorbitol, sucrose, potassium sorbate, pH 4.2. Ora-Sweet SF® constituents: citric acid, flavouring, glycerin, methylparaben, propylparaben, potassium sorbate, purified water, sodium saccharin, sodium citrate, sorbitol, xanthan gum, pH 4.2. Sugar-free.

Table 2. Stability of different drugs in oral suspension extemporaneously prepared from commercially available sodium carboxymethylcellulose (CMC-Na) in Ora-Plus® and/or other vehicle constituents.

Most of the suspensions prepared presented no substantial changes in pH, odour or physical appearance, showing that CMC-Na base usually provides products with a satisfactory safety and stability (drug content equal or superior to 90% of the original concentration) over a period of 8 weeks.

The use of Ora-Plus® extemporaneous oral suspension has provided satisfactory stability for aminophylline, cyclophosphamide, domperidone, granisetron, itraconazole, ursodiol and tramadol hydrochloride associated with acetaminophen [14,78]. However, for captopril, hydralazine hydrochloride and tetracycline hydrochloride, the suspensions were reported as not having enough stability. The problem of stability with captopril is due to its oxidative degradation, which can be solved by the addition of EDTA disodium [79]. Although, these studies are important, most of them have not evaluated the microbiological stability, an essential criterion for liquid dosage forms.

In addition to MC and CMC-Na, HPMC has also been employed in extemporaneous oral suspension. For instance, nifedipine tablets or drug powder, prepared with HPMC 1% solution in order to obtain a suspension of 1 mg/mL concentration, was stable for at least 4 weeks when stored at room or refrigerated temperatures and protected from light [80].

There is no doubt that these drug stability results are important for providing formulations for both paediatric patients and the geriatric populations that have difficulty in swallowing capsules or tablets.

3.2. Cellulose and its derivatives in oral solid dosage forms

3.2.1. Mycrocrystalline cellulose in immediate release

Microcrystalline cellulose is reported as an excipient (diluent) in oral powder and capsules extemporaneously compounded for paediatric use. Capsules and powders were prepared from commercial tablets containing 10 mg of nifedipine, which was mixed with different amounts of lactose or microcrystalline cellulose in a mortar with pestle using standard geometric dilution. Capsules were filled by a hand-operated capsule-filling machine. The oral powders and capsules containing extemporaneously prepared nifedipine showed acceptable quality regarding content uniformity, but considerable loss of the active ingredient occurred during the compounding process for both preparations. The authors demonstrated that oral powders of nifedipine (a light sensitive drug) can be replaced by capsules, which were adequately safe with either lactose monohydrate or microcrystalline cellulose as excipients for delivering a paediatric medication [81].

Recently, microcrystalline cellulose and two other common pharmaceutical excipients (starch and lactose) were investigated with regards to the choice of the best diluent for *Gymnema sylvestre* extract (a plant used as an adjuvant in the treatment of diabetes mainly in China) used to compound capsules. The *Gymnema sylvestre* extract is available as powder that presents low flowability due to its small particle size which causes problems in the filling of the hard capsules. An evaluation of these excipients was also performed in the presence of different lubricants (magnesium stearate or talc). The study showed that microcrystalline cellulose is a better diluent than lactose or starch, because it produces the most uniform particle size distribution when added to *Gymnema sylvestre* extract and also reduces the percentage of fine particles resulting in acceptable variation of the weight among the capsules (RSD< 4%). On the other hand, starch and lactose increase the number of small particles that worsen the flowability of the powder mixture. Furthermore, microcrystalline cellulose associated with 1% lubricant renders a powder mixture ready for encapsulation of *Gymnema sylvestre* extract in hard gelatin capsules, since flow agents optimize the capsule filling in the compounding routine practices [82]. The foregoing suggests that microcrystalline cellulose can be an appropriate diluent in formulating similar flowable plant extracts.

3.2.2. Cellulose ether derivatives in sustained/controlled release

3.2.2.1. Carboxymethylcellulose and hydroxypropylmethylcellulose

Some studies reveal improper use of cellulosic excipients. For example, using a high percentage (30% w/w) of CMC-Na (anionic polymer) as a diluent in the compounding of capsules of simvastatin has a deleterious effect. These capsules showed serious drug release problems in pharmaceutical tests because they did not disintegrate or dissolve at all [83]. In this case, CMC-Na should have been used as a capsule disintegrating agent at a much inferior concentration (< 6%) [12].

CMC-Na and HPMC (nonionic polymer) were evaluated by *in vitro* release studies with regards to ibuprofen (non-steroidal anti-inflammatory) extended-release from hard gelatin capsules. The study showed that different grades of CMC-Na and HPMC could control ibuprofen release to a substantial degree when used as diluents. Furthermore, the molecular weight of the polymer group that is directly related to the viscosity grade affects the drug release: the higher the molecular weight is, the slower the drug release is [40]. One year later, these researchers evaluated ibuprofen bioavailability (healthy volunteers) from hard gelatin capsules containing different grades of HPMC (K100 and K15M) and CMC-Na (low, medium, high viscosity). These capsules were prepared by filling the shells with the simple mixture of the powders (drug and polymer). The study showed that different viscosities of HPMC can modify the absorption rate of ibuprofen from hard gelatin capsules, in close correlation with a previous *in vitro* study. In particular, a higher viscosity HPMC (K15M) was a better diluent in sustained-release. On the other hand, the use of the CMC-Na with different viscosity grades did not allow for the control of the absorption rate of ibuprofen and did not correspond to *in vitro* results. However, none of the polymers seemed to have any effect on the bioavailability of the ibuprofen from hard gelatin capsules [39].

Slow-release morphine (opioid analgesic used for the relief of pain) capsules extemporaneously prepared were investigated regarding their dissolution profile. Three batches of capsules prepared by a pharmacist were compared with each other and with tablets acquired in the market. The authors describe how similar slow-release profiles were found for tablets and compounded capsules, though the latter showed a faster release-rate for morphine sulphate. Despite small variations from batch to batch, the authors describe that compounded capsules showed a remarkably consistent slow-release profile in *in vitro* studies [34].

Another study of compounded capsules containing 300 mg of morphine sulphate (a dosage unavailable in the market) reported the use of HPMC in sustained-release. There has been considerable controversy about the advisability of this practice. Release studies, performed according to the United States Pharmacopeia (USP) using a dissolution apparatus of type III, showed that almost half of the morphine was released in the first hour and that the release of the remainder was not adequately sustained. As verified in other studies, the increase of HPMC prolonged release and reduced drug release in the first hour. Other formulations prepared by placing compressed pellets in capsules showed a sustained release significantly beyond that of the pellets' original formulation. Considering that the medication can be taken after a meal, the agitation of the gastrointestinal tract would have increased, resulting in the reduction of the sustained release period and in a slight increase of the drug amount during the first hour after administration. In the first formulation, the capsules did not exhibit sustained-release that could be adequate for most applications. Formulations with a greater percentage of the HPMC are preferred. Furthermore, the pelleted formulation was superior, but it may not be feasible because it is too labour-intensive [35].

Slow-release capsules of morphine sulphate (15, 60, 200 mg) and oxycodone hydrochloride (10, 80, 200 mg) were evaluated *in vitro* by USP dissolution apparatus II. All capsules (three batches of each) were compounded in a local pharmacy employing 40% HPMC (E4M Premium CR) and lactose as excipients and a specific machine for capsule-filling. The authors observed that the release of the active ingredients from the compounded capsules after 0.5, 4 and 12 h were less than 23%, 85% and 98%, demonstrating that HPMC is an adequate excipient for preparing slow-release capsules of morphine sulphate and oxycodone hydrochloride. The authors recommend that the ratio of active ingredient to polymer should remain constant regardless of the capsule size in order to achieve similar release rates, provided there is some degree of compression within the capsule shell. *In vitro* performance showed small intra-batch variations as well as inter-batch variations which were not statistically significant. Thus the compounding of slow release capsules yielded reproducible formulations. However, the authors mention that clinical evaluation is needed in order to determine whether the small differences are significant [37].

A variety of drugs, especially natural bioidentical hormones, have been exploited in compounding using matrix systems since 2002. Hydrophilic matrix systems were mentioned as being successfully used in slow-release capsules. The authors report that a good response of patients in a dose-related manner was observed in response to all micronized hormones administered in slow-release capsules [36].

Polymers of HPMC (K100MPRCR, K15MPRCR and E4MCR) in different proportions from 15 to 35% w/w were also used as extended-release excipients in the compounding of capsules containing 100 mg of theophylline. The polymers of HPMC were employed to prepare capsules by volumetric method for powder filling in a manual encapsulator. The extended drug release was evaluated using USP apparatus I for industrially-produced batches and for those obtained by compounding process. The dissolution profile obtained for the higher ratio (35% w/w) of HPMC (E4MCR) met USP specifications. Furthermore, reproducibility was observed with ten other compounded batches. HPMC was efficient in controlling the release of theophylline from the matrix of the capsules prepared by compounding. However, extended-release capsules containing 100 mg of theophylline (pellets) available in the market did not show prolonged release when submitted to the same test conditions [38].

To summarize, CMC-Na does not seem to be an adequate slow-releasing agent for preparing the capsules regardless of its viscosity. On the other hand, HPMC is a promising agent for prolonging the release of drugs, since it has been used before with success. The results suggest a relationship between degree of viscosity of HPMC and slow release (ibuprofen, morphine). However, reproducibility is an important requisite, and may not be assured for all formulations. In addition, studies of therapeutic efficacy are also scarce for such compounded products.

3.2.3. Cellulose ester derivatives in delayed release

Cellulose ester derivatives are used for enteric coatings of capsules, making them resistant to dissolution in low pH environments, such as the stomach, but allowing for their rapid disintegration in higher pH environments, such as the intestine. The efficiency of the coating is limited by the smooth and nonporous surface of the hard capsules. Studies into anti-inflammatory and anti-secretory (H⁺ pumps inhibitors) drugs, such as diclofenac and pantoprazole sodium salts, respectively, show that these drug formulations must be coated because they can irritate the stomach walls or degrade in acid environments [84,85].

3.2.3.1. Cellulose acetate phthalate

CAP and other agents (formaldehyde, methacrylic acid copolymer) have been used to compound delayed-release capsules of diclofenac using specific small-scale machinery or manual immersion in order to evaluate the efficiency of these enteric coating processes [32]. Capsules coated with CAP (using acetone as solvent) prepared with either small machinery or twofold manual immersion showed adequate gastro-resistance, for which the release of the drug was less than 10% in acid and greater than 75% in buffered conditions. However, the capsules coated by machinery had a poorer visual aspect than those coated by the manual process [32]. In spite of this difference, the authors did not suggest which method was the most adequate to compound delayed-release capsules of diclofenac.

A simple, quick and easily reproducible method for compounding enteric-release capsules containing diclofenac has also been described. Twenty-two batches of diclofenac sodium capsules (n=60) were divided into three groups, which were submitted to different processes of coating. A small-scale machine and an enteric coating by atomization (spraying) of organic solutions of polymers (5% CAP in a mix of acetone and alcohol) were employed for ten and six batches, respectively. Before coating, the capsules' hemi receptacles were sealed by treatment with 50% v/v hydroalcoholic solution. The dissolution test results were statistically compared inter-batch and also with reference commercial product (Voltaren® DR). Most of the batches (>75%) met the pharmacopeial requirements for enteric release, in both acid (less than 10%) and buffered (greater than 80%) conditions [33]. Results confirmed that CAP is an effective enteric coating agent in compounding practice and that the application of adequate techniques in pharmacies is important.

Delayed release capsules obtained by compounding and coating with organic solutions of CAP have been evaluated for pro-drug sodium pantoprazole, a proton pump inhibitor that undergoes degradation in the acid environment of the stomach [31,84,85]. Quality control tests were performed on capsules locally acquired in compounding pharmacies. Dissolution studies for gastro-resistance evaluation were performed with granules of pantoprazole coated with CAP and encapsulated, as well as with capsules coated with CAP or other agents (formaldehyde, shellac, methacrylic acid copolymer). However, all the samples prepared by coating with CAP (capsules or granules) released their content in an acid environment and did not show adequate gastro-resistance [31]. These results reveal the need for suitable coating techniques for compounding gastro-resistant capsules, since CAP is admittedly an effective agent for enteric coating.

4. Conclusion

Cellulose and its derivatives are very important excipients in compounded medicines. Many compounded preparations containing such excipients have been investigated since 1992, especially for extemporaneous use; however, a great effort is still necessary in this field in order to assure quality, safety and efficacy for several other drugs. This aspect is even more relevant when these products require specific pharmaceutical features (such as delayed release) and, consequently, adequate techniques for achieving drug therapy success. It points towards the need for more research into ways to properly disseminate the appropriate use of cellulose derivatives in compounding pharmacies. A greater attention should be paid to this field because compounding is a growing practice in many countries as a result of pharmaceutical care that prioritizes the person in his/her individuality, as opposed to the average population usually targeted by companies. For all these reasons, cellulose derivatives and their applications in compounding practice were reviewed, with an emphasis on their use in solid dosage forms with modified release. Addressing the use of the cellulose derivatives, such as cellulose acetate phthalate, can be critical in the coating of capsules by hand.

Author details

Flávia Dias Marques-Marinho* and Cristina Duarte Vianna-Soares
Department of Pharmaceutical Products, Federal University of Minas Gerais, Belo Horizonte, MG, Brazil

Acknowledgement

To CAPES for the fellowship to Marques-Marinho FD, to Lima AA and Reis IA for the important initial collaboration.

5. References

[1]   Anderson S. Making Medicines- A Brief History of Pharmacy and Pharmaceuticals. Great Britain: Pharmaceutical Press; 2005.

[2]   Kremers E., Sonnedecker, G., editors. Kremers and Urdang' s History of Pharmacy. Madison: American Institute of the History of Pharmacy; 1986.

[3]   Buurma H, De Smet PA, van den Hoff OP, Sysling H, Storimans M, Egberts AC. Frequency, Nature and Determinants of Pharmacy Compounded Medicines in Dutch Community Pharmacies. Pharm. World Sci. 2003;25(6): 280-7.

[4]   Galston SK. Federal and State Role in Pharmacy Compounding and Reconstitution. In: US Food and Drug Administration; 2003. Available: http://www.fda.gov/NewsEvents/Testimony/ucm115010.htm   (accessed   26   February 2013)

[5]   Jew RK, Soo-hoo W, Erush SC. Extemporaneous Formulations for Pediatric, Geriatric, and Special Needs Patients. Bethesda: American Society of Health-System Pharmacists; 2010.

---

* Corresponding Author

[6]   da Costa PQ, Rey LC, Coelho HL. Lack of Drug Preparations for Use in Children in Brazil. J. Pediatr. 2009;85(3): 229-35.

[7]   Nahata MC, Allen LV Jr. Extemporaneous Drug Formulations. Clin. Ther. 2008;30(11): 2112-9.

[8]   Standing JF, Tuleu C. Paediatric Formulations- Getting to the Heart of the Problem. Int. J. Pharm. 2005;300: 56-66.

[9]   Prot-labarthe S, Bussières JF, Brion F, Bourdon O. Comparison of Hospital Pharmacy Practice in France and Canada: Can Different Practice Perspectives Complement each Other? Pharm. World Sci. 2007;29: 526-33.

[10] Brion F, Nunn AJ, Rieutord A. Extemporaneous (magistral) preparation of oral medicines for children in European hospitals. Acta Paediatr. 2003;92: 486-490.

[11] Allen LV Jr. The Art, Science, and Technology of Pharmaceutical Compounding. Washington: American Pharmacists Association; 2008.

[12] Rowe RC, Sheskey PJ, Quinn ME., editors. Handbook of Pharmaceutical Excipients. London: Pharmaceutical Press; 2009.

[13] Pifferi G, Santoro P, Pedrani M. Quality and functionality of excipients. Il Farmaco 1999;54: 1-14.

[14] Glass BD, Haywood A. Stability Considerations in Liquid Dosage Forms Extemporaneously Prepared from Commercially Available Products. J. Pharm. Pharm. Sci. 2006;9(3): 398-426.

[15] Edgar K J. Cellulose Esters in Drug Delivery. Cellulose 2007;14(1): 49-64.

[16] Kamel S, Ali N, Jahangir K, Shah SM, El-Gendy AA. Pharmaceutical Significance of Cellulose- A Review. eXPRESS Polymer Letters 2008;2(11): 758-78.

[17] Nobles D, Romanovicz DK, Brown RM Jr. Cellulose in Cyanobacteria. Origin of Vascular Plant Cellulose Synthase? Plant Physiol. 2001;127: 529-42.

[18] Brown RM Jr. Algae as Tools in Studying the Biosynthesis of Cellulose Nature' s Most Abundant Macromolecule. In Wiessner G, Robinson DG, Starr RC, editors. Experimental Phycology. Cell Walls and Surfaces, Reproduction, Photosynthesis. Berlin:Springer-Verlag. 1990; 20-39.

[19] Kimura S, Itoh T. A New Cellulosic Structure, the Tunic Cord in the Ascidian Polyandrocarpa misakiensis. Protoplasma 1998;204: 94-102.

[20] Brown RM Jr. Cellulose Structure and Biosynthesis: What is in Store for the 21 st Century? J. Polym. Sci. A Polym. Chem. 2004;42: 487-95.

[21] Klemm D, Heublein B, Fink HP, Bohn A. Cellulose: Fascinating Biopolymer and Sustainable Raw Material. Angew. Chem. Int. Ed. 2005;44: 3358-93.

[22] Weimer PJ, French AD, Calamari TA Jr. Differential Fermentation of Cellulose Allomorphs by Ruminal Cellulolytic Bacteria. Appl. Environ. Microbiol. 1991;57(11): 3101-6.

[23] Kuga S, Brown RM Jr. Silver Labeling of the Reducing Ends of Bacterial Cellulose. Carbohydr. Res. 1988;180: 345-50.

[24] Saxena IM, Brown RM Jr. Cellulose Biosynthesis: Current Views and Evolving Concepts. Ann. Bot. 2005;96: 9-21.

[25] Wada M, Chanzy H, Nishiyama Y, Langan P. Cellulose IIII Crystal Structure and Hydrogen Bonding by Synchrotron X-ray and Neutron Fiber Diffraction. Macromolecules 2004;37: 8548-55.

[26] Ishikawa A, Okano T, Sugiyama J. Fine Structure and Tensile Properties of Ramie Fibres in the Crystalline Form of Cellulose I, II, IllI and IVI. Polymer 1997;38(2): 463-8.

[27] United States Pharmacopeial Convention. The United States Pharmacopeia-National FormularyRockville: United States Pharmacopeia; 2011. p. 1416-19.

[28] Béchard SR, Levy L, Clas SD. Thermal, Mechanical and Functional Properties of Cellulose Acetate Phthalate (CAP) Coatings Obtained from Neutralized Aqueous Solutions. Int. J. Pharm. 1995;114(2): 205-13.

[29] Prista LN, Alves AC, Morgado R, Lobo JS. Tecnologia Farmacêutica. Lisboa: Fundação Calouste Gulbekian; 2002. p. 562-7.

[30] Prista LN, Alves AC, Morgado R. Tecnologia Farmacêutica. Lisboa: Fundação Calouste Gulbekian; 1995. p. 2027-8.

[31] Marques-marinho FD, Vianna-soares CD, Carmo VA, Campos LM. Avaliação da Qualidade de Pantoprazol Cápsulas Manipuladas Gastro-Resistentes. Lat. Am. J. Pharm. 2009;28(6): 899-906.

[32] dos Santos L, Guterres SS, Bergold AM. Preparação e Avaliação de Cápsulas Gastro-Resistentes de Diclofenaco de Sódio. Lat. Am. J. Pharm. 2007;26(3): 355-61.

[33] Ferreira AO, Holandino C. Pharmaceutical Development of Enteric-Release Hard Gelatin Capsules in the Compounding Setting. Int. J. Pharm. Compound. 2008;12(2): 163-9.

[34] Webster KD, Al-achi A, Greenwood R. In Vitro Studies on the Release of Morphine Sulfate from Compounded Slow-Release Morphine-Sulfate Capsules. Int. J. Pharm. Compound. 1999;3(5): 409-11.

[35] Bogner RH, Szwejkowski J, Houston A. Release of Morphine Sulfate from Compounded Slow Release Capsules: the Effect of Formulation on Release. Int. J. Pharm. Compound. 2001;5(5): 401-5.

[36] Timmons ED, Timmons SP. Custom-Compounded Micronized Hormones in a Slow-Release Capsule Matrix. Int. J. Pharm. Compound. 2002;6(5): 378-9.

[37] Glowiak DL, Green JL, Bowman BJ. In Vitro Evaluation of Extemporaneously Compounded Slow-Release Capsules Containing Morphine Sulfate or Oxycodone Hydrochloride. Int. J. Pharm. Compound. 2005;9(2): 157-64.

[38] Pinheiro VA, Kaneko TM, Velasco MV, Consiglieri VO. Development and In Vitro Evaluation of Extended-Release Theophylline Matrix Capsules. Braz. J. Pharm. Sci. 2007;43(2): 253-61.

[39] Ojantakanen S, Marvola M, Hannula AM, Klinge E, Naukkarinnen T. Bioavailability of Ibuprofen from Hard Gelatin Capsules Containing Different Viscosity Grades of Hydroxypropylmethylcellulose and Sodium Carboxymethylcellulose. Eur. J. Pharm. Sci. 1993;1: 109-14.

[40] Ojantakanen S. Effect of Viscosity Grade of Polymer Additive and Compression Force on Dissolution of Ibuprofen from Hard Gelatin Capsules. Acta Pharm. Fennica 1992;101(33): 119-26.

[41] United States Pharmacopeial Convention. USP pharmacists´ pharmacopeia Rockville: United States Pharmacopeia; 2008. p. 333.

[42] Roy JJ, Besner JG. Stability of Clonazepam Suspension in HSC Vehicle. Int. J. Pharm. Compound. 1997;6(5): 378-9.

[43] Nahata MC. Development of Two Stable Oral Suspensions for Gabapentin. Pediatr. Neurol. 1999;2(3): 195-7.

[44] Nahata MC, Morosco RS, Willhite, EA. Stability of Nifedipine in Two Oral Suspensions Stored at Two Temperatures. J. Am. Pharm. Assoc. 2002; 42(6): 865-7.

[45] Metras JI, Swenson CF, Mcdermott MP. Stability of Procainamide Hydrochloride in an Extemporaneously Compounded Oral Liquid. Am. J. Health Syst. Pharm. 1992;49(7): 1720-4.

[46] Nahata MC, Morosco RS, Trowbridge JM. Stability of Propylthiouracil in Extemporaneously Prepared Oral Suspensions at 4 and 25 °C. Am. J. Health Syst. Pharm. 2000;57(12): 1141-3.

[47] Nahata MC, Morosco RS, Peritore SP. Stability of Pyrazinamide in Two Suspensions. Am. J. Health Syst. Pharm. 1995;52(14): 1558-60.

[48] Nahata MC, Morosco RS, Hipple TF. Stability of Pyrimethamine in a Liquid Dosage Formulation Stored for Three Months. Am. J. Health Syst. Pharm. 1997;54(23): 2714-6.

[49] Nahata MC, Morosco RS, Brady MT. Extemporaneous Sildenafil Citrate Oral Suspensions for the Treatment of Pulmonary Hypertension in Children. Am. J. Health Syst. Pharm. 2006;63(3): 254-7.

[50] Sidhom MB, Rivera N, Almoazen H, Taft DR, Kirschenbaum HL. Stability of Sotalol Hydrochloride in Extemporaneously Prepared Oral Suspension Formulations. Int. J. Pharm. Compound. 2005;9(5): 402-6.

[51] Nahata MC, Morosco RS. Stability of Sotalol in Two Liquid Formulations at Two Temperatures. Ann. Pharmacother. 2003;37(4): 506-9.

[52] Nahata MC, Morosco RS. Stability of Tiagabine in Two Oral Liquid Vehicles. Am. J. Health Syst. Pharm. 2003;60(1): 75-7.

[53] Alexander KS, Haribhakti RP, Parker GA. Stability of Acetazolamide in Suspension Compounded from Tablets. Am. J. Hosp. Pharm. 1991;48(6): 1241-4.

[54] Allen LV Jr, Erickson MA 3rd. Stability of Acetazolamide, Allopurinol, Azathioprine, Clonazepam, and Flucytosine in Extemporaneously Compounded Oral Liquids. Am. J. Health Syst. Pharm. 1996;53(16): 1944-9.

[55] Allen LV Jr, Erickson MA 3rd. Stability of Alprazolam, Chloroquine Phosphate, Cisapride, Enalapril Maleate, and Hydralazine Hydrochloride in Extemporaneously Compounded Oral Liquids. Am. J. Health Syst. Pharm. 1998;55: 1915-20.

[56] Allen LV Jr, Erickson MA 3rd. Stability of Baclofen, Captopril, Diltiazem Hydrochloride, Dipyridamole, and Flecainide Acetate in Extemporaneously Compounded Oral Liquids. Am. J. Health Syst. Pharm. 1996;53(18): 2179-84.

[57] Allen LV Jr, Erickson MA 3rd. Stability of Bethanechol Chloride, Pyrazinamide, Quinidine Sulfate, Rifampin, and Tetracycline Hydrochloride in Extemporaneously Compounded Oral Liquids. Am. J. Health Syst. Pharm. 1998;55(17), 1804-9.

[58] Nahata MC, Morosco RS, Trowbridge JM. Stability of Dapsone in Two Oral Liquid Dosage Forms. Ann. Pharmacother. 2000;34(7): 848-50.

[59] Johnson CE, Wagner DS, Bussard WE. Stability of Dolasetron in Two Oral Liquid Vehicles. Am. J. Health Syst. Pharm. 2003;60(21): 2242-4.

[60] Dentinger PJ, Swenson CF, Anaizi NH. Stability of Famotidine in an Extemporaneously Compounded Oral Liquid. Am. J. Health Syst. Pharm. 2000;57(14): 1340-2.

[61] Allen LV Jr, Erickson MA 3rd. Stability of Ketoconazole, Metolazone, Metronidazole, Procainamide Hydrochloride and Spironolactone in Extemporaneously Compounded Oral Liquids. Am. J. Health Syst. Pharm. 1996;53(17): 2073-8.

[62] Allen LV Jr, Erickson MA 3rd. Stability of Labetalol Hydrochloride, Metoprolol Tartrate, Verapamil Hydrochloride, and Spironolactone with Hydrochlorothiazide in Extemporaneously Compounded Oral Liquids. Am. J. Health Syst. Pharm. 1996;53(19): 2304-9.

[63] Nahata MC, Morosco RS, Hipple TF. Stability of Lamotrigine in Two Extemporaneously Prepared Oral Suspensions at 4 and 25 Degrees C. Am. J. Health Syst. Pharm. 1999;56(3): 240-2.

[64] Nahata MC, Morosco RS, Leguire LE. Development of Two Stable Oral Suspensions of Levodopa-Carbidopa for Children with Amblyopia. J. Pediatr. Ophthalmol. Strabismus. 2000;37(6): 333-7.

[65] VandenBussche HL, Johnson CE, Fontana EM, Meram JM. Stability of Levofloxacin in an Extemporaneously Compounded Oral Liquid. Am. J. Health Syst. Pharm. 1999; 56(22): 2316-8.

[66] Hutchinson DJ, Johnson CE, Klein KC. Stability of Extemporaneously Prepared Moxifloxacin Oral Suspensions. Am. J. Health Syst. Pharm. 2009;66(7): 665-7.

[67] Ensom MH. Stability of Mycophenolate Mofetil in a 1:1 Mixture of Ora-Sweet and Ora-Plus. Can. J. Hosp. Pharm. 2002;55(1): 63-5.

[68] Zhang YP, Trissel LA, Fox JL. Naratriptan Hydrochloride in Extemporaneously Compounded Oral Suspensions. Int. J. Pharm. Compound. 2000;4(1): 69-71.

[69] Johnson CE, Price J, Hession JM. Stability of Norfloxacin in an Extemporaneously Prepared Oral Liquid. Am. J. Health Syst. Pharm. 2001;58: 577-9.

[70] Haslam JL, Egodage KL, Chen Y, Rajewski RA, Stella V. Stability of Rifabutin in Two Extemporaneously Compounded Oral Liquids. Am. J. Health Syst. Pharm. 1999;56(4): 333-6.

[71] Nahata MC, Morosco RS, Hipple TF. Stability of Spironolactone in an Extemporaneously Prepared Suspension at Two Temperatures. Ann. Pharmacother. 1993; 27(10): 1198-9.

[72] Navid F, Christensen R, Minkin P, Stewart CF, Furman WL, Baker S. Stability of Sunitinib in Oral Suspension. Ann. Pharmacother. 2008;42(7): 962-6.

[73] Jacobson PA, Johnson CE, West NJ, Foster JA. Stability of Tacrolimus in an Extemporaneously Compounded Oral Liquid. Am. J. Health Syst. Pharm. 1997;54(2): 178-80.

[74] Han J, Beeton A, Long PF, Wong I, Tuleu C. Physical and Microbiological Stability of an Extemporaneous Tacrolimus Suspension for Paediatric Use. J. Clin. Pharm. Ther. 2006;31(2): 167-72.

[75] Stefano VD, Cammarata SM, Pitonzo R. Paediatric Oral Formulations: Comparison of Two Extemporaneously Compounded Suspensions from Tacrolimus Capsules. Eur. J. Hosp. Pharm. Pract. 2011;17(6): 70-2.

[76] Abdel-rahman SM, Nahata MC. Stability of Terbinafine Hydrochloride in an Extemporaneously Prepared Oral Suspension at 25 and 4 Degrees C. Am. J. Health Syst. Pharm. 1999;56(3): 243-5.

[77] Johnson CE, VanDeKoppel S, Myers E. Stability of Anhydrous Theophylline in Extemporaneously Prepared Alcohol-Free Oral Suspensions. Am. J. Health Syst. Pharm. 2005;62(23): 2518-20.

[78] Kennedy R, Groepper D, Tagen M, Christensen R, Navid F, Gajjar A, Stewart CF. Stability of Cyclophosphamide in Extemporaneous Oral Suspensions. Ann. Pharmacother. 2010;44(2): 295-301.

[79] Brustugun J, Lao YE, Fagernæs C, Brænden J, Kristensen S. Long-Term Stability of Extemporaneously Prepared Captopril Oral Liquids in Glass Bottles. Am. J. Health Syst. Pharm. 2009;66(19): 1722-5.

[80] Helin-tanninen M, Naaranlahti T, Kontra K, Ojanen T. Enteral Suspension of Nifedipine or Neonates. Part 2. Stability of an Extemporaneously Compounded Nifedipine Suspension. J. Clin. Pharm. Ther. 2001;26(1): 59-66.

[81] Helin-tanninen M, Naaranlahti T, Kontra K, Savolainen K. Nifedipine Capsules May Provide a Viable Alternative to Oral Powders for Paediatric Patients. J. Clin. Pharm. Ther. 2007;32(1): 49-55.

[82] Carbinatto FM, Castro AD, Oliveira AG, Silva AA Jr. Preformulation Studies of Gymnema sylvestre Extract Powder Formulation for Hard Gelatin Capsules. J. Basic Appl. Pharm. Sci. 2011;32(2): 175-80.

[83] Marques-marinho FD, Zanon JC, Sakurai E, Reis IA, Lima AA, Vianna-soares CD. Quality Evaluation of Simvastatin Compounded Capsules. Braz. J. Pharm. Sci. 2011;47(3): 495-502.

[84] Jungnickd PW. Pantoprazole: A New Proton Pump Inhibitor. Clin. Ther. 2000;22(11): 1268-93.

[85] Brunton LL, Lazo JS, Parker KL. Goodman & Gilman' s The Pharmacological Basis of Therapeutics. New York: McGraw-Hill; 2006. 2021 p.

Chapter 9

Bioactive Bead Type Cellulosic Adsorbent for Blood Purification

Shenqi Wang and Yaoting Yu

Additional information is available at the end of the chapter

1. Introduction

Cellulose, a natural polymer, has been widely used in blood purification due to its good biocompatibility, and excellent processing which can be easily formulated into beads, membranes and hollow fibers. Sorbent-perfusion is a novel approach of blood purification which can specifically remove endogenous and exogenous pathogenic toxins from the blood of patients [1]. The technique involves passing whole blood or plasma of the patient through a cartridge filled with an adsorbent which can easily adsorb the toxin molecules, see **Figure 1 a,b**. According to selectivity, generally adsorbents can be classified as broad spectrum, affinity adsorbents and immuno-adsorbents, of which the latter has the highest selectivity [2-5]. Materials, most commonly used are activated charcoal [6], porous resins and fibers. The pathogenic substances in the blood of patients are adsorbed by the adsorbent via hydrophilic (electro-static forces) or hydrophobic interactions. Macroporous resins usually show high adsorption capacities especially for the removal of high molecular weight or "middle molecules" toxins [7-9].

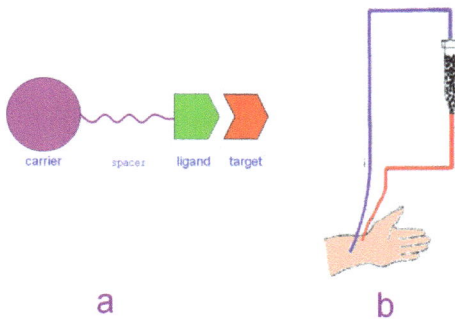

Figure 1. Schematic diagram of sorbent-perfusion

2. Preparation and activation of cellulosic beads

2.1. Preparation of cellulosic beads [10, 11]

One hundred grams of cotton (medical grade) was soaked in a flask containing 19% NaOH solution for 3 h at room temperature. The cotton was squeezed and weighed then placed in a 1500 ml conical flask at 25°C for 3 days. Fifty milliliters of carbon disulfide was added to the conical flask which was then sealed and aged for 5 h to convert the cellulose into a viscose solution, which was then diluted to 1000 ml with 6% NaOH solution to make a 10% viscous solution of cellulose. In a reactor equipped with a stirrer, a mixture of 800 ml chlorobenzene, 200 ml carbon tetrachloride and 2.0 g of potassium oleate was stirred for 30 min at 300 rpm under room temperature. Then 300.0 ml of 10% cellulose viscose solution was added to the reactor slowly and continued stirring for 30 min until the liquid particles were dispersed uniformly. Thereupon, the temperature was slowly raised to 90°C and kept for 2.5 h, after which it was cooled to room temperature to solidify the liquid particles into resin beads. Cellulosic beads were filtered (20–40 mesh) and washed thoroughly with alcohol and distilled water to remove all the impurities.

Compared to the gel type cellulosic beads, macroporous beads can greatly enhance the adsorption capacity for middle and high molecules in the therapeutic embolization of meningiomas [12-15]. It can be synthesized according to reference [16, 17]. In brief, a certain amount of pore-forming agent such as calcium carbonate granules, with an average diameter of about 0.2mm was added to a 10% viscous solution of cellulose, then mixed and dispersed to form cellulosic beads. After washing with dilute HCl to remove the pore-forming agent, various kinds of porous adsorbents could be prepared. Alternatively, macroporous cellulose beads could also be prepared from cellulose solution in ionic liquid by double emulsification [18, 19].

Recently, cellulosic microspheres with a particle size below 5μm have been widely adopted in blood purification [20, 21], which can be an excellent matrix for the preparation of adsorbent.

Bead porosity and density are calculated by the following equations: [10, 17]

$$P = \frac{\rho_s \times Q}{\rho_s Q + (1-Q)\rho_{H_2O}} \times 100\% \qquad (1)$$

$$D_p = \frac{wt_w}{V_W} \qquad (2)$$

$$Q = \frac{wt_w - wt_d}{wt_w} \times 100\% \qquad (3)$$

where P stands for porosity percentage; r_s stands for skeleton density; Q stands for water content; ρ_{H2O} stands for density of water; wt_w stands for weight of wet beads; wt_d stands for weight of dried beads; V_W stands for volume of wet beads; D_p stands for packing density.

Adsorption percentage and capacity can be calculated by the following equations,

$$AP = \frac{[C]_B - [C]_A}{[C]_B} \times 100\% \qquad (4)$$

$$AC = ([C_B] - [C]_A) \times V_P \qquad (5)$$

Where *AP* and *AC* stand for adsorption percentage and adsorption capacity respectively; *[C]*B is the concentration before adsorption, *[C]*A is the concentration after adsorption, *V*P is the volume of plasma used during adsorption.

2.2. Activation of cellulose beads

Cellulose can be easily activated by reaction with epichlorohydrin which is frequently used for the preparation of cellulosic adsorbent [10,22,23].Briefly, 10 grams of cellulosic beads was activated with 10ml epichlorohydrin in 20ml 2mol/l sodium hydroxide solution. The mixture was stirred at 40°C for 4 h. Then the epoxy-activated cellulosic beads was washed thoroughly with distilled water and further reacted with amino acids or proteins, see **Figure 2**. The concentration of sodium hydroxide solution used in the condensation reaction plays an important role on the amount of activated expoxy groups linked onto cellulose. This is attributed to the condensation and ring opening reaction of epichlorohydrin molecule that competes in the reactions, see **Figure 3**.

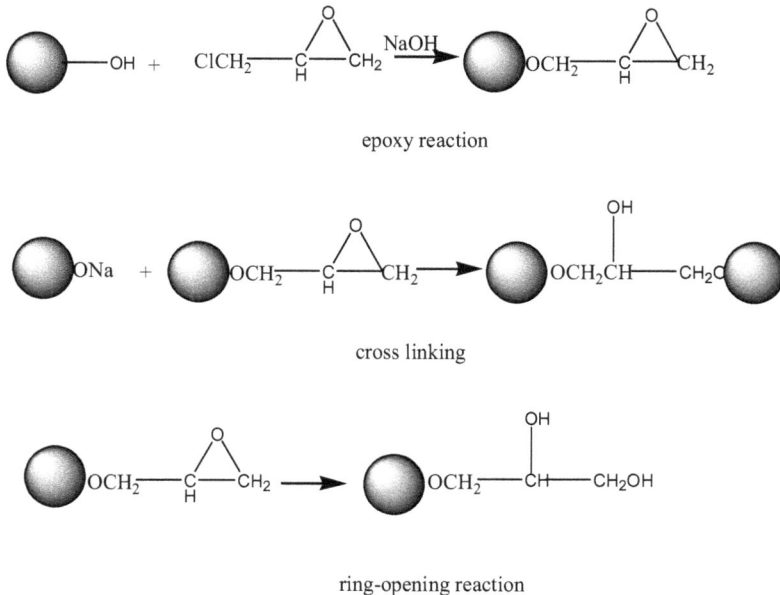

epoxy reaction

cross linking

ring-opening reaction

Figure 2. Activation reaction of cellulosic beads

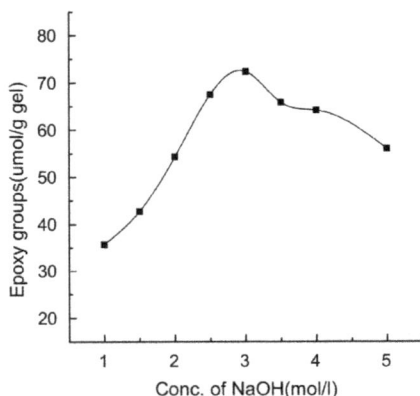

Figure 3. Amount of epoxy groups on cellulose versus concentration of NaOH

3. Mechanism study of molecular recognition between the ligand and the pathogenic toxic molecule.

3.1. Molecular recognition

To understand the interaction mechanism of pathogenic toxins with different ligands is essential, since it not only provides fundamental insight to biomaterial science, but also can lead to the discovery of more efficient ligands for the removal of pathogenic toxins in human blood. Chemical modification of proteins has been frequently used in the studies of structure-function relationships of proteins, especially in the determination of the active sites in biologically active proteins [23,24]. In the present study, we selectively modified the arginine, tryptophan, lysine residues and carboxyl terminus on the protein for the molecular recognition studies.

Lianyong Wang et al [25] investigated the interaction between ss- DNA and IgGRF by selectively modification of the arginine, tryptophan, lysine residues and carboxyl terminus on IgGRF, which was purified from patients' serum. It is well known that the density of negative charge is high on the surface of ss-DNA molecule, due to the large amount of phosphate groups. After the ss-DNA was covalently attached to the cellulose carrier, the immunoadsorbent is negatively charged, so it has a high adsorption capacity for the positively charged N-bromosuccinimide (NBS) modified IgGRF. The same situation occurred when N-Ethyl-N'-[3-(dimethylamino)propyl]carbodiimide(EDC) modified IgGRF because of its decrease in negatively charged density. The low adsorption capacity for 1 ' 2-cyclohexanedion（CHD） and pyridoxal 5-phosphate (PP) modified IgGRF may be attributed to the reduction of positively charged density after modification. From all the experimental results, it is assumed that there is an ionic bond formed between the modified IgGRF and the ss-DNA immobilized immunoadsorbent.

Shenqi Wang and Yaoting Yu et al [24] studied the mechanism of recognition and interactions of low density lipoprotein cholesterol (LDL-C) with different charged ligands on the adsorbents. Tryptophan, lysine residues and carboxyl terminus on LDL were chemically modified by PP, EDC and NBS respectively. Due to the effectiveness of L-lysine in the removal of LDL-C, it was selected to study the interaction of ligand with the modified LDL. Experimental results show that positive charge on the surface of LDL interacted with the negatively charged carboxyl groups of L-lysine by electrostatic force, thus resulting in the adsorption of LDL by the absorbent. We also found that increasing the positive charge on the surface of LDL could enhance the adsorption capacity of the adsorbent. On the contrary, increasing the negative charge could decrease the adsorption ability. Thus, different adsorbents containing sulfonic groups, phosphoric groups, L-lysine and carboxyl groups as the ligand were synthesized for investigating the effect of electric charge on their adsorption capacity. Results show that the adsorption capacity increases with the increase of the electro-negativity of the ligand on the adsorbent. See **Table1**

| | TC | | LDL-C | |
Terminus group of adsorbent	Adsorption percentage (%)	Adsorption capacity (mg /ml)	Adsorption percentage (%)	Adsorption capacity (mg/ ml)
$-SO_3^{2-}$	52.58	1.998	60.9	1.432
$-PO_4^{3-}$	43.86	1.667	44.25	1.039
$PP-PO_4^{3-}$	40.39	1.535	39.51	0.928
$DNA-PO_4^{3-}$	34.94	1.328	33.14	0.778
$L-lysine-COO^-$	31.68	1.203	27.98	0.657
$-COO^-$	26.02	0.989	13.75	0.323

Source: Wang S Q et al, Reactive & Functional Polymers (2008), 68: 261-267

Table 1. Adsorption capacity and percentage of total cholesterol(TC), LDL-C by cellulosic beads having different terminus groups

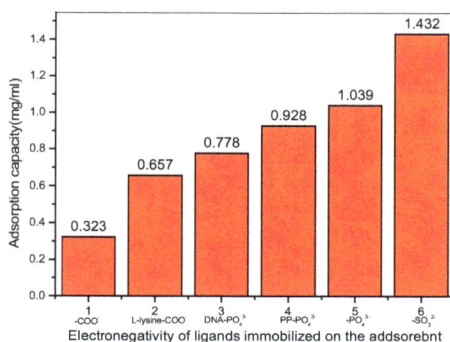

(From Wang S Q et al, Reactive & Functional Polymers (2008), 68: 261-267, adapted)

Figure 4. The relationship of absorption capacity versus electronegativity of ligands immobilized on the adsorbent

Experimental results show that the adsorption capacity (mg/ ml or percentage) for TC and LDL-C decreased with decreasing of electro-negativity of ligands on the adsorbents $(-SO_3^{2-}>-PO_4^{3-}>-COO^-;\ -PO_4^{3-}>PP-PO_4^{3-}>DNA-PO_4^{3-})$, which demonstrate that the electro-negativity of ligand on adsorbent plays an important role in adsorbing TC and LDL-C. This relationship of the adsorption capacity to its electro-negativity is shown in Figure. 4

3.2. Spacer effect

Spacers have a significant effect on the adsorption property of the resin adsorbents. It can reduce the steric hindrance between the ligand and the large toxic molecules, resulting in an increase of adsorption capacity of the adsorbent. Different spacers have an obvious effect on the adsorption properties of adsorbents. The density of ligands on the carrier and the effect of steric hindrance are both important factors in specific adsorption. When the target substance is a small molecule, there may be no steric hindrance, see **Figure 5 a,** so the enhancement of the density of ligands can improve the adsorption capacity. But when the target substance is a large molecule, due to the presence of steric hindrance [26-33], a high density of ligands linked may display a low adsorption capacity of target protein, see **Figure 5b.** In theory, a flexible spacers can reduce the steric hindrance, see **Figure 5c.** In order to study flexible spacers play the role in reducing steric hindrance between the target protein and immobilized ligands, Xinji Guo et al [34]designed and prepared cellulosic adsorbents with L-lysine acid as ligands and PEG having different molecule weights as spacers.

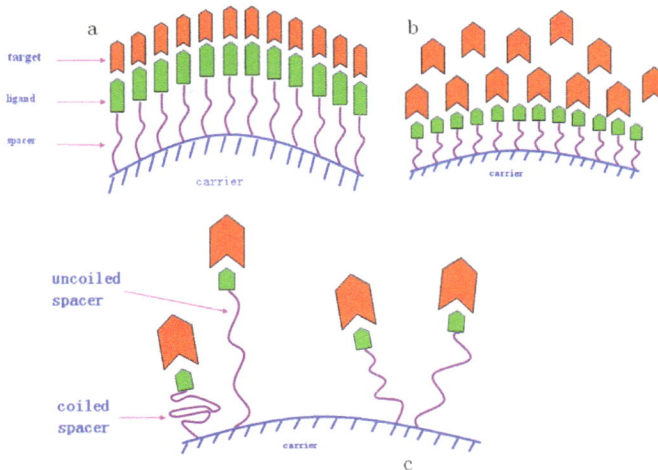

Figure 5. Schematic diagram of interaction between the ligand and target toxins

Note: a, Interaction between small target molecules and the immobilized ligands; b, Large target molecules having a steric hindrance to the immobilized ligands; c, Flexible spacer can reduce the steric hindrance,

In this study, carboxyl modified PEG spacer was synthesized and linked covalently to cellulose beads. L-lysine ligand was coupled to the spacer and its selective affinity for low-density lipoprotein-cholesterol (LDL-C) was determined. It was found that the adsorption capacity and the efficiency of the ligand for adsorption of LDL-C were increased when PEG spacer was used. Experimental results showed that by increasing the molecular weight of PEG spacers from 1000Da to 6000Da, the average adsorption capacity of LDL-C was enhanced from 0.242mg/ml to 0.903mg/ml. According to the analytical data of cellulose adsorbents, the amount of L-lysine ligand could be calculated. Although the amount of L-lysine linked to the adsorbent with PEG spacers (10.5, 9.8, 9.0, 8.6 mg per ml cellulose adsorbent respectively) was lower than those without PEG spacers (121.6mg per ml cellulose adsorbent), see **Table 2**, the average adsorption capacity for LDL-C per ml cellulose adsorbent increased from 0.130 mg/ml to 0.903 mg/ml. After the introduction of PEG spacers, (see **Table 3**) and consequently the adsorption capacity for LDL-C per unit ligand increased significantly from 0.001 mg/mg L-lysine to 0.105 mg/mg L-lysine, see **Table 4**, the adsorption capacity of LDL-C per unit L-lysine ligand (0.027mg LDL-C/mg L-lysine) was much higher than that without PEG spacer (0.001mg LDL-C/mg L-lysine). This result indicated that in the presence of PEG spacer, the adsorption efficiency of L-lysine ligands was enhanced significantly, see **Table 5**. It is postulated that appropriate increasing the amount of the L-lysine ligands and the use of PEG spacers can enhance the adsorption capacity for LDL-C.

Molecular weight of PEG spacers (Da)	TC		LDL-C	
	Removal capacity (%)	Removal amount (mg/ml)	Removal capacity (%)	Removal amount (mg/ml)
1000	11.45±0.35	0.351±0.011	12.01±0.79	0.242±0.017
2000	14.90±0.69	0.458±0.020	13.04±0.71	0.263±0.013
4000	28.94±0.33	0.889±0.011	35.13±0.69	0.708±0.017
6000	33.48±0.33	1.028±0.011	44.76±0.36	0.903±0.003

Source: Wang S Q et al, Reactive & Functional Polymers (2008), 68: 261-267

Table 2. Adsorption capacity and adsorption percentage of TC, and LDL-C by cellulosic beads with different molecular weight of PEG as a spacer

Mol. weight of PEG spacers (Da)	L-Lysine amount (mg/ml)	Average removal amount of LDL−C (mg/ml)	Stoichiometric capacity (mg LDL-C/mg L-lysine)	Efficiency of active site (%)*
1000	10.5	0.242	0.023	22
2000	9.8	0.263	0.027	26
4000	9.0	0.708	0.079	75
6000	8.6	0.903	0.105	100

Source: Wang S Q et al, Reactive & Functional Polymers (2008), 68: 261-267

Table 3. Adsorption capacity of LDL-C per mg L-lysine

Mol .weight of PEG spacers (Da)	L-lysine amount (mg/ml)	Average removal capacity of LDL (mg/ml)	Stoichiometric capacity (mg LDL/mg L-lysine)	Efficiency of active site (%)*
—	121.6	0.130±0.013	0.001	1.018±0.098
1000	10.5	0.242±0.017	0.023	21.947±1.469
2000	9.8	0.263±0.013	0.027	25.556±1.178
4000	9.0	0.708±0.017	0.078	74.917±1.549
6000	8.6	0.903±0.003	0.105	100.000

Source: Wang S Q et al, Reactive & Functional Polymers (2008), 68: 261-267

Table 4. Adsorption capacity of LDL –C by L-lysine ligand

	CPS-lysine		CPS-PEG1000-lysine	
	Adsorption capacity(mg/g)	Adsorption percentage (%)	Adsorption capacity(mg/g)	Adsorption percentage (%)
TC	0.664	8.769	0.402	5.31
LDL-C	1.493	33.559	4.158	93.29
TG	0.419	6.213	0.132	1.97
HDL-C	0.204	15.170	0.341	24.48

Source: Wang S Q et al, Reactive & Functional Polymers (2008), 68: 261-267

Table 5. Adsorption capacity of TC and LDL by CPS beads with and without PEG as spacer

4. Typical bioactive bead type cellulosic adsorbent for blood purification

4.1. Cellulosic adsorbents for removing low density lipoprotein –cholesterol (LDL-C)

Familial hypercholesterolemia is characterized by a high concentration of plasma cholesterol in the form of low-density lipoprotein-cholesterol (LDL-C). In order to decrease the LDL-C level in patients, drugs and surgical intervention were reported [35]. Sorbent-perfusion treatment is currently employed when the reduction of LDL-C level appears impossible to be achieved by drug administration [36-38]. Since the late 1970s, scientists have been engaged in developing different kinds of adsorbents to remove pathogenic substances [39-41].

Yaoting Yu and Shenqi Wang et al [42,43] have developed cellulosic adsorbent with amphiphilic ligands for the adsorption of (LDL-C) which was prepared by the following procedure: Cellulose beads were reacted with cholesterol N-(6-isocyanatohexyl) carbamate in the presence of pyridine in DMSO at 80°C in order to introduce the hydrophilic moiety. It was then reacted with chlorosulfonic acid in dimethyl formamide to introduce the sulfonic group see **Figure 6**

The effects of sulfonation and grafting time of cholesterol on the swelling property of adsorbent were studied. Results showed that sulfonation and grafting time of cholesterol

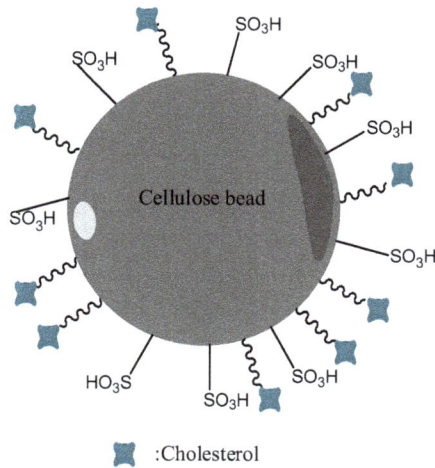

:Cholesterol

(From Wang S Q et al, Artif Cells Blood Sub (2002), 30: 285-292, adapted)

Figure 6. Schematic structure of amphiphilic cellulose adsorbent

was 3 and 5 h, respectively. The amphiphilic adsorbent had a high adsorption capacity for LDL-C without significantly adsorbing high-density lipoprotein. Rabbit model was constructed according to the following method [44]. In brief, Japanese white male rabbits were purchased from local experimental animal institute and housed in a standard facility. After feeding with standard chow and water *ad libitum* for one week, the healthy rabbits were divided into control group (group 1, n=6) and hyperlipidemia group. Rabbits in the control group consumed standard chow from 120-150g/d and water *ad libitum*. In the hyperlipidemia model group, the rabbits were fed with standard chow supplemented with 0.5-1% cholesterol, 15% egg yolk and 5% animal oil. After 8 weeks, the rabbits in the hyperlipidemia group were further divided into two groups, that was group No.2 (n=6), (without any treatment) and group No.3 (n=6), (treated by sorbent-perfusion.). Experimental results showed that the LDL-C levels decreased significantly after 2 h perfusion indicating the adsorbent could effectively remove LDL-C, see **Table 6.** Furthermore, sorbent-perfusion also reduced all the subfractions of LDL-C, therefore decreased the risk for the development of atherosclerosis and myocardial infarction, see **Table 7.**

Parameter	Before (mmol/l)	After (mmol/l)	Reduction (%)
TC	8.54±1.01	3.33±0.63*·	61.20±2.81
TG	1.845±0.191	1.05±0.153*·	43.09±2.43
LDL-C	3.619±0.354	0.724±0.07*·	78.56±0.147
HDL-C	0.216±0.06	0.205±0.057	5.09±0.042

Note: n=6,
Source: Wang S Q et al, Biomaterials (2003), 24: 2189-2194

Table 6. Removal of lipoproteins by amphiphilic adsorbent

	Before perfusion		After perfusion		Change=(A-B)
	mmol/l	%	mmol/l	%	
LDL-1	0.51±0.061	13.95±2.1*	0.12±0.046	15.1±2.3	8.24±2.6
LDL-2	1.65±0.019	45.4±5.3*	0.38±0.043	53.4±5.8*	17.36±3.4
LDL-3	1.45±0.052	40.1±4.7*	0.23±0.024	31.6±3.6*	-21.20±3.9

Note: n=6
Source: Wang S Q et al, Biomaterials (2003), 24: 2189-2194

Table 7. Removal of LDL-C subfraction by amphiphilic adsorbent

The authors also developed the adsorbent with lysine and phosphate groups as ligand for the treatment of hyperlipidemia [34, 45]. Comparing to the amphiphilic adsorbent, the adsorbent with lysine as ligand has a lower adsorption capacity for LDL-C, total cholesterol and higher HDL-C. On the other hand, the phosphalated cellulosic adsorbent has a higher adsorption capacity for LDL-C and lower for HDL-C, see **Table 8.**

Sample		1	2	3	4
Amount of coupled phosphate(μmol/ml)		0	68.4	94.3	128.6
Adsorption capacity (mg/ml)	TC	0.893	0.735	0.939	1.586
	LDL	1.235	1.515	1.707	2.721
	HDL	0.134	0.039	0.058	0.063

Source: Wang T et al, Chinese Journal of Biomedical Engineering (2008), 27: 132-136

Table 8. Effect of amount of coupled phosphate on adsorption capacity

Haofeng Yu et al [46] synthesized PAA-grafted cellulosic adsorbent for the removal of LDL-C from human plasma. In-vitro studies showed that this adsorbent could remove total cholesterol (TC), LDL-C, and triglyceride (TG) at levels of 5.55, 4.46, and 2.48 mg /ml respectively. Unfortunately, it removed 30% HDL-C in the plasma.

4.2. Cellulosic adsorbents for removal of rheumatoid factors

Rheumatoid Arthritis (RA) is a rather wide-spread immune disease. Spector reported that there is about 1% of the world's population is infected by this disease. Lianyong Wang and Yaoting Yu et al [25] covalently linked ss-DNA to cellulosic beads for the removal of rheumatoid factors. In vitro and in vivo studies showed that the adsorbent had a high adsorption capacity for IgMRF, IgGRF and IgARF, see **Figure 7**. Furthermore, the adsorbent attained good blood compatible properties.

Sorbent-perfusion using the above mentioned adsorbent was conducted on 35 RA patients clinically. Clinical protocol was designed as follows: 46 patients were hospitalized from Dec. 1998 until Nov.2000 and diagnosed as RA sickness. The extent of joint pain, swelling, morning stiffness, nodules under the skin, titer of rheumatoid arthritis factor (RF), value of immune-globulin and X-ray diagnosis of sick ankle etc. were all performed. The patients were divided into several groups from the above 46 patients. 35 of them matched the American Rheumatoid Arthritis Standard of 1987[47].

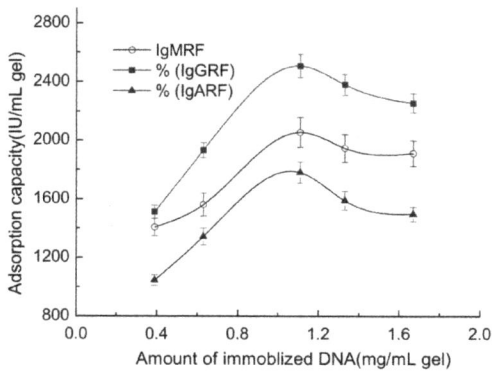

Note:IU is abbreviation of international unit
(Source: Wang L Y et al, Chemical Research in Chinese Universities(2004),20: 795-800)

Figure 7. Adsorption capacity versus amount of DNA immobilized

RA patients were graded into 4 grades according to the functional ability of the sick joint. 11cases having grade II joint sickness, 20 cases having grade III and 1 case having gradeIV. Conventional method of treatment was performed on 30 cases. The control group compared to treated group with a significance of P>0.05.

After perfusion the levels of rheumatoid factor was reduced and platelet count showed no significant changes in 35 RA patients (P<0.05) during treatment. After one week joint pain, swelling, tenderness, morning stiffness disappeared in 90 percent patients; 80 percent abnormal indexes recovered to normal value. After two to three weeks, joint function of 82 percent patients were improved and reached grade II or I. Two to three months later, X-ray examinations showed 80 percent bone matrix destruction was restored. Results of follow-up (0.67-2.5 years) proved that effective rate (97.14%, 81.25%) and total remission rate (82.86%, 59.38%) all had a significant improvement and recovery when compared to routine therapy.

In conclusion, clinical results show that sorbent-perfusion by cellulosic adsorbents is an effective approach for the therapy of rheumatoid arthritis. The treatment is safe, without hemolysis or pyrogenic side reactions, the adsorbent is easy to sterilize and cost effective.

4.3. Cellulosic adsorbents for removal of endotoxin

Bacterial endotoxins (ET), frequently named as lipopolysaccharides (LPS), are components of the outer cell wall of gram-negative bacteria and supposed to be a key factor in the pathogenesis of endotoxemia and septic shock [48,49] . Sorbent-perfusion is one of the best methods to remove endotoxin. Hui Fang and Yaoting Yu et al [10] synthesized a new type adsorbent for the removal of bacterial endotoxins by immobilizing lysine covalently onto cellulosic beads. Results showed that the adsorbent has good biocompatibility, see **Table 9**. In order to evaluate the adsorbent's properties, rabbit models were constructed by the following method [10]: Thirteen New Zealand white rabbits (weight 2.0-2.5 Kg) were injected intravenously with 0.20 mg LPS (*E.coli O55: B5, sigma*) to induce endotoxemia after being anaesthetized. The rabbits were classified into two groups, one was the treated group (n=8) perfused through adsorbents while the other was the control group (n=5) without undergoing perfusion. Sorbent-perfusion was conducted 1.5 hours after LPS administration and conventional equipments for perfusion were used. Blood was drawn from the artery and returned to the vein by peristaltic pump (Pharmacia-LKB). The perfusion was carried out at a rate of 5 ml/min for 2 hours and the adsorbent showed a strong ET-binding capacity. After perfusion, the blood ET level was decreased from 5.56±0.54 EU/ml to 0.41±0.26 EU/ml, see **Figure 8**. Liver function and renal function tests as well as SOD, malondialdehyde (MDA) assays were conducted. Results all showed that the septic symptoms were ameliorated with the removal of large amounts of ET in the blood which obviously prevented further damage to the organs, see **Table 10, Figure 9** and **10**.

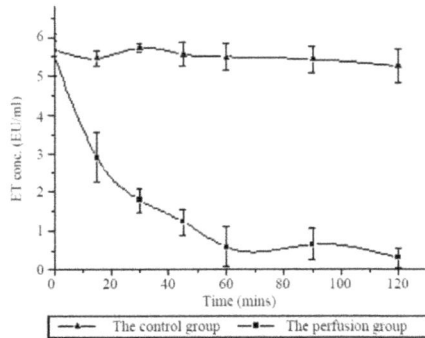

Note: EU is the abbreviation of Endotoxin unit, which is used to indicate the content of endotoxin (From Fang H and Yu Y T,et al, Biomaterials(2004), 25: 5433-5440)

Figure 8. Removal of ET from rabbit's blood by perfusion using Lys immobilized cellulosic beads

Parameters tested	Results
Hemolysis	1.1% (standard <5%),(n=3)
Platelet adhesion	18.76% (n=3) (10% is excellent; 10-30% is good)
Whole body toxicology	Wt. of rat before test Wt. of rat after test 21.5±0.61 (g) 24.4±2.04 (g) (n=5)
Allergic reaction	Grade 1, not higher than negative control (n=10), none allergic
Cytotoxicology	R^* index of sample=0/0 (n=6), none toxic
Skin stimulation	PII^{**}=0.0-0.4, (N=15), very mild toxic

R^* index of sample=R index of extract − R index of sample tested=0.5/0- 0.5/0=0/0, PII^{**} is the skin stimulation index
Note: The biocompatibility and toxicology tests of the adsorbent were conducted by the Testing and Evaluation Research Centre of Biomedical Materials in Tianjin, China according to the Criteria of GB/T16886.5-1997 in correlation to ISO 10993.4:2002, Source: Fang H and Yu Y T et al, Biomaterials (2004), 25: 5433-5440

Table 9. Biocompatibility and toxicology properties of the adsorbent

Parameter	Groups	0 min	60 min	120 min
SA	Perfusion	26.7±2.4	24.5±1.8	23.2±2.3
(g/L)	Control	26.1±3.7	22.0±2.2	24.1±4.6
TBL	Perfusion	0.04±0.01	0.05±0.02	0.05±0.01
(mg/ml)	Control	0.05±0.01	0.08±0.03	0.09±0.02
AST	Perfusion	15.3±4.0	12.1±3.2	14.0±4.2
(U/L)	Control	17.9±5.7	27.6±6.9	35.7±8.0
ALT	Perfusion	34.4±9.5	25.9±9.1	24.8±8.5
(U/L)	Control	35.3±9.9	34.5±6.2	41.4±1.8
AKP	Perfusion	113.1±8.5	116.9±19.2	118.1±25.7
(U/L)	Control	109.4±10.3	153.0±17.3	195.5±31.2
BUN	Perfusion	5.2±0.6	4.9±1.3	4.6±0.6
(mmol/L)	Control	5.6±1.1	7.3±0.8	7.8±0.7
CRK	Perfusion	61.8±4.8	59.3±7.5	70.9±2.0
(umol/L)	Control	62.7±4.2	80.4±3.7	94.3±4.7

Note: n=3

Source: Fang H and Yu Y T ,et al, Biomaterials (2004), 25: 5433-5440

Table 10. Liver and renal function tests

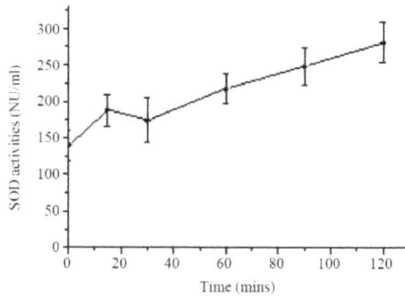

Note: the NU is a unit that is used to indicate the activity of SOD, (From Fang H and Yu Y T et al, Biomaterials (2004), 25: 5433-5440)

Figure 9. Improved activities of serum superoxide dismutase (SOD) versus ET removal by perfusion

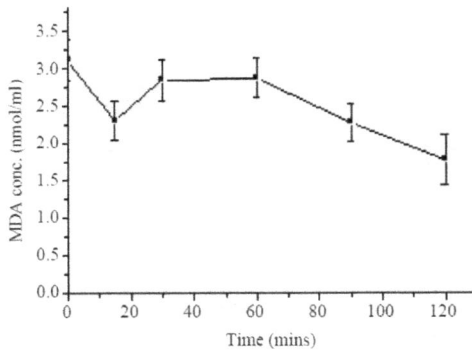

(From Fang H and Yu Y T et al, Biomaterials (2004), 25: 5433-5440)

Figure 10. Decrease in malondialdehyde (MDA) concentrations versus ET removal by perfusion

4.4. Cellulosic adsorbents for removal of anti-DNA antibody in treatment of systematic lupus erythematosus

The abnormally high levels of anti-DNA antibodies and immune complex in the sera of systemic lupus erythematosus (SLE) patients can be removed by sorbent- perfusion. Yaoting Yu and Deling Kong [50] synthesized cellulose adsorbents with DNA as ligand for the removal of anti-DNA antibodies. The activation of cellulosic beads were conducted according to reference [50]: In brief, 2 ml of activated cellulose beads was added to 4.0ml buffer solution containing 4.0mg DNA in a flask and stirred at 25°C for 20 h on a shaker. Then the immobilized DNA beads were washed consecutively with buffer solution and water, until no leakage of DNA in the rinse water was detected at 260nm by UV spectrometer. The immunoadsorbent thus obtained was stored at 4°C. In vitro adsorption tests showed that the DNA immuno-adsorbent could remove 40%-70% of anti-DNA antibody from plasma [29]. The maximum decrease of anti-DNA level was 80% after 60 min in a dynamic experiment. This high adsorption capacity shows a high potential for clinical application.

4.5. Cellulosic adsorbent for the treatment of myasthenia gravis

Myasthenia gravis (MG) is an autoimmune disorder characterized by a disturbance in neuromuscular transmission that results in muscle weakness. Yaoting Yu and Li Yang et al [51, 52] synthesized immobilized tryptophan cellulosic adsorbent and evaluated its adsorption capacity for binding acetylcholine receptor in the plasma of MG patients. Experimental autoimmune myasthenia gravis (EAMG) rabbits were induced by Ta183-200 peptide according to the following method: Briefly, Female rabbits weighing approximately 2 kg were injected intradermally at multiple sites with 500µg of Ta183-200, which was emulsified with an equal volume of Freund's complete adjuvant. A booster injection of 500 µg Ta183-200 with Freund's incomplete adjuvant was administered after 4 weeks.

The rabbits underwent extracorporeal whole blood perfusion for 2 h. Results showed no significant damages on blood cells and changes in the concentration of electrolytes. Whole blood sorbent-adsorption improved clinical manifestation and neuromuscular function of the EAMG rabbits, see **Table 11**

No. of rabbits	Pre WBIA	5th day after WBIA
1	+	0 - +
2	+	0
3	+	0
4	+	0
5	+	0
6	+	0
7	++	+
8	+	0

Note : 0 stands for the rabbits recovered normally after therapy
Source: Yang L et al, Artif Cell Blood Sub(2004), 32: 519-528

Table 11. Grading of clinical manifestation pre and 5th day after WBIA

The neuromuscular transmission function was evaluated by the stimulation of the deep peroneal nerve. The mean decrement of potentials evoked from the anterior tibial muscle, at three stimulation frequencies in the therapeutic rabbit group were determined. At 3Hz, the potential decreased from 21.87% to 17.87%, at 5Hz, decreased from 22.25% to 18.75% and at 10 Hz, decreased from 24.37% to 23.25%. Table 12 shows the changes of the electro-physiological features of EAMG rabbits after WBIA which was on the 5th (D5) and the 8th (D8) day after passive transfer. The same RNS was performed in the control rabbit group, but no decrement was found, see **Table 12**

Frequency	Therapeutic group (%)		Control group (%)	
	D5	D8	D5	D8
3Hz	21.875±3.226	17.875±1.642	21.375±2.615	21.125±2.416
5Hz	22.250±2.815	18.750±1.388	21.875±2.232	22.000±1.772
10Hz	24.375±1.685	23.250±1.388	23.875±1.126	24.125±2.696

Source: Yang L et al, Artif Cell Blood Sub(2004), 32: 519-528

Table 12. Comparison of the decrement of RNS between the therapeutic and the control group

The quantity of neuromuscular junction can reflect the neurotransmission function. On the 3rd day after WBIA (D8), the quantity of neuromuscular junction per unit area ($25mm^2$) of the therapeutic rabbit group was determined. **Figure 11** shows the quantity of neuromuscular junction per unit area increased from 9.825±3.401 to 10.90±2.879(P<0.05) after WBIA, which was higher than that of the control group (P<0.01)

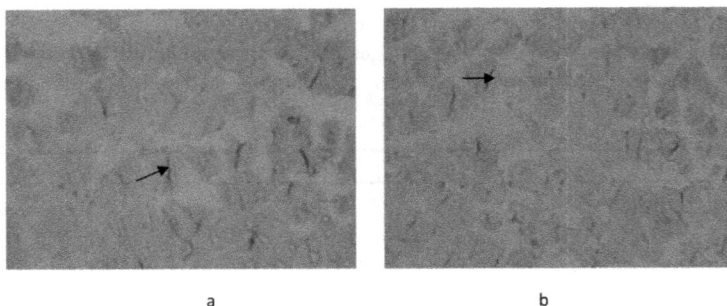

a   b

NOTE: a and b stand for the results before and after WBIA
(From Li X H et al, Artif. Cell Blood Sub (2010), 38:186-191)

Figure 11. Neuromuscular junctions of the therapeutic rabbit group before and after WBIA on the 3rd day（×200）

In conclusion, extracorporeal whole blood sorbent-adsorption is an effective and safe approach in treating the passive experimental autoimmune myasthenia gravis by improving clinical manifestation, neuromuscular transmission function, enhancing the quantity of neuromuscular junction and antibody titer.

5. Conclusion and future perspectives

Intensive research and development of cellulosic bead type adsorbents by sorbent-perfusion in blood purification have paved a path for the treatment of patients with autoimmune diseases, hyperlipidemia and inflammatory disorders. Animal experiments and clinical trials have proved that it is safe, efficient and cost effective. Therefore, it is highly potential to be used clinically on patients for upgrading the quality of their living standard and prolonging their survival rates.

Author details

Shenqi Wang
College of Life Science and Technology & Advanced Biomaterials and Tissue Engineering Center, Huazhong University of Science and Technology, 1037 Luoyu Road, Wuhan, P.R. China

Yaoting Yu
The Key Laboratory of Bioactive Materials, Ministry of Education, Nankai University, Tianjin, P.R. China

Acknowledgement

The authors highly appreciate the financial support of Key project of National Basic Science and Development (G1999064707), the Project of 863（N0.2002AA326060）and Tianjin-Nankai University Co-Construction Foundation.

6. References

[1] Behim E, Klinkmann H (1989) Selective and specific adsorbents for medical therapy. Int J Artif Organs. 12: 1-10.

[2] Lupien P J, Moorjani S, Awad J (1976) A new approach to the management of familial hypercholesterolaemia: removal of plasma cholesterol based on the principle of affinity chromatography, Lancet.1:1261-1265.

[3] Sinitsyn V V, Mamontova A G, Konovalov G A, Kukharchuk V V (1990) Apheresis of low density lipoproteins using a heparin-based sorbent with low antithrombin III binding capacity. Atherosclerosis. 84:55-59.

[4] Maaskant N, Bantjes A, Kempen H J M (1986) Removal of low density lipoprotein from blood plasma using cross-linked, sulfated polyvinylalcohol. Atherosclerosis. 62:159-166.

[5] Kojima S, Harada-Shiba M, Toyota Y, Kimura G, Tsushima M, Kuramochi M, Sakata T, Uchida K, Yamamoto A, Omae T (1992) Changes in coagulation factors by passage through a dextran sulfate cellulose column during low-density lipoprotein apheresis. Int J Artif Organs. 15:185-190.

[6] Yatzidis H (1964) Research on extrarenal purification with the aid of activated charcoal. Nephron. 72: 310-312.

[7] Terman D S, Petty D, Harbeck, Carr R I, Buffaloe G (1977) Specific removal of DNA antibodies *in vivo* by extracorporeal circulation over DNA immobilized in collodion charcoal. Clin. Immunol. Immunopathol. 8: 90-96.

[8] Falkenhagen D; Schima H; Loth F (1999) Arrangement for removing substances from liquids, in particular blood. United States Patent, #5,855,782.

[9] Von Appen K, Weber C, Losert U, Schima H, Gurland H J, Falkenhagen D (1996) Microspheres based detoxification system: a new method in convective blood purification. Artif Organs 20: 420-425.

[10] Fang H, Wei J, Yu Y T (2004) In vivo studies of endotoxin removal by lysine-cellulose adsorbents. Biomaterilas. 25: 5433-5440.

[11] Zhao Y P, Huang M S, Wu W, Jin W (2009) Synthesis of the cotton cellulose based Fe(III)-loaded adsorbent for arsenic(V) removal from drinking water. Desalination. 249:1006-1011.

[12] Kai Y, Hamada J-I, Morioka M, Yano S, Nakamura H, Makino K, Mizuno T, Takeshima H, Kuratsu J-I (2006) Clinical evaluation of cellulose porous beads for the therapeutic embolization of meningiomas. *AJNR Am J Neuroadiol*.27: 1146-1150.

[13] Hamada J, Ushio Y, Kazekawa K, Tsukahara T, Hashimoto N, Iwata H (1996) Embolization with cellulose porous beads I An experimental study. *AJNR Am J Neuroradiol* .17:1895-1899.

[14] Hamada J, Kai Y, Nagahiro S, Hashimoto N, Iwata H, Ushio Y (1996) Embolization with cellulose porous beads II Clinical trial. *AJNR Am J Neuroradiol* .17: 1901-1906.

[15] Kai Y, Hamada J-I, Morioka M, Yano S, Nakamura H, Makino K, Mizuno T, Takeshima H, Kuratsu J-I (2007) Preoperative cellulose porous beads for therapeutic embolization of meninggioma:provocation test and technical considerations. Neuroradiology. 49:437-443.

[16] Wang D M, Hao G, Shi Q H, Sun Y (2007) Fabrication and characterization of superporous cellulose bead for high-speed protein chromatography. J Chromatogr A. 1146:32-40.

[17] Bai Y X, Li Y F (2006) Preparation and characterization of crosslinked porous cellulose beads. Carbohyd Polym. 64:402-407.

[18] Du K F, Yan M, Wang Q Y, Song H (2010) Preparation and characterization of novel macroporous cellulose beads regenerated from ionic liquid for fast chromatography. J Chromatogr A. 1217:1298-1304.

[19] Wang H F, Li B, Shi B L (2008) Preparation and surface acid-base properties of porous cellulose. BioResources. 3: 3-12.

[20] Thummler K, Fischer S, Feldner A, Weber V, Ettenauer M, Locth F, Faulkenhagen D (2011) Preparation and characterization of cellulose microspheres. Cellulose.18:135-142.

[21] Weber V, Ettenauer M, Linsberger I, Loth F, Thummler K, Feldner A, Fisher S, Faukenhagen D (2010) Functionalization and application of cellulose microparticles as adsorbents inextracorporeal blood purification. Macomol.Symp. 294- II :90-95.

[22] Weber V, Linsberger I, Ettenauer M, Loth F, Hoyhtya M, Faulkenhagen D (2005) Development of specific adsorbents for human tumor necrosis factor-α: Influence of antibody immobilization on performance and biocompatibility. Biomacromolecules. 6:1864-1870.

[23] Zhou H M, Wang H R (1998) Chemical Modification of Proteins. Beijing: Tsinghua University Press.19 p.

[24] Wang S Q, Li C M, Guo X J, Yu Y T, Wang W C (2008) Interaction Study of LDL with Charged Ligands for Effective LDL-C Removing Adsorbents. React Funct Polym. 68: 261-267.

[25] Wang L Y, Lu J, Yu Y T (2004) Interaction of Rheumatoid Factor with Immobilized ss-DNA. Chem Res Chinese U. 20: 795-800.

[26] Yang Y (2002). Agarose resin for removal of rheumatoid arthritis factors. Master thesis.

[27] Hou K C, Zaniewski R (1991) Protein A immobilized cartridge for immunoglobulin purification. Biotechnol. Appl. Bioc. 13: 257-268.

[28] Adrian M, Roberto F-L, Jose' M G (2000) Essential role of the concentration of immobilized ligands in affinity chromatography: Purification guanidinobenzoatase on an ionized ligand. J Chromatogr B. 740: 211-218.

[29] Chu J Q, Zhang W H, Yu Y T, Zhu B R, Chen C Z (2004) In vitro studies of the immunoadsorbent for removal of IgA-nephropathy(III). Chem J Chinese U. 25: 1454-1457.

[30] Leckband D E, Kuhl T, Wang H K, Herron J, Muller W, Ringsdorf H (1995) 4-4-20 anti-fliorescyl IgG Fab' recognition of membrane bpund hapten: direct evidence for the role of protein and interfacial structure. Biochemistry-US. 34: 11467-11478.

[31] Zalipsky S (1995) Chemistry of polyethylene glycol conjugates with biological active molecules. Adv Drug Deliver Rev.16: 157-182.

[32] Harris M J, Struck C E, Case G M, Paley S M, Yalpanic M, Van Alstine M J, Brooks E D (1984) Synthesis and characterization of poly(ethylene glycol)derivatives. J Polym Sci. 22: 341-352.

[33] Soltys J P, Etzel R M (2000) Equilibrium adsorption of LDL and gold immunoconjugates to affinity membranes containing PEG Spacers. Biomaterials. 21: 37-48.

[34] Wang S Q, Guo X J, Wang L Y, Wang W C, Yu Y T (2006) Effect of PEG Spacer on Cellulose Adsorbent for the Removal of Low Density Lipoprotein Cholesterol. Artif Cell Blood Subs. 34: 101-112.

[35] Armstrong V W, Windisch M, Wieland H, Fuchs C, Rieger J, Kostering H, Nebendahl K, Scheler F, Seidel D (1983) Selective continuous extractacorporeal ekimination of low-density lipoprotein with heparin at acidic pH . Trans. Am Soc. Artif. Intern Organs. 29: 323-327.

[36] Cao N N, Yu Y Y, Wang M Y, Chen C Z. In vitro study of a novel low density lipoprotein adsorbent (2002), Artif Cell Blood Sub. 30:53-61.

[37] Sato Y, Agishi T (1996) Low –density lipoprotein adsorption for arteriosclerotic patients. Artif. Organs. 20: 324-327.

[38] Chen B, Pan J L, Tong M R, Yu Y T (1993) A study of adsorbents for the adsorption of low density lipoprotein (LDL). Ion Exch Adsorption. 9: 330-334.

[39] Sreenivasan K (1997) Synthesis and evaluation of cyclodextrin 2 hydroxyethy methacrylate copolymer as anovel adsorbent. Polym Int. 42:22-24.

[40] Bosch T (1996) Low density lipoprotein hemoperfusion using a modified polyacrylate adsorber:in vitro, exvitro and first clinical results. Artif Organs. 204: 344-345.

[41] Bosch T, Schmidt B, Kleophas W, Gillen C, Otto V, Passlick-Deetjen J, Gurland H J (1997) LDL hemoperfusion-a new procedure for LDL apheresis: first clinical application of an LDL adsorber compatible with human whole blood. Artif Organs. 209: 977-982.

[42] Wang S Q, Yu Y T, Cui T, Cheng Y (2002) Cellulose amphiphilic adsorbent for the removal of low density lipoprotein. Artif Cell Blood Subs. 30: 285-292.

[43] Cheng Y, Wang S Q, Yu Y T, Yuan Y (2003) In vitro, in vivo studies of a new amphiphilic adsorbent for the removal of low-density lipoprotein. Biomaterials. 24: 2189-2194.

[44] Chapman M J (1980) Animal lipoproteins: chemistry, structure and comparative aspects. J Lipid Res. 21: 789-853.

[45] Wang T, Wang W C, Wang L Y, Wei D, Wang S Q, Yu Y T, Kong D L (2008) Phosphate-based affinity adsorbents for the removal of low density lipoprotein: Preparation and In vitro Adsorption capability test. Chinese J Biomed Eng. 27:132-136.

[46] Yu H F, Fu G Q, He B L (2007) Preparation and adsorption properties of PAA-grafted cellulose adsorbent for low density lipoprotein from human plasma. Cellulose. 14: 99-107.

[47] Arnett F C, Edworthy S M, Bloch D A, Mcshane D J, Fries J M, Cooper N S, Healey L A, Kaplan L A, Liang M H, Luthra H S, Medsger T A, Mitchell D A, Neustadt D H, Pinals R S, Schaller J G, Sharp J T, Wilder R, Hunder G G (1988) The 1987 revised ARA criteria for rheumatoid arthritis Rheum. Arthritis and Rheumatism. 31: 315-324.

[48] Raetz C R H. Biochemistry of endotoxins (1990) Annu Rev Biochem. 59:129-170.

[49] Callery M P ,Kamei T, Mangino M J, Flye M W (1991) Organ interactions in sepsis. Host defense and the hepatic-pulmonary macrophage axis. Arch Surg-Chicago. 126: 28-32.

[50] Kong D L, Chen C Z, Lin E F, Yu Y T (1998) Clinical trials of type I and invitro studies of type II immunoadsorbents for systemic lupus erythematosus therapy. Artif Organs. 22: 644-650.

[51] Yang L, Cheng Y, Yan W R, Yu Y T (2004) Extracorporeal whole blood immunoadsorption of autoimmune myasthenia gravis by cellulose tryptophan adsorbent. Artif Cell Blood Sub. 32: 519-528.

[52] Li X H, Cheng Y, Yang L, Wang S Q, Yu Y T (2010) Extracorporeal whole blood immune-adsorption of passively transferred myasthenia gravis rabbits by cellulose-tryptophan column. Artif Cell Blood Sub. 38:186-191.

Chapter 10

Plasma Induced Hydrophilic Cellulose Wound Dressing

Zdenka Persin, Miran Mozetic, Alenka Vesel, Tina Maver,
Uros Maver and Karin Stana Kleinschek

Additional information is available at the end of the chapter

1. Introduction

The wound care industry is a highly diverse and competitive arena - including standard products such as dry bandages to sophisticated hydrogels and alginate dressings, and encompassing artificial skin and anti-infective used in wound care. A broad range of available modern dressings composed of layers possessing specific functionality has been placed on the market. Waddings and gauzes were replaced by new natural, synthetic and semi-synthetic biomaterials, which depending on their physical, chemical and technological properties are capable to absorb the exudate excess by maintaining the proper moisture environment on/in the wound. The use of textiles in medicine has a long tradition; specially woven textiles are mostly used.

Cellulose and its derivatives are very often used as a functional part of different wound dressing material e.g. as hydrogels, fibre, non-woven. Cellulose micro-fibrillar structure control the accessibility of the surface, while the bulk polar groups (-OH groups) ensure an attraction of the water molecule to the cellulose fibre (hydrogen bonds). Generally different standard chemical pre-treatments (i.e. alkaline washing, bleaching, and slack-mercerization) are used to improve the sorption capacity [1-4], but besides the ecological pollution, they also worsen fibres mechanical properties to a large extent. On the subject of the latter, plasma treatments are known to be rather harmless surface activation procedures leading to enhancement of hydrophilic properties; the best results are obtained using highly non-equilibrium oxygen plasma. Reactive particles created in plasma readily interact with cellulose resulting in formation of surface functional groups, as well as etching. The properties of plasma treated materials depend on the processing parameters and are measured using surface sensitive techniques. Numerous authors worldwide addressed modification of cellulose and other suitable materials by gaseous plasma treatment [5-16].

Unfortunately, however, only few authors are dared to measure plasma parameters so the results obtained by different authors cannot be directly compared. Many authors addressed surface properties such as the surface free energy [17-18], and surface functional groups. The surface free energy is usually determined from measurements of the contact angle of a suitable liquid or by more sophisticated techniques [19-20], while the appearance of specific functional groups has been determined by X-ray Photoelectron Spectroscopy (XPS) [6,21-23], Secondary Ion Mass Spectroscopy (SIMS) [24] and Furrier Transform Infrared Spectroscopy (FTIR) [21,25], just to mention the most commonly used ones. The modification of surface morphology of plasma treated textiles for wound dressings has been often studied by Scanning Electron Microscopy (SEM) [23-25] and Atomic Force Microscopy (AFM) [22, 26-27].

A variety of gases has been used for modification of surface properties in order to achieve better hydrophilicity. Many authors just use noble gases [9,21,28-29], but more oxidizing gases such as oxygen, air, carbon dioxide and water vapour often give better results, especially when the process speed is the merit. Traditionally, gaseous plasma was always created at low pressure in vacuum compatible chambers. In the past couple of decades, however, plasmas which run at atmospheric pressure are becoming more and more popular. In fact, many resent works on modification of organic materials by gaseous plasma treatment report application of a kind of atmospheric gaseous plasma. As mentioned earlier, the richness of available plasmas as well as the lack of experimental techniques used by different authors prevents both direct comparison of reported result as well as good insight in the phenomena taking place during plasma treatment of textiles and similar materials. An important scope of this chapter is to explain some basic considerations about plasma textile interaction and give hints for optimal utilization of this advanced technique for modification of materials.

2. Plasma treatment of cellulose wound dressing

The hydrophilicity of organic materials is improved dramatically by incorporation of oxygen rich functional groups on to the surface, so a natural choice of gas used in plasma treatment of materials is oxygen. In thermodynamic equilibrium and at room temperature oxygen gas contains only neutral oxygen molecules. According to the basic law of statistical mechanics the molecules move randomly, collide with each other and the basic kinetic properties are described well taking into account the approximation of an ideal gas. The distribution of molecules over the kinetic energy is given by Maxwell-Boltzmann law [30] and the average random velocity is calculated as:

$$v = \sqrt{8kT / (\pi m)}, \tag{1}$$

The random velocity only depends on the gas temperature. At room temperature the average random velocity of oxygen molecules is about 440 m/s, while at 1000 K it is about 810 m/s. Oxygen molecules reach the surface of any material with different velocities taking into account the Maxwell-Boltzmann distribution, but as long as the gas temperature is close to the room temperature the velocity is too low to bring enough energy for chemical reactions on the surface of cellulose materials. Increasing gas temperature or the temperature of the cellulose material the available energy can become large enough to overcome the potential barrier and chemical interaction between oxygen molecules and cellulose is observed. The chemical interaction is highly exothermic so the material is locally heated due to a chemical reaction, which results in even higher chemical affinity resulting in more exothermic reactions etc. so finally the control of oxidation mechanism is lost and the chemical reactions leading to the oxidation of cellulose become very intensive. A common expression for such experimental conditions is burning. Treatment of textiles by oxygen gas at equilibrium conditions is therefore not useful because the oxidation is difficult if not impossible to control. One could use oxygen at very low concentration in order to avoid too extensive chemical reactions, but the results are usually not optimal. This is the reason why equilibrium gaseous treatment has not been used for introduction of highly polar oxygen-rich functional groups onto the surface of organic materials. Instead, non-equilibrium gases should be used.

A typical example of non-equilibrium gas is gaseous plasma. It can be created by different means but the most common method is application of a gaseous discharge. In a gaseous discharge an electric field is applied and the gaseous molecules ionize to positive ions and free electrons. Charged particles are accelerated in electric fields and collide with neutral gaseous molecules as well as surfaces. If the kinetic energy of charged particles is above the ionization threshold the collisions will result in ionization of a neutral molecule and thus multiplication of charged particles. Particles with opposite charges attract each other and may or may not neutralize upon collisions. The neutralization may occur either in the gas phase or on surfaces. After turning on the electric field the multiplication of charged particles is quickly balanced by the loss due to neutralization so rather steady concentration of charged particles is established. The concentration depends on numerous parameters and the discharge power is only one of them.

Ionization is definitely the most important collision event since it allows for sustaining the discharge and thus transformation of oxygen gas into the state of plasma. In practical applications, however, it has a minor role in modification of organic materials. Other collisions are often much more important in order to make oxygen a suitable medium for surface functionalization of organic materials. Such collisions include excitation and dissociation of oxygen molecules as well as excitation of oxygen atoms. Namely, oxygen molecules are rich in excited states and so are oxygen atoms. Many excited states are metastable which means that do not decay in a short time by electrical dipole radiation. Such metastable molecules or atoms are not only more reactive than the counterparts in the ground state, but also play a very important role in formation of oxygen particles in highly excited states. Table 1 presents some most important excited states of oxygen molecules, while appropriate values for oxygen atoms are summarized in Table 2.

Examination of Table 1 reveals important data which should be taken into account when dealing with application of oxygen plasma for modification of cellulose materials. The ground state has the potential energy 0 eV by definition. A variety of vibrational and rotational excited states are not shown in Table 1 due to the clearness. Also, the population of oxygen molecules in highly excited vibrational states is usually negligible in plasma suitable for modification of organic materials due to super elastic collisions between vibrational excited molecules and atoms which are very likely to appear in oxygen plasma [31]. At such super-elastic collisions a part of molecular vibrational energy is transferred to the kinetic energy of colliding particles. Since the cross-section for super-elastic collisions between vibrational excited oxygen molecules and neutral oxygen atoms is very large, this event effectively reduces the number of oxygen molecules in highly vibrational excited states so any interaction between such molecules and organic materials is regarded not important. The rotational states are usually coupled rather well with translational ones so the rotational population is similar to that calculated using equations of equilibrium thermodynamics. In fact, the rotational population detected by high resolution optical emission spectroscopy is often used for estimation of the neutral gas kinetic temperature in gaseous plasmas.

Excited state	Optical symbol	Excitation energy
Ground	$X^3\Sigma_g^-$	0 eV
First excited	$a^1\Delta_g$	1 eV
Second excited	$b^1\Sigma_g^+$	2 eV
	$B^3\Sigma_u^-$	6 eV
Ground ionized	$X^2\Pi_g$	12 eV
First ionized	$a^4\Pi_u$	16 eV
Second ionized	$A^2\Pi_u$	17 eV
Dissociation energy		4.5 eV

Table 1. Most important excited states of oxygen molecules and their excitation energies

Excited state	Optical symbol	Excitation energy
Ground	$2s^2\,2p^4\,^3P_2$	0 eV
First excited	$2s^2\,2p^4\,^1D$	2 eV
Second excited	$2s^2\,2p^4\,^1S$	4 eV
Metastable highly excited	$2s^2 2p^3(^4S^\circ)3s\,^5S^0$	9.1 eV
	$2s^2 2p^3(^4S^\circ)3s\,^3S^0$	9.5 eV
	$2s^2 2p^3(^4S^\circ)5s\,^5S^0$	12.7 eV

Table 2. Most important excited states of oxygen atoms and their excitation energies

More important that vibrational states are electronically excited states. Oxygen molecules are famous of two metastable states which appear at the excitation energies of about 1 and 2 eV, respectively (Table 1). Especially the first excited state has a very long radiative life-time of nearly one hour. The life-time of second excited state is much shorter at about ten seconds, but still very long for the atomic world. The molecules in excited states are definitely chemically more reactive than those found in the ground state, but very little work has been done on explanation of interaction mechanisms between electronically excited oxygen molecules and organic materials. Many authors neglect such an interaction since they claim other particles found in oxygen plasma are more suitable for modification of surface properties. All authors, however, agree that metastable oxygen molecules play an important role in dissociation events. Table 1 reveals that the dissociation energy of oxygen molecules is about 4.5 eV. An electron colliding with the molecule in the ground state will thus need kinetic energy higher than this value to be capable of a dissociation collision. The dissociation energy of excited molecule, however, is much lower than for a molecule in the ground state: for a molecule in the first excited state the dissociation energy is about 3.5 eV, and for a molecule in the second excited state it is about 2.5 eV. The difference in excitation energies of molecules found in the ground or excited states does not seem extremely important, but in fact it is due to two important considerations: (i) the high-energy tail of the electron energy distribution function in gaseous plasma is exponential, and (ii) the dissociation cross-section just above the threshold increases extremely steep with increasing kinetic energy of an electron. These two considerations indicate that the dissociation of molecules in oxygen plasma will be a very probable event due to step-like excitation. The result of the oxygen molecule dissociation is formation of neutral oxygen atoms in the ground state. Taking into account the upper considerations the production of atoms will be extensive in oxygen plasma created by any suitable electrical discharge. In spite of this fact the density of neutral oxygen atoms in technological plasmas vary enormously between different configurations. Neutral oxygen atoms tend to recombine (associate to form a molecule). The recombination event may take place either in the gas phase or on surfaces. The recombination event is simply described by reaction $O + O \rightarrow O_2$. Such a simple denotation is, however, wrong since it does not take into account the conservation of energy and momentum. According to Table 1 a large amount of energy is released at a recombination event. If the recombination results in formation of a molecule in the ground state the amount of energy released is 4.5 eV. If a molecule in an excited state is formed the appropriate amount of released energy is accordingly lower. In any case, the released energy cannot be transferred into the kinetic energy of newly born oxygen molecule due to the conservation of momentum. This is why simple recombination as mentioned above does not occur in oxygen at low pressure. A possible recombination mechanism which can occur in the gas phase is radiative recombination $O + O \rightarrow O_2 + \gamma$, where γ is a newly born light quantum (a photon) which takes the excessive energy. In this case two particles are formed at the recombination event so there are no problems with conservation of the momentum. The probability of such reaction depends on the quantum characteristics of particles involved. For the case of recombination of neutral oxygen atoms such reactions are highly improbable so neutral oxygen atoms are regarded stable in vacuum.

Oxygen atoms may recombine in the gas phase at so called three-body collisions $O + O + O_2 \rightarrow O_2 + O_2$. In such a case, the released energy is shared between the two molecules. The molecules may gain the released energy either as kinetic or excitation energy. The sharing of released energy again depends on the quantum characteristics of particles involved, and very typically the newly born molecule is found in an excited state. Such a reaction is actually an important source of excited molecules in plasma afterglows [32]. Since a three body collision is required the loss of atoms through this channel depends on the pressure (or the density of gaseous molecules). As long as the pressure is below several mbar the probability of three-body collisions is very low and this channel is regarded a marginal one at low-pressure plasmas. The opposite is true at atmospheric pressure where the probability for such collision is very high and practically any collision regardless the kinetic energy and momentum is highly probable. This is an important reason for observed large gradients on reactive gaseous particles in atmospheric plasmas.

Low-pressure plasma is therefore characterized by an absence of atomic loss mechanisms in the gas phase. The only channel for atom recombination in such plasmas is heterogeneous surface recombination. The reaction is written as $O + O_{(surf)} \rightarrow O_{2(surf)}$. In the case of surface recombination the excessive energy released is shared between the atoms in the solid material and the newly born molecule. An important consideration about surface recombination is the presence of oxygen atoms on the solid material surface. Namely, the recombination is unlikely to occur if atoms are not stuck on the surface. The recombination mechanisms have been explained neglecting details many decades ago. A couple of recombination mechanisms have been proposed: (i) Langmuir-Hinshelwood and (ii) Eley-Rideal mechanisms. The first one predicts adsorption of at least two atoms on the solid state surface, migration to an appropriate site and association of two atoms into a molecule. Since both atoms involved are already thermalized on the surface the newly born molecule leaves the surface in the ground state. In the case of Eley-Rideal mechanism oxygen atom is adsorbed on the surface where it waits for an atom from the gas phase and recombines to a molecule immediately (before the atom from the gas phase manages to thermalize on the solid surface). In this case the newly born molecule is likely to leave the surface in an excited state.

Whatever the recombination mechanism is, the reaction probability and thus the loss of atoms on surfaces depends on the ability of adsorption and the life-time of an adsorbed atom. Many materials tend to chemisorb atoms. Chemisorption stands for formation of a chemical bond between an oxygen atom and atoms on the surface. A good example is interaction between a metal and an oxygen atom. The oxygen atom is chemically bonded to the metal surface where it waits for other atoms to interact with them and form a molecule. The recombination probability on such surfaces is therefore very large. Appropriate literature reports experimentally determined values which are of the order of 0.1 for many metals. The consequence of such extensive surface recombination is heating of metals in atom rich plasma. This effect is used by catalytic probes - the thermal load onto a selected catalyst is a measure of the O atom density in the vicinity of the probe [33-36]. The exception to highly recombinant metals is a group of metals which form extremely stable oxides such as aluminium, titanium and alike. In such cases the atoms form very stable compounds and are thus not any more available for recombination. Recombination coefficients for such materials are orders of magnitude lower that for metals mentioned above.

Materials which do not chemisorb oxygen atoms are famous of having low coefficients for heterogeneous surface recombination of oxygen atoms. Apart from oxide ceramics the group of materials inert to chemisorption of oxygen atoms includes glasses. A glass is often a mixture of metal oxides which are chemically extremely inert and thus unlikely to attract neutral oxygen atoms. Such inertness is preserved also to elevated temperatures so different glasses are often used as building materials for plasma reactors where high density of neutral oxygen atoms is required. The recombination coefficient for neutral oxygen atoms on glasses depends on the type of material as well as surface roughness and is often as low as 10^{-4}. A typical low-pressure plasma reactor made from glass is shown in Figure 1. In such glass chambers it is common to achieve very high dissociation fractions of oxygen molecules even at moderate discharge power. Values exceeding 10% have been reported by different authors [37-39]. The reasons are obvious from upper considerations: a very low recombination coefficient on plasma facing materials, negligible loss in the gas phase and favourite production by dissociation of molecules taking into account their metastable excited states.

Figure 1. A photo of plasma created in a glass discharge chamber

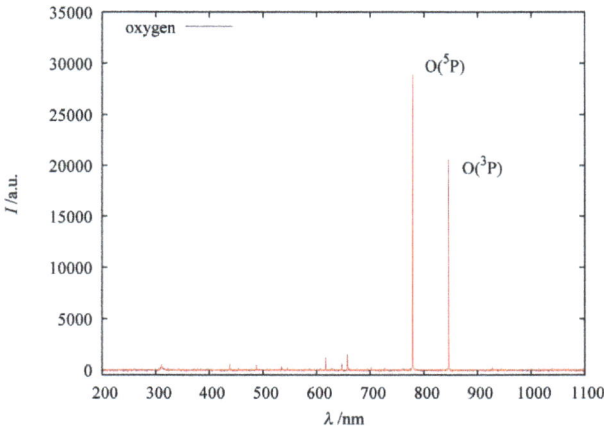

Figure 2. An optical spectrum acquired in oxygen plasma created by an electrode-less discharge in a glass chamber

Not only oxygen molecules, but also neutral atoms have a lot of excited states. Table 2 represents the most important ones. The first excited state is at the excitation energy of about 2 eV and is metastable with the characteristic radiation life-time of about 1 second. As mentioned earlier this is a very long life for atomic world. Excited states which are not metastable de-excite by electrical dipole radiation in a fraction of a microsecond. The second excited state is also metastable, and the radiation life-time is about 100 s. Many other excited states are not metastable which in practice means that an atom found in such an excited state de-excites quickly by emission of a light quantum. An important metastable excited state is found at very high excitation energy of about 9 eV. The existence of such a state allows for excitation of oxygen atoms from this metastable state to very high levels and even ionization. The ionization energy of an oxygen atom is about 13.6 eV. Taking into account that one has to dissociate molecule first and then ionize the atom in order to get a positively charged oxygen atomic ion it is rather improbable that such ions would abound in plasma created by an electrical discharge of a low or moderate power. Still, the existence of metastable states definitely makes ionization of oxygen atoms easier. In practical applications it does not really matter whether the positive ions are atomic or molecular, but the distribution of atoms over excited states is important since the chemical reactivity of atoms depends largely on the excitation energy. Due to the high energies involved one would think that it is unlikely to excite atoms above the level of the $2s^2 2p^3(^4S^\circ)3s$ $^5S^0$ using low power discharges, but such an assumption is not justified experimentally. Experiments with a simple optical emission spectrometer clearly show abundance of atoms in highly excited states. Figure 2 represents an optical spectrum of plasma created in oxygen by high-frequency discharge in a glass discharge chamber. In this spectrum atomic oxygen lines arising from transitions of highly excited atoms prevail indicating the presence of such atoms. Since the excitation energy of radiative excited states exceeds 11 eV it is clear that excitation is a step-like process: after successful dissociation of a molecule an atom is excited to a metastable state of low energy, such a state is excited to a metastable state of high energy and finally an atom is excited to a radiative state. After emission of a photon the atom does not de-excite to the ground state, but rather to the metastable state so another excitation to a radiative state is likely to occur. The spectra such as one presented in Figure 2 therefore represent a good fingerprint of the existence of highly excited metastable atoms in oxygen plasma.

Plasma is usually defined as an ionized gas. Taking into account upper considerations, such a simple definition is hardly useful in practical applications such as activation of polymer materials. Namely, the concentration of neutral reactive gaseous particles is often orders of magnitude larger than the concentration of charged particles. From this point of view it may be more convenient to define plasma as a state of gas which is essentially close to room temperature, but the chemical reactivity is as high as the parent gas at thousands of degrees if not higher. A huge difference between the gas temperature and chemical reactivity steams from the fact that, unlike the case of equilibrium gases, high temperature is not a necessary condition for achieving an appropriate concentration of highly reactive gaseous particles. Instead of high temperatures, non-equilibrium gases take advantage of the fact that the electrons are found at extremely high temperatures while the rest of the gas practically remains at room temperature. The reason for a huge difference between the electron temperature which easily exceeds 10,000 K and the gas temperature comes from two facts: (i) electrons are easily accelerated to high energy in electrical fields, and (ii) the kinetic energy coupling between electrons and other particles is negligible. In a simple DC (direct-current) discharge the electrons are born on the cathode by secondary electron emission due to bombardment of the cathode by positive ions. All charged particles accelerate in the potential fall near the cathode. Electrons are accelerated towards plasma while positively charged particles are directed towards the cathode and thus cannot contribute to gas heating. The energy positive ions gain by passing the cathode sheath is therefore used for emission of free electrons, but the major part is simply lost for heating of the cathode material. Electrons born by secondary emission are accelerated towards the plasma. They gain a lot of kinetic energy which they slowly loose at elastic collisions with other electrons from plasma. Since the energy transfer between electrons is very intensive, fast electrons born on the cathode quickly thermalize in plasma thus heating other electrons. The electrons gain Maxwell-Boltzamann distribution throughout the volume of gaseous plasma. Since they are continuously heated by fast electrons from the cathode their temperature remains very high despite of the loss of kinetic energy at inelastic collisions with gaseous molecules. This is why electrons are usually at very high temperature as compared to other particles present in non-equilibrium gaseous plasma. Similar observations are valid for capacitively coupled high-frequency discharges except that the electrons are accelerated in DC self-bias sheath next to the smaller (powered) electrode. The physics of such discharges has been studied to detail in [40-42]. The heating of electrons follows a different mechanism in inductively coupled discharges, especially those created in pure H-mode [43]. In such cases the electrons follow oscillations of the electrical field induced by oscillating magnetic field. Since the magnetic field oscillates in the entire volume of the discharge the electrons are accelerating in the entire volume, too. No acceleration across sheath is required so plasma can be created in an absence of any electrode. This is a favourite property since electrodes often represent a major sink for neutral oxygen atoms. As mentioned above the recombination of neutral oxygen atoms on metallic surfaces is always orders of magnitude more important than on glass surfaces. Another advantage of inductively coupled discharge is an absence of ion acceleration. If the frequency of the oscillating field is large enough (typically above 1 MHz) the ions cannot follow oscillations of the local electric field since the frequency is well above the resonant one. Massive ions therefore gain a negligible energy from a high-frequency electric field so they remain at room temperature. On the other hand, electrons are light enough to be able to follow the oscillations. The kinetic energy they gain from the electric filed is spent for their multiplication at ionizing collisions as well as for excitation and dissociation of oxygen molecules. Inductively coupled plasmas are therefore very suitable sources of highly excited oxygen atoms which are chemically extremely reactive.

In the first part of this chapter we clearly stressed an importance of avoiding high temperature during oxidation of materials. Although gaseous plasma is basically at room (or little elevated) temperature the heating of materials exposed to gaseous plasma is not always negligible. We already mentioned heating of materials with a high coefficient for heterogeneous surface recombination. A variety of surface physical and chemical reactions are exothermic. Among physical reactions one should stress weak bombardment of a material by positively charged ions, neutralization of charged particles, relaxation of metastables and recombination of neutral oxygen atoms. The thermal load obviously depends on the flux of reactive particles on to the surface, which in turn depends on the density of different particles in the gas phase.

Since the electron temperature (T_e) in plasma is much larger than the positive ion temperature and since they have a favourable random velocity the electrons escape from plasma towards the surfaces of the discharge vessel much faster than the positive ions. The surface of plasma facing components thus becomes negatively charged against gaseous plasma. The potential difference between plasma (V_p) and the surface (V_f) is created in such a way to repel most electrons except the fastest and accelerate ions. In steady state conditions the flux of electrons on to the surface is equal to the flux of positive ions. Obviously, the potential difference between unperturbed plasma and the solid surface increase with increasing electron temperature. In the first approximation the potential difference is calculated using equation:

$$V_p - V_f = \frac{kT_e}{2e_o}\ln\left(\frac{m_+}{2m_e}\right),$$ (2)

Positively charged ions are therefore accelerated towards the surface. They impinge the surface with kinetic energy which corresponds to the potential difference as illustrated by equation (2). Their kinetic energy when impinging the surface therefore does not depend on the ion temperature in unperturbed plasma but rather on the potential difference between plasma and floating potentials. A typical value for the kinetic energy of ions impinging the surface is about 10 eV taking into account the collision-less sheath approximation. This energy is not large enough to cause implantation of ions into a solid material, but causes heating of the solid material. Furthermore, positively charged ions are very likely to neutralize on solid surfaces. As mentioned before the surfaces are negatively charged against plasma so electrons abound on the surface. The observed probabilities for neutralization of slow ions on solid materials are actually very close to 1. This means that an average ion reaching the surface of a plasma facing component brings energy equal to the sum of ionization and kinetic energies. Table 1 reveals that the ionization potential of oxygen molecule is about 12 eV so the contribution of both energies is practically the same. In any case, interaction of positively charged ions with solid materials causes heating so highly ionized plasmas are not very suitable for treatment of delicate materials such as cellulose for wound dressings.

Not only charge particles but many other particles created in oxygen plasma are suitable for heating of cellulose materials. Highly excited atoms de-excite on the surfaces of practically any material at very high probability so they directly contribute to the heating. Their role, however, is difficult to quantify since the density of atoms in highly excited states is difficult to measure [44]. Theoretical simulations indicate that the density of oxygen atom metastables depends largely on the excitation energy and not so much on the radiative life-time [32]. This finding is explained by a very high probability for quenching of highly excited states. Quenching stands for de-excitation by mechanisms other than radiation. The concentration of excited oxygen atoms therefore depends on the electron density and the electron temperature. As long as the density of charged particles is rather low it is reasonable to assume a rather low density of atoms in excited states and thus negligible contribution to the heating of plasma facing materials.

Another source of heating of cellulose materials is heterogeneous surface recombination of neutral atoms in the ground state. As mentioned earlier the density of atoms in plasma created by electrode-less discharge in the glass tube is always high. The available energy dissipated at surface recombination is 4.5 eV. As explained above this energy is shared between the solid material and the newly born molecule. If the surface recombination was intensive, the heat load would have been high enough to cause immediate burning of cellulose materials. The experimental observations, however, show that cellulose materials do not burn in plasma containing large quantities of neutral oxygen atoms providing the density of charged particles is very low. This observation leads to the qualitative conclusion that the probability for heterogeneous recombination of neutral oxygen atoms in the ground state is rather low. The lack of data on reaction mechanisms is unfortunately severe. While recombination coefficients have been determined rather accurately on many other materials especially inorganic, a lot of work will have to be performed on recombination of neutral oxygen atoms in the ground state on organic materials in order to get a good insight into this mechanism and thus to estimate the corresponding thermal load. The same applies for excited oxygen molecules. In any case, these reactions are much less important for heating of cellulose materials than interaction of highly excited gaseous particles, so weakly ionized, highly dissociated oxygen plasma is suitable for modification of surface properties of cellulose materials.

The first effect of exposure of practically any organic material to oxygen plasma is functionalization of surface with oxygen rich functional groups. A usual technique for determination of functional groups is X-ray Photoelectron Spectroscopy (XPS) which is a very surface sensitive method with detection depth of only few nanometers. Therefore, it is a powerful technique for studying elemental composition and chemical state of the surfaces. In this chapter we present results obtained with our XPS device. The cellulose samples were analysed by TFA XPS instrument from Physical Electronics. The base pressure in the XPS analysis chamber was about 6×10^{-10} mbar and the samples were excited with X-rays over a 400 μm area using monochromatic Al $K_{\alpha1,2}$ radiation at 1486.6 eV. The photoelectrons were detected by a hemispherical analyser, positioned at an angle of 45° with respect to the surface normal. Survey-scan spectra were measured at a pass energy of 187.85 eV, while for C1s individual high-resolution spectra were taken at pass energy of 23.5 eV and a 0.1-eV energy step. An additional electron gun was used for surface neutralization of samples during the measurements. The measured spectra were analysed using MultiPak v7.3.1 software from Physical Electronics, which was supplied with the spectrometer. The high-resolution carbon C 1s peaks were fitted with symmetrical Gauss-Lorentz functions. A

Shirley-type background subtraction was used. Both the relative peak positions and the relative peak widths (FWHM) were fixed in the curve fitting process.

Cellulosic fibres, in a form of woven or non-woven textiles are one of the most used wound dressing base materials. Therefore, the cellulose material as viscose fibres in its non-woven form was studied, as produced by Kemex, the Netherlands. The surface mass was 175 g/m^2 (SIST ISO 3801), the thickness under normal conditions was 1.7 mm (SIST EN ISO 5084), and the air permeability 650 l/m^2 s (DIN 53 887).

A typical survey spectrum of photoelectrons on cellulose material is presented in Figure 3. The spectrum is dominated by carbon and oxygen peaks. By far the largest carbon peak is marked as C1s and results from photoelectrons ejected from carbon 1s orbital. A much weaker peak is observed at the energy of about 25 eV. Several oxygen peaks are observed in the spectrum as well and the O1s predominates. The survey spectrum is suitable for estimation of the material composition in the surface layer with the thickness of few nanometres but tells nothing about the functional groups. In order to reveal the chemical structure one should acquire a high-resolution spectrum of C1s. Such a spectrum is shown in Figure 4 for the case of a very flat cellulose material while a corresponding spectrum for cellulose material used for wound dressing is shown in Figure 5. The curve presented in Figure 4 reveals three different functional groups. The major peak at about 286.4 eV corresponds to C-O functional group which abounds in cellulose materials. A well distinguished peak at 284.8 eV corresponds to the C-C bonds, while the third one which appears as a shoulder at about 287.9 eV corresponds to the O-C-O bond. The spectrum presented in Figure 5, on the other hand, does not reveal well expressed peaks due to the artefacts typical for this sort of surface characterization. The major reason for unclear peaks is a rich roughness of the material. Detailed explanation of such deviations is presented elsewhere [45].

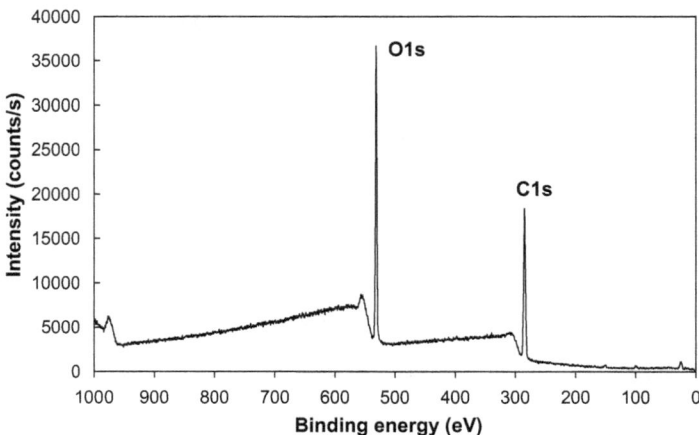

Figure 3. A survey spectrum of cellulose material

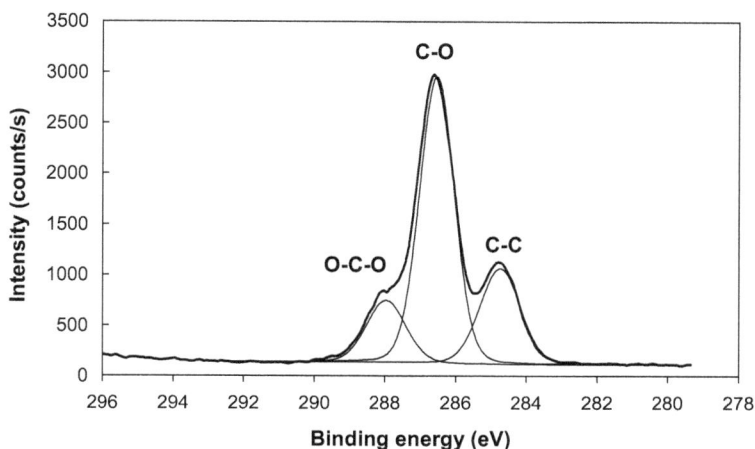

Figure 4. A high-resolution XPS spectrum obtained on a cellulose film

The survey spectrum of cellulose used for wound dressings treated by oxygen plasma is not much different from the one presented in Figure 3 so it is not shown in this chapter. Instead, a comparison of atomic composition between non-treated and oxygen plasma treated cellulose is presented in Table 3. Of particular importance is the ratio between oxygen and carbon atoms as detected by XPS and presented in Table 3. One could observe a large difference of almost 20 %. Here, it is worth stressing that the escape depth of photoelectrons is several nm so the experimental technique give somehow averaged value over the few nm thick surface layer. Taking into account gradients of oxygen concentration (it is the largest on the very surface) one could qualitatively argue that the actual oxygen concentration on the very surface is much larger than shown in Table 3. Any gradients in the elemental composition in a thin surface film could be estimated by tilting the sample during XPS characterization but such procedure is efficient only in the case of very flat samples. Cellulose materials used for wound dressings definitely do not fulfil this condition.

Sample treatment	Surface composition (at. %)		
	C	O	O/C ratio
Non-treated	57.2	42.2	0.73
Oxygen plasma treated	52.1	47.9	0.91

Table 3. Surface composition of non-treated and oxygen plasma treated cellulose fabric

Figure 6 represents a typical high resolution C 1s peak of cellulose material which has been treated by oxygen plasma. Unlike the non-treated material the peak now contains four different sub-peaks. The newly formed sub-peak at high binding energies in Figure 6 corresponds to a new functional group which is formed on the cellulose material during treatment with oxygen plasma and it corresponds to the carboxyl functional group

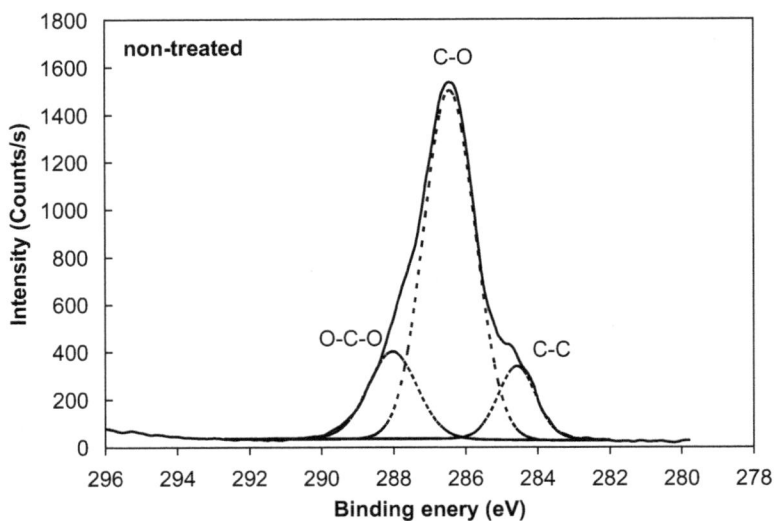

Figure 5. A high-resolution XPS spectrum obtained on cellulose material used for wound dressing

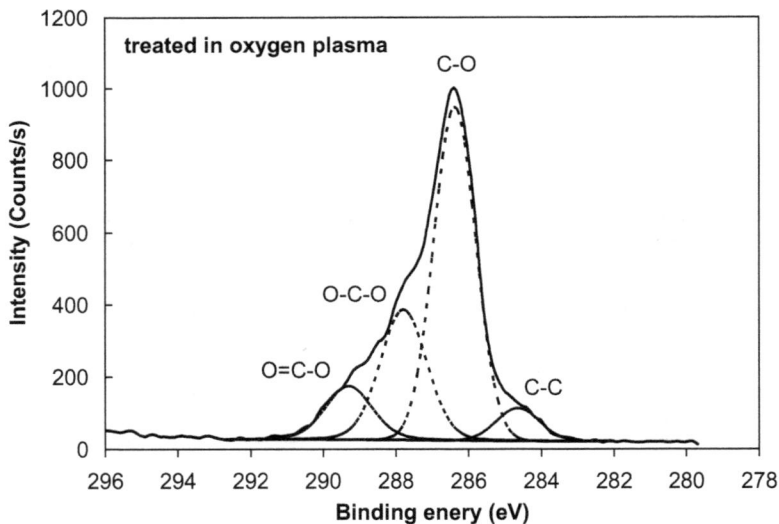

Figure 6. A high-resolution XPS spectrum obtained on oxygen plasma treated cellulose material used for wound dressing

O=C-OH. Furthermore, the sub-peak C-C in Figures 5 and 6 is much smaller after plasma treatment, while the O-C-O sub-peak is enlarged somehow and it can contain also contribution of C=O bonds. Taking into account upper considerations one can easily speculate that the C-C bonds completely vanish from the very surface of cellulose material upon oxygen plasma treatment. Plasma treatment thus facilitates formation of a functional group which is extremely polar and thus increases the polar component of the surface energy which in turn leads to improved soaking properties of the cellulose material.

XPS results indicate information of surface functional groups which should influence sorption properties of cellulose materials. One of the best techniques for studying the sorption properties is contact angle measurements. This is actually one of the oldest experimental techniques, and probably the simplest. In its basic configuration it is just a drop of water placed onto a surface of the material for which the surface properties are to be determined. Large contact angles of a water drop indicate hydrophobic character of a material, while small contact angles are typical for hydrophilic materials. Such a simple technique is applicable for rapid estimation of the surface character but it is not accurate and usually fails at attempt to use it on porous materials such as cellulose for wound dressing. In order to overcome the limitations of the method in its original form more sophisticated versions have been developed and applied by different authors [46-51]. Although large progress has been obtained in the last decades the technique is still regarded semi-quantitative due to complexity of interaction between liquid drops and solid materials. In order to give a comprehensive explanation of this technique we present different models currently used for quantification of measured results.

Contact angle measurements, which can be used to characterize the hydrophobic/hydrophilic properties of surfaces and from which changes in adhesion can be calculated [52], are frequently used to evaluate the effect of chemical changes on the adhesive properties of polymer surfaces. Various approaches for further analyses of these properties have appeared. It is generally agreed that the surface free energy contains the non-polar and polar components. In this sense, the dispersive (Lifshitz van der Waals, LW) and polar (acid-base, AB) interactions are known. The most common mathematical models used for determining the surface free energy (SFE) are the ones suggested by Fowkes [53-54], Zisman [55], Kwok and Neumann [56], denoted "Equation of state", Rabel [57], Kaelble [58-59], Owens and Wendt [60], denoted "OW", and van Oss, Chaudhury and Good [61-63], denoted "vOGC".

The determination of the solid SFE is a complex procedure, and it cannot be measured directly. As such, the evaluation technique consists of three stages i) characterisation of test liquids using Wilhelmy plate (platinum and PTFE) and Du-Nuoy ring methods, ii) contact angle determination by bringing various standard liquids in contact with the solid surface, and iii) evaluating the SFE by using an appropriate mathematical model. Prior to contact angle determination, standard liquids must be investigated in order to determine their surface tensions (SFT) and their dispersive and polar components. Test liquids with different surface tensions and different polar and dispersive components must be chosen in order to determine the SFE.

The surface tensions of all liquids, except water, are determined using the Wilhelmy plate method according to DIN 53 914. The Du-Nuöy ring method according to DIN 53 993 is applied for water. The pulling force of a liquid, when it is brought into contact with the lower edge of the platinum plate or the ring, is a linear measure of the liquid surface tension, γ, i.e. for the plate:

$$\gamma_l = \frac{F_w - F_b}{L \cdot \cos\theta} \tag{3}$$

and for the ring

$$\gamma_l = \frac{F_{max} - F_r}{L \cdot \cos\theta}, \tag{4}$$

where F_w is the force on the plate, F_{max} is the maximum pull on the ring, and F_b and F_r are the buoyancy corrections for the plate and the ring, respectively, L is the wetted length (mm) i.e. for the plate $L=2(a+b)$ where a is the thickness and b is the width of the plate, and for the ring $L=4\pi R$ where R is the major radius of the ring, and θ is the contact angle (°). Both the ring and the plate are made of platinum and it is assumed that the contact angle of all liquids on platinum is 0°.

The dispersive component of used test liquid is determined by measuring the contact angle between standard poly(tetrafluoroethylene) (PTFE) plate and liquid. According to Good et al. [64] PTFE is assumed to be capable only of dispersive interactions ($\gamma_S = \gamma_S{}^D = 18.0$ mJ/m^2 and $\gamma_S{}^P = 0$ mJ/m^2). The polar component of the liquid can be then calculated by the difference as:

$$\gamma_l^p = \gamma_l - \gamma_l^d \tag{5}$$

where γ_l is the surface tension of the wetting liquid, γ_l^d is the dispersive surface tension component of the wetting liquid, γ_l^p is the polar surface tension component of the wetting liquid.

Contact angle between a cellulose sample and a liquid drop is determined using a capillary rise technique [65], applicable for porous materials. The samples weight is measured as a function of time during the adsorption of the liquid phase. The wetting rate curve (m^2/t) is used to determine the contact angle using modified Washburn equation [66-67]:

$$\cos\theta = \frac{m^2}{t} \cdot \frac{\eta}{\rho^2 \cdot \gamma \cdot c}, \tag{6}$$

where θ is the contact angle between the solid and liquid phases, m is the weight of the liquid that penetrates into sample, t is the penetration time, η is the liquid viscosity, ρ is the liquid density, γ is the surface tension of the liquid, and c is a material constant. N-heptane or n-hexane (for which the contact angle on the solid sample is zero) is used as a liquid (for which the $\cos\theta \approx 1$), in order to determine the material constant.

Using water as the test liquid, hydrophilic/hydrophobic properties of solid surfaces are described by contact angle measurement results i.e. the hydrophilic ($\theta<$ 90°) or hydrophobic ($\theta > 90°$). Contact angle measurement results, using various standard liquids, the samples surface energy can be determined using one among several common models for evaluating.

Fowkes SFE model [53-54] takes advantage of the theory that the molecular interactions contributing to surface tension are divided into two groups: dispersion (non-polar) interactions and polar interactions, i.e.:

$$\gamma = \gamma^d + \gamma^p \tag{7}$$

and that the adhesion (W) between two materials 1 and 2 that interact only through dispersive forces ($\gamma_p = 0$) are estimated from:

$$W_{12} = 2\sqrt{\gamma_1^d \gamma_2^d}. \tag{8}$$

The thermodynamic work of adhesion W_a of a liquid with surface tension γ_l on a solid with surface tension γ_s is defined by:

$$W_a = \gamma_s + \gamma_l - \gamma_{sl}, \tag{9}$$

where γ_{sl} is the solid-liquid interfacial tension.

The Young equation for the equilibrium contact angle on a solid surface is:

$$\gamma_l \cos\theta = \gamma_{s,0} - \gamma_{sl} - \pi_s, \tag{10}$$

where $\gamma_{s,0}$ is the surface energy of the pure solid and π_s is the surface pressure due to adsorption on the solid. On low-energy surfaces π_s is usually assumed to be insignificant, i.e. $\gamma_{s,0} = \gamma_s$. Combination with Equation (9) then yields the Young-Dupré equation:

$$W_a = \gamma_l (\cos\theta + 1). \tag{11}$$

The combination of Equations 8, 9 and 11 gives:

$$\gamma_{sl} = \gamma_s + \gamma_l - 2\sqrt{\gamma_s^d \gamma_l^d}, \tag{12}$$

and

$$W_a = \gamma_l (1 + \cos\theta) = 2\sqrt{\gamma_s^d \gamma_l^d} \tag{13}$$

or

$$\gamma_l^d = \frac{\gamma_l^2 (1 + \cos\theta)}{4\gamma_d^s}. \tag{14}$$

Therefore, when only dispersive interactions are significant, the dispersive component of the surface tension of the liquid can be determined by measuring the contact angle of a liquid on a solid with a known γ_s^d. The polar component of the surface tension is then determined by Equation (5).

The Owens-Wendt-Rabel-Kaelble (OW) SFE model [59-60] assumes that the polar interaction between solid and liquid can be estimated by a combination of γ_l^p and γ_s^p. Thus, in the model it is proposed that the polar component can be described by a geometric mean (GM) approximation i.e.:

$$\gamma_{sl} = \gamma_s + \gamma_l - 2\sqrt{\gamma_s^d \gamma_l^d} - 2\sqrt{\gamma_s^p \gamma_l^p}, \tag{15}$$

$$W_a = \gamma_l(1 + \cos\theta) = 2\sqrt{\gamma_s^d \gamma_l^d} + 2\sqrt{\gamma_s^p \gamma_l^p}, \tag{16}$$

where γ_s^p and γ_l^p are the polar components of the surface free energies of the solid and the liquid, respectively. Equation 15 can be used to determine γ_s^p by measuring the contact angle of two liquids with known γ_l^d and γ_l^p (usually one with only dispersive and one with both dispersive and polar components). This approach has been criticized for being too simple because a polar, but purely acidic surface will show only dispersive interactions with acid solvents, and the same will be the case for the interaction of basic surfaces with basic solvents [68-69]. However, it has received widespread use as a method to characterise the adhesive properties of polymers, because at least two various liquids are necessary to measure the contact angle and then solve the system of equations. One of these liquids should be characterized by high value of γ_l^d and low value of γ_l^p and the second liquid - inversely. The most often, water and diiodomethane were used as the pair of testing liquids.

In order to average the interactions with several solvents, Owens and Wendt [60] wrote Equation 16 in the linear form:

$$y = ax + b, \tag{17}$$

where

$$x = \sqrt{\gamma_s - \gamma_s^d}\sqrt{\frac{\gamma_l^p}{\gamma_l^d}}; \quad a = \sqrt{\gamma_s^p}, \tag{18}$$

$$y = \frac{1 + \cos\theta}{2}\frac{\gamma_l}{\sqrt{\gamma_l^d}}; \quad b = \sqrt{\gamma_s^d}. \tag{19}$$

Thus, in the OW approach, the contact angles of several liquids with known values of γ_l, γ_l^d and γ_l^p are measured after which data are fitted to straight lines (Equation 17). The slope and intercept of this line yields values of γ_l^d and γ_l^p (Equations 18 and 19).

Several professional devices are currently available for characterisation of surface properties by contact angle methods. The work described in this contribution has been performed using a professional device Kruss K12. The KRÜSS Tensiometer of the K12 series is the standard for the measurement of the surface and interfacial tension of liquids and solids. This instrument enables measuring surface and interfacial tension measurement of liquids, dynamic contact angle measurements, wetting behaviour of tablets, pharmaceutical active substances, auxiliaries, lacquers, paints, and printing inks, of fibre bundles, textiles, wetting and adhesion of coatings. The K12 Tensiometer has been specially designed for versatile and demanding applications in research of surface-active agent development, wetting properties of tablets, pharmaceutical active ingredients and auxiliary products, pigments, as well as in textiles, and monitoring surface optimization in the production of polymers and foils, lacquers, dyes, inks.

In order to evaluate the SFE of cellulose materials for wound dressing 6 different test liquids: water, ethylene glycol, ethanol, chloroform, tetrahydrofuran, and diiodomethane have been used. The results of surface tensions and contact angles of water and diiodomethane were used for geometric mean (GM) calculations (Equation 16). In OW calculations (Equation 18 and 19) the data for all six solvents were used.

The total surface tension and its dispersive and polar components have been measured and are found as follows: water ($\gamma = 72$ mJ/m^2; $\gamma^d = 28$ mJ/m^2; $\gamma^p = 44$ mJ/m^2); Ethylene glycol ($\gamma = 48$ mJ/m^2; $\gamma^d = 26$ mJ/m^2; $\gamma^p = 21$ mJ/m^2); Ethanol ($\gamma = 22$ mJ/m^2; $\gamma^d = 20$ mJ/m^2; $\gamma^p = 2$ mJ/m^2); Chloroform ($\gamma = 27$ mJ/m^2; $\gamma^d = 26$ mJ/m^2; $\gamma^p = 1$ mJ/m^2); Tetrahydrofuran ($\gamma = 27$ mJ/m^2; $\gamma^d = 12$ mJ/m^2; $\gamma^p = 15$ mJ/m^2), and Diiodomethane ($\gamma = 51$ mJ/m^2; $\gamma^d = 51$ mJ/m^2; $\gamma^p = 0$ mJ/m^2). These measured values were used in evaluation of SFE using the mathematical model described above.

The cellulose samples were cut into rectangular pieces (2 x 5 cm^2) and hung up onto the sample holder in the Tensiometer Krüss K12 apparatus. Non-treated and plasma treated cellulose materials for wound dressing have been characterised by all six liquids. The wetting rise curves for water as the test liquid are presented in Figure 7. The results were statistically processed (a set of parallel measurements of wetting rise curves were performed until the standard deviation of calculated contact angle was less than 2°) and represent the average value of at least ten measurements of each sample.

The slope of curves presented in Figure 7 characterise the rate of water sorption. One can observe a significant difference in the rate of water sorption when comparing non-treated and plasma treated cellulose samples. In the case of plasma treated cellulose sample, the rate of water sorption was very fast at the beginning, while after a few seconds (e.g. 10 s) the sorption slows and curve riches plateau indicating complete wetting of the material.

The wetting rise curves obtained using test liquids allow for determination of the contact angles which are used for evaluating the surface energies of non-treated and plasma treated materials, their dispersive and polar components. The contact angles are shown in Figure 8, while the corresponding values for surface energy, polar and dispersive components (as obtained by using Owens-Wendt-Rabel-Kaelble (OW) model) are summarised in Figure 9. As shown in Figure 8 the highest contact angles by all samples were determined using diiodomethane as a test liquid. The lowest values were obtained by ethylene glycol. The

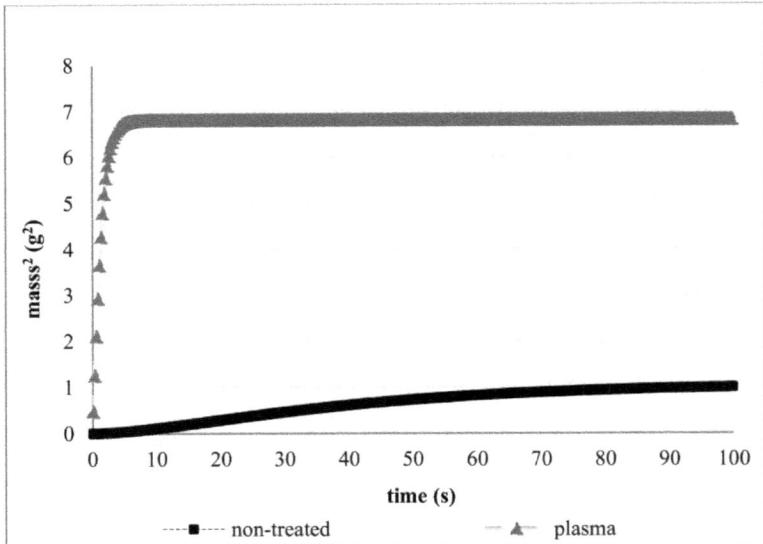

Figure 7. The wetting rise curves for non-treated and oxygen plasma treated cellulose wound dressing obtained with water

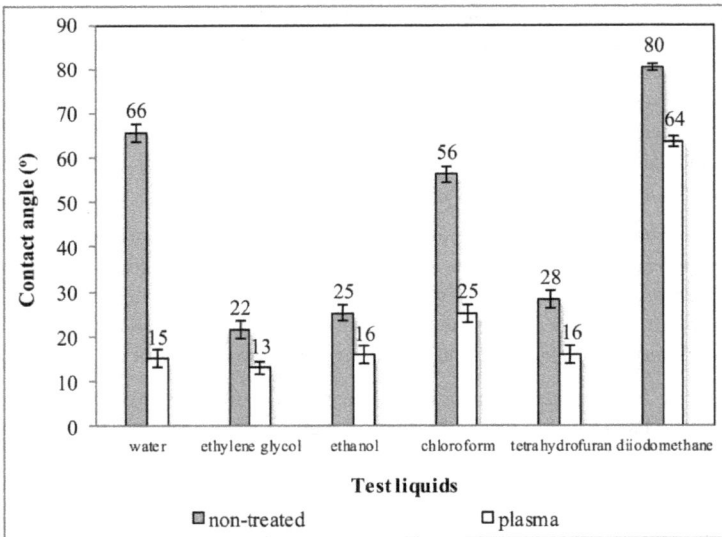

Figure 8. Contact angles of test liquids on non-treated and plasma treated cellulose

water contact angle determined using water was 66°. As well, similar results were obtained by Hsieh et al., where the water contact angle varied from 56° to 60° depending on measuring the contact angle on a single fibre or on a woven fabric [70]. Simoncic et al. obtained even higher values, i.e. 73° for a plane-wave cotton sample [71], and 80° for a desized and alkaline scoured cotton samples [72]. In addition, modified Washburn equation as applied procedure for porous samples leads to overestimated contact angle values compared to those measured directly on smooth surfaces of the same solid, if such surface can be obtained at all [69, 73]. The results indicate a decrease of contact angles for all liquids on plasma treated samples; most effect is evident when using water as the test liquid. Contact angle decrease is explained by increased polarity due to plasma activation.

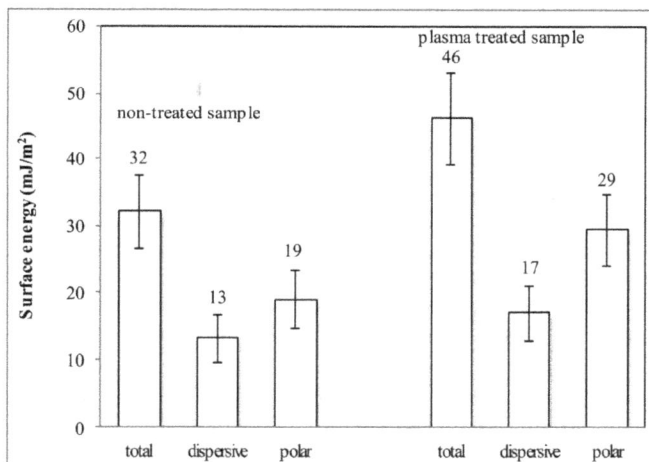

Figure 9. The total surface energy, their dispersive and polar components of non-treated and plasma treated cellulose, as evaluated using OW model

As discussed earlier by other authors [68-69, 74-75], the values of SFEs and their components depend on the choice of solvents and on the combination of the test liquids chosen. In addition, it should be kept in mind that independently of the applied approaches, the obtained values for surface free energy and its components are relative ones and regarded as indicators of the adhesive properties of the surfaces and not as measures of their actual SFEs. Therefore studies on new approaches and experimental procedures to solve the problems are still needed.

Both XPS and SFE results indicate important modifications of the surface properties of cellulose as a consequence of plasma treatment. Formation of oxygen rich functional groups on the cellulose surface causes an increase of the surface energy, especially the polar component. Although the techniques cannot reveal the mechanisms involved in interactions of oxygen plasma with cellulose materials it is possible to explain the results by taking advantage of the known properties of oxygen plasma. Experiments on modification of cellulose materials presented in this work have been performed using weekly ionised oxygen plasma. As mentioned earlier, such plasma created in a glass-discharge chamber is rich in neutral reactive particles. The density of charged particles in our plasma is about 10^{15} m^3. The corresponding flux of ions onto the surface is of the order of 10^{17} m^{-2} s^{-1}. This flux is rather low specially when taking into account a very large surface of cellulose materials used in wound dressings. The modification of surface properties thus cannot be explained by interaction of positively charged oxygen ions with solid material. The functionalization of material is rather explained by interaction with particles which abound in our plasma. The non-equilibrium gas is rich in neutral oxygen molecules excited to the first state, and the dissociation fraction is also very high. The excited molecules and neutral oxygen atoms are therefore appropriate candidates for explanation of the observed surface properties. Unfortunately, very little work has been reported on interaction of excited oxygen molecules with solid materials. Many authors agree that probability for surface de-excitation of molecules with the excitation energy of only 1 eV is always low and practically independent from characteristics of solid materials. On the other hand, the literature on interaction between neutral oxygen atoms and solid materials abounds [76-80].

Different authors reported moderate interaction probabilities. The chemical reactivity definitely depends on the type of organic material, but it seems that all authors observed rapid functionalization of practically all organic materials upon exposure to oxygen atoms. The improved surface properties of cellulose material are therefore explained by chemical interaction between neutral oxygen atoms in the ground state and cellulose materials. The interaction leads to formation of oxygen rich functional groups. The etching of cellulose material upon treatment with oxygen atoms also occurs but it can be considered as a minor effect as long as the temperature of cellulose materials remains low. The etching is illustrated in Figure 10, which presents high resolution scanning electron microscope images of non-treated and plasma treated cellulose materials.

Figure 10. SEM images of non-treated (a) and plasma treated (b) cellulose fibres

Figure 10 reveals a richer surface morphology of cellulose fibres treated by oxygen plasma. Obviously, some etching occurred during plasma treatment. This observation is sound with effects reported in appropriate literature. Namely, many authors have shown that a treatment of organic materials with gaseous plasma leads to increased surface roughness [77, 81-83]. While the effect is not yet fully understood, a plausible explanation is selective interaction with amorphous and crystalline segments of cellulose materials. Namely, the interaction between neutral oxygen atoms and organic materials is extremely selective. It has to be pointed out that even chemically identical materials are etched at different rates upon exposure to oxygen plasma [84]. In the cited reference [84] major differences in the behaviour of PET polymer with different degrees of crystallinity were reported.

An important consideration associated with plasma treatment is modification of bulk properties of plasma treated materials. A suitable method for estimation of such modifications is measurement of the breaking force and elongation. Figure 11 represents typical results of such measurements. The results are rather surprising: the breaking force does not decrease after plasma treatment but even a small increase is observed, while the elongation is reduced. Since the breaking force and elongation should not depend much on plasma treatment, the results are difficult to explain. Still, the observed improve of the mechanical properties might be explained by increased surface wettability. Namely, the attraction force between particular fibres in the cellulose material may increase as a result of the surface functionalization. Therefore the shrinkage of the origin emptiness between longitudinal and horizontal placed fibres occurred resulting in more compact structure of the material. In any case, the modification of the mechanical properties is not very relevant for the application of cellulose materials in wound dressings since they are designed for regular changing and disposal (and not washing).

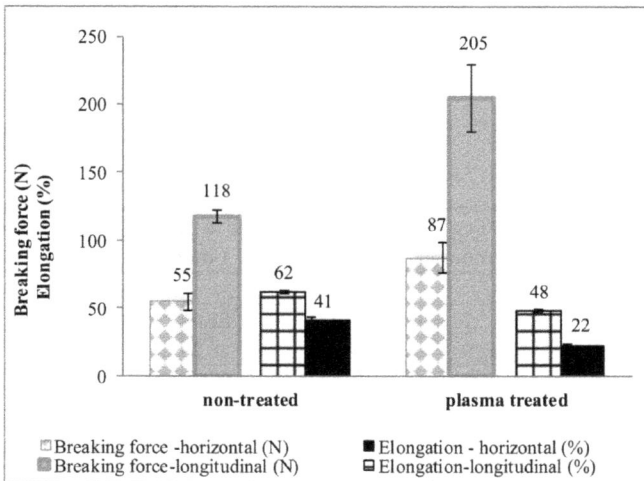

Figure 11. The breaking force and elongation for non-treated and plasma treated cellulose materials

Finally is worth mentioning that cellulose materials for wound dressings are usually not used directly after surface modifications. An important consideration is ageing of plasma treated materials [77, 85-87]. The surface functional groups are found in non-equilibrium state and any thermodynamically non-equilibrium system tends to approach the equilibrium value. Oxygen rich functional groups may decay by spontaneous release of excessive oxygen, may re-orient from the surface towards the bulk, or may be simply blocked by adsorption of gaseous molecules present in the atmosphere. In order to address the ageing properly, a set of experiments have been performed. Table 4 reveals time dependency of plasma activation effects presented as the loss of oxygen, as determined by XPS and water contact angle results on cellulose measured 1 and 4 days after activation treatment was performed. It is found that the loos rates are within the limits of the experimental error, so it can be concluded that functional groups formed on the surface of cellulose upon exposure to weakly ionized highly dissociated oxygen plasma are rather stable. Figure 12 presents the results of breaking force and elongation of cellulose materials after 10 days since plasma treatment. Again, any modification of these parameters is within the limits of the experimental error. Still, it can be observed that the water contact angle slowly increases with increasing ageing time (see Table 4), but the increase is not dramatic, which indicates that plasma modification of cellulose materials for wound dressings is suitable for practical applications. To demonstrate this, sorption properties evaluated by two different solutions (i.e. synthetic exudate and blood) to simulate real system, were used. As presented in Figure 13, plasma treated samples revealed as effective absorbent for wound fluids i.e. synthetic exudate and blood. The results evident a 3-times faster up-taking of blood and a 15-times faster sorption of exudate by plasma treated samples. One can conclude that the plasma effect was not as significant as by water sorption (see Figure 7), but one should keep in mind that chemical composition of wound fluids are different. Wound exudate may be beneficial, but it can also be problematic, especially in chronic wounds such as leg ulcers. In leg ulcer patients, exudate levels may be high even from small ulcers. Nevertheless, the obtained results illustrate the value of using plasma modified cellulose material as a dressing in order to control exudating by facilitating and supporting the healing of the wound.

	Time dependency of plasma activation performance		
	Immediately after	1 day after	4 days after
Surface composition			
C (at. %)	52.1	51.9	52.8
O (at. %)	47.9	48.2	47.3
O/C ratio	0.91	0.92	0.89
Hydrophilicity			
Water contact angle (°)	15±2	17±1	16±2

Table 4. The elemental surface compositions and water contact angles of plasma treated cellulose as a function of ageing

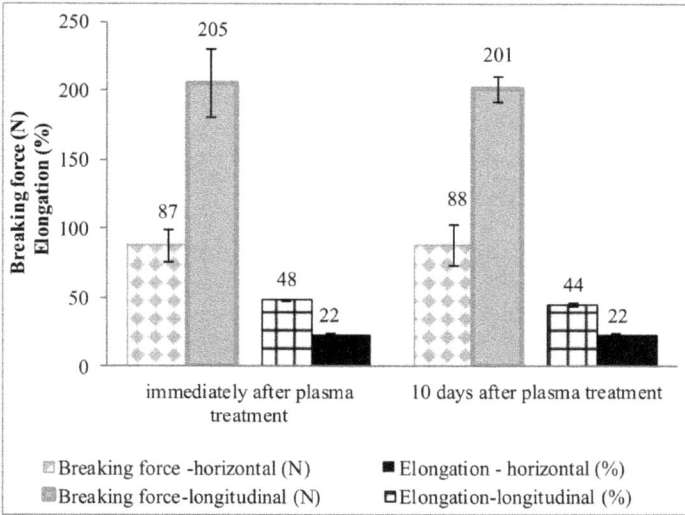

Figure 12. Breaking force and elongation of plasma treated cellulose samples as a function of 10-days ageing

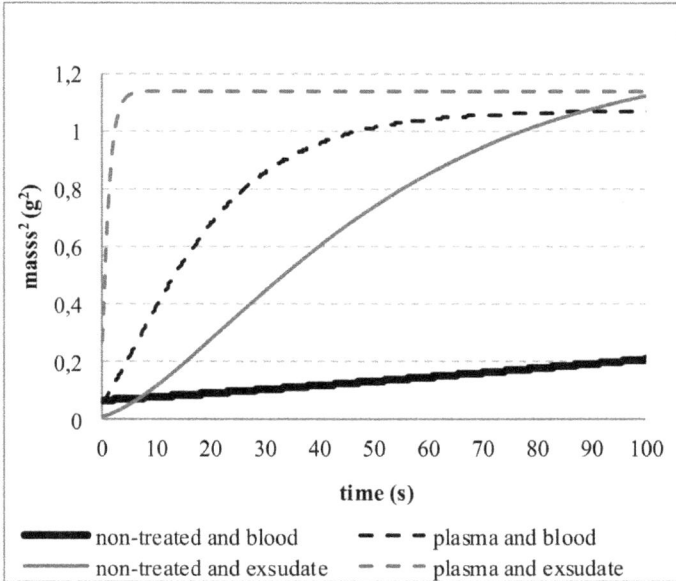

Figure 13. The wetting rise curves for non-treated and oxygen plasma treated cellulose obtained during synthetic exudate and blood sorption

3. Conclusion

Oxygen plasma treatment is a suitable method to improve the quality of cellulose materials. The drastically improvement in sorption capacity far outweigh the minor changes in mechanical properties; while no changes in morphology was observed. Ageing of plasma activation effect was insignificant present within a few days concerning hydrophilicity and tensile properties. Since oxygen plasma contains a variety of different reactive particles, the proper choice of plasma parameters is the key issue. Plasma containing rather large concentration of charged particles is better to be avoided since it usually leads to excessive thermal effects and thus degradation of the material. The available energy dissipated on the cellulose surface at the interaction between charged particles and solid material is simply too large to assure for modification of cellulose material at low temperature. Instead of moderately ionised plasma it is better to use extremely weekly ionised plasma with a large density of neutral reactive particles. The best species in term of reasonable chemical reactivity and low thermal load are neutral oxygen atoms in the ground state. Plasma rich with these particles, but poor in charged particles, is created using an electrodeless high frequency discharge. The absence of electrodes in such discharges is favourable since it prevents substantial drain of neutral oxygen atoms by heterogeneous surface recombination. Best materials for construction of a suitable plasma reactor are different glasses and some ceramics. The plasma facing components should be smooth and should not allow for chemisorption of neutral oxygen atoms. If this requirements are fulfilled the dissociation fraction of oxygen molecules in gaseous plasma at pressures up to several mbar, exceeds the ionisation fraction for at least five orders of magnitudes. Such plasma is therefore safe to use since it does not lead to excessive exothermic reactions and thus thermal degradation of treated materials. Based on that, plasma treatment is advantage surface activation procedure in order to obtain super-hydrophilic matric, used as a potential layer in wound dressings.

Author details

Zdenka Persin, Tina Maver and Karin Stana Kleinschek*
Laboratory for Characterisation and Processing of Polymers, Faculty of Mechanical Engineering, University of Maribor, Maribor, Slovenia

Zdenka Persin, Alenka Vesel, Tina Maver, Uros Maver and Karin Stana Kleinschek
Centre of Excellence for Polymer Materials and Technologies, Ljubljana, Slovenia

Alenka Vesel
Jozef Stefan International Postgraduate School, Ljubljana, Slovenia

Miran Mozetic
Jozef Stefan Institute, Ljubljana, Slovenia

---

* Corresponding Author

Acknowledgement

This work was supported by the Ministry of Higher Education, Science and Technology of the Republic of Slovenia [Grant number 3211-10-000057].

4. References

[1] Freytag R., Donzé J.J (1983) Alkali treatment of cellulose fibres. In Lewin M, Sello SB, editors. Handbook of fiber science and technology: Volume I. Chemical processing of Fibers and Fabrics, Fundamentals and Preparation, Part A. New York and Basel: Marcel Deckker, Inc. pp. 91-121.

[2] Lewin M. (1984) Bleaching of Cellulosic and Synthetic Fabrics. In Lewin M, Sello SB, editors. Handbook of fiber science and technology: volume I. Chemical processing of Fibers and Fabrics, Fundamentals and Preparation, Part B. New York and Basel: Marcel Deckker, Inc. pp. 91-243.

[3] Durso D.F (1978) Chemical modification of cellulose – A historical review. In Rowell RM, Young RA, editors. Modified Cellulosics. NY, ZDA: Academic Press. pp. 23-37.

[4] Fengel D., Wegener G. (1984) Reactions in alkaline medium. In: Fengel D, Wegener G, editors. Wood, Chemistry, Ultrastructure, Reactions. Berlin, New York: Walter de Gruyter. pp. 66-105.

[5] Gorjanc M, Bukosek V, Gorensek M, Vesel A (2010) The influence of water vapor plasma treatment on specific properties of bleached and mercerized cotton fabric. Tex. Res. J. 80: 557-567.

[6] Vesel A, Mozetic M, Strnad S, Stana-Kleinschek K, Hauptman N, Persin Z (2010) Plasma modification of viscose textile. Vacuum 84: 79-82.

[7] Gorensek M, Gorjanc M, Bukosek V, Kovac J, Petrovic Z Lj, Puac N (2010) Functionalization of polyester fabric by Ar/N2 plasma and silver. Tex. Res. J. 80:1633-1642.

[8] Radetic M, Jovancic P, Puac N, Petrovic Z Lj (2005) Environmental impact of plasma application to textiles. J. Phys.: Conf. Series, 71: 12017 doi:10.1088/1742-6596/71/1/012017. Available: http://iopscience.iop.org/1742-6596/71/1/012017. Accessed 2012 April 6th.

[9] Mihailovic D, Saponjic Z, Molina R, Puac N, Jovancic P, Nedeljkovic J, Radetic M (2010) Improved properties of oxygen and argon RF plasma-activated polyester fabrics loaded with TiO2 nanoparticles. ACS applied materials and interfaces. 2 (6): 1700-1706.

[10] Canal C, Gaboriau F, Villeger S, Cvelbar U, Ricard A (2009) Studies on antibacterial dressings obtained by fluorinated post-discharge plasma. Int. J. Pharm. 367 (1-2): 155-161.

[11] Canal C, Villeger S, Cousty S, Rouffet B, Sarrette JP, Erra P, Ricard A (2008) Atom-sensitive textiles as visual indicators for plasma post-discharges. Appl. Surf. Sci. 254 (18): 5959-5966.

[12] Ristic N, Jovancic P, Canal C, Jocic D (2009) One-bath one-dye class dyeing of PES/cotton blends after corona and chitosan treatment. Fiber Polym. 10 (4): 466-475.

[13] Gorensek M, Gorjanc M, Bukosek V, Kovac J, Jovancic P, Mihailovic D (2010) Functionalization of PET fabrics by corona and nano silver. Tex. Res. J. 80/3: 253-262.

[14] Guimond S, Hanselman B, Amberg M, Hegemann D (2010) Plasma functionalization of textiles: Specifics and possibilities. Pure Appl. Chem. 82 (6): 1239-1245.

[15] Morent R, De Geyter N, Verschuren J, De Clerck K, Kiekens P, Leys C (2008) Non-thermal plasma treatment of textiles. Surf. Coat. Technol. 202: 3427-3449.

[16] Canal C, Erra P, Molina R, Bertran E (2007) Regulation of surface hydrophilicity of plasma treated wool fabrics. Tex. Res. J. 77 (8): 559-564.

[17] Devetak M, Skoporc N, Rigler M, Persin Z, Drevensek-Olenik I, Copic M, Stana-Kleinschek K (2012) Effects of plasma treatment on water sorption in viscose fibres. Mater. Tehnol. 46 /1 69–73.

[18] Ferrero F (2003) Wettability measurements on plasma treated synthetic fabrics by capillary rise method. Polym. Test. 22 (5): 571-578.

[19] Persin Z, Stenius P, Stana-Kleinschek K (2011) Estimation of the Surface Energy of Chemically and Oxygen Plasma-Treated Regenerated Cellulosic Fabrics Using Various Calculation Models. Tex Res J. 81/16: 1673-1685.

[20] Persin Z, Stana-Kleinschek K, Sfiligoj-Smole M, Kreze T, Ribitsch V (2004) Determining the Surface Free Energy of Cellulose Materials with the Powder Contact Angle Method. Tex Res J. 74/1: 55-62.

[21] Hua ZQ, Sitaru R, Denes F, Young RA (1997) Mechanisms of oxygen- and argon-RF-plasma-induced surface chemistry of cellulose. Plasmas Polym. 2: 199-224.

[22] Alfredo Calvimontes, Peter Mauersberger, Mirko Nitschke, Victoria Dutschk and Frank Simon, Effects of oxygen plasma on cellulose surface, Cellulose, 18 (2011) 803-809.

[23] Inbakumar S, Morent R, De Geyter N, Desmet T, Anukaliani A, Dubruel P, Leys C (2010) Chemical and physical analysis of cotton fabrics plasma-treated with a low pressure DC glow discharge. Cellulose 17: 417-426.

[24] Vander Wielen LC, Ragauskas AJ (2004) Grafting of acrylamide onto cellulosic fibers via dielectric-barrier discharge. Eur. Polym. J. 40/3: 477–482.

[25] Jun W, Fengcai Z, Bingqiang C (2008) The Solubility of Natural Cellulose After DBD Plasma Treatment. Plasma Sci. Technol. 10/743. Available: http://iopscience.iop.org/1009-0630/10/6/18 (doi:10.1088/1009-0630/10/6/18). Accessed: 2012 April 6th.

[26] Ercegovic Razic S, Cunko R, Svetlicic V, Segota S (2011) Application of AFM for identification of fibre surface changes after plasma treatments. Mater. Technol.: Adv. Perform. Mater. 26: 146-152.

[27] Mahlberg R, Niemi HEM, Denes FS, Rowell RM (1999) Application of AFM on the Adhesion Studies of Oxygen-Plasma-Treated Polypropylene and Lignocellulosics. Langmuir. 15: 2985–2992.

[28] Ward TL, Jung HZ, Hinojosa O, Benerito RR (1978) Effect of rf cold plasmas on polysaccharides. Surf. Sci. 76: 257–273.

[29] Wong KK, Tao XM, Yuen CWM, Yeung KW (1999) Low Temperature Plasma Treatment of Linen. Tex. Res. J. 69: 846-855.

[30] Strnad J (1977) Fizika 1. Ljubljana: DZS. 256 p.

[31] Ricard A (1996) Reactive Plasmas, 1 edn. Paris: SFV.

[32] Kutasi K, Guerra V, Sá PA (2011) Active species downstream of an Ar–O2 surface-wave microwave discharge for biomedicine, surface treatment and nanostructuring. Plasma Sources Sci. Technol. 20/3: doi:10.1088/0963-0252/20/3/035006. Available: http://iopscience.iop.org/0963-0252/20/3/035006/. Accessed: 2012 April 6th.

[33] Mozetic M, Cvelbar U, Vesel A, Ricard A, Babic D, Poberaj I (2005) A diagnostic method for real-time measurements of the density of nitrogen atoms in the postglow of an Ar–N2 discharge using a catalytic probe. J. Appl. Phys. 97/ 10: 103308-1-103308-7. Available: http://dx.doi.org/10.1063/1.1906290. Accessed: 2012 April 6th.

[34] Mozetic M, Vesel A, Cvelbar U, Ricard A (2006) An iron Catalytic Probe for Determination of the O-atom Density in an Ar/O2 Afterglow. Plasma Chem. Plasma P. 26/ 2: 103-117.

[35] Zaplotnik R, Vesel A, Mozetic M (2012) Fiber Optic Catalytic Sensor for Neutral Atom Measurements in Oxygen Plasma. Sensors. 12: 3857-3867.

[36] Drenik A, Cvelbar U, Ostrikov K, Mozetic M (2008) Catalytic probes with nanostructured surface for gas/discharge diagnostic: a study of a probe signal behaviour. J. Phys. D Appl. Phys. 41/11: 115201-1-115201-7.

[37] Balat-Pichelin M, Vesel A (2006) Neutral oxygen atom density in the Mesox air plasma solar furnace facility. Chem. Phys. 327/1: 112-118.

[38] Vesel A, Mozetic M, Balat-Pichelin M (2007) Oxygen atom density in microwave oxygen plasma. Vacuum. 81/9: 1088-1093.

[39] Primc G, Zaplotnik R, Vesel A, Mozetic M (2011) Microwave discharge as a remote source of neutral oxygen atoms. AIP Advances. 1/2: 022129-1-022129-11. Available: http://dx.doi.org/10.1063/1.3598415. Accessed 2012 April 6th.

[40] Denysenko I, Ostrikov K, Azarenkov NA (2009) Dust charge and ion drag forces in a high-voltage, capacitive radio frequency sheath. Phys. Plasmas 16/11: 113707-1 - 113707-10.

[41] Kratzer M, Brinkmann RP, Sabisch W, Schmidt H (2001) Hybrid model for the calculation of ion distribution functions behind a direct current or radio frequency driven plasma boundary sheath. J. Appl. Phys. 90/5: 2169-2179.

[42] Schuengel E, Zhang QZ, Iwashita S, Schulze J, Hou LJ, Wang YN, Czarnetzki U (2011) Control of plasma properties in capacitively coupled oxygen discharges via the electrical asymmetry effect. J. Phys. D Appl. Phys. 44/28: 285205-1 - 285205-14.

[43] Zaplotnik R, Vesel A, Mozetic M (2011) Transition from E to H modes in inductively coupled oxygen plasma: behaviour of oxygen atom density. Europhys. Lett. 95: 55001-1 – 55001-5.

[44] Takeda K, Takashima S, Ito M, Hori M (2008) Absolute density and temperature of O(^{1}D2) in highly Ar or Kr diluted O2 plasma. Appl. Phys. Lett. 93/2: 021501-1 - 021501-3.

[45] Mozetic M (2012) Application of X-ray photoelectron spectroscopy for characterization of PET biopolymer. Mater. Tehnol. 46/1: 47–51.

[46] Morent R, De Geyter N, Leys C, Vansteenkiste E, De Bock J, Philips W (2006) Measuring the wicking behavior of textiles by the combination of a horizontal wicking experiment and image processing. Rev. Sci. Instrum. 77. Available: http://dx.doi.org/10.1063/1.2349297. Accessed 2012 March 3rd.

[47] Grundke K, Bogumil T, Gietzelt T, Jacobasch HJ, Kwok DY, Neumann AW (1996) Wetting measurements on smooth, rough and porous solid surfaces. Prog. Colloid Polym. Sci. 101:58-68.

[48] Chibowski E, Perea-Carpio R (2001) A Novel Method for Surface Free-Energy Determination of Powdered Solids. J Colloid Interf Sci. 240/2: 473-479.

[49] Dang-Vu T, Hupka J (2005) Characterization Of Porous Materials By Capillary Rise Method. Physicochem Probl Mi. 39: 47-65.

[50] Carroll B.J (1993) Direct measurement of the contact angle on plates and on thin fibres: some theoretical aspects. In: Mitall KL, editor. Contact angle, wetting, and adhesion. The Netherlands: VSP BV. pp. 235-246.

[51] Chan C.M (1994) Contact angle measurements. In Chan CM, editor. Polymer surface modification and characterization. Münich Vienna New York: Carl Hanser Verlag. pp. 35-76.

[52] Good R.J (1993) Contact angle, wetting, and adhesion: a critical review. In: Mitall KL, editor. Contact angle, wetting, and adhesion. The Netherlands: VSP BV.pp.3-37.

[53] Fowkes FM (1962) Determination of interfacial tensions, contact angles, and dispersion forces in surfaces by assuming additivity of intermolecular interactions in surfaces. J Phys Chem. 66: 382-382.

[54] Fowkes FM (1964) Attractive forces at interfaces. Ind Eng Chem. 56: 40-52.

[55] Zisman W.A (1964) Relation of the equilibrium contact angle to liquid and solid constitution. In: Fowkes FM editor. Contact angle, wettability and adhesion. Washington, DC: American Chemical Society. pp. 1-51.

[56] Kwok DY, Neumann AW (1999) Contact angle measurement and contact angle interpretation. Adv Colloid Interface Sci. 81: 167-249.

[57] Rabel W (1977) Flüssigkeitsgrenzflächen in Theorie und Anwendungstechnik. Physkalische Blätter. 33: 151-161.

[58] Kaelble DH (1969) Peel adhesion: Influence of surface energies and adhesive rheology. J Adhes. 1: 102-123.

[59] Kaelble DH (1972) Physical chemistry of adhesion. New York: Wiley Interscience. Chapter 9.

[60] Owens DK, Wendt RC (1969) Estimation of the surface free energy of polymers. J Appl Polym Sci. 13: 1741-1747.

[61] van Oss CJ, Chaudhury MK, Good RJ (1988) Additive and nonadditive surface tension components and the interpretation of contact angles. Langmuir. 4: 884-891.

[62] van Oss CJ (1993) Acid-base interfacial interactions in aqueous media. Colloids Surf. A. 78: 1-49.

[63] van Oss CJ, Chaudhury MK, Good RJ (1986) The role of van der Waals forces and hydrogen bonds in "hydrophobic interactions" between biopolymers and low energy surfaces. J Colloid Interface Sci. 111: 378-390.

[64] Good RJ, Girifalco LA (1960) A theory for estimation of interfacial energies III. Estimation of surface energies of solids from contact angle data. J Phys Chem. 64: 561-565.

[65] Jakobasch H.J, Grundke K, Mäder E, Freitag K.H, Panzer U (2003) Application of the surface free energy concept in polymer processing. In: Mittal KL editor. Contact angle, wettability & adhesion. The Netherlands: Zeist, VSP BV. pp. 921–936.

[66] Washburn EW (1921) The dynamics of capillary flow. Phys Rev. 17/3: 273-283.

[67] Grundke K, Boerner M, Jacobasch HJ (1991) Characterization of fillers and fibres by wetting and electrokinetic measurements. Colloid Surface. 58/1-2: 47–59.

[68] Fowkes FM, Riddle FL, Pastore WE, Weber AA (1990) Interfacial interactions between self-associated polar liquids and squalane used to test equations for solid−liquid interfacial interactions. Colloids Surf. 43: 367–387.

[69] Chibowski E, Perea-Carpio R (2002) Problems of contact angle and solid surface free energy determination. Adv Colloid Interface Sci. 98: 245–264.

[70] Hsieh YL, Yu B (1999) Liquid Wetting, Transport, And Retention Properties Of Fibrous Assemblies, Part I: Water Wetting Properties Of Woven Fabrics And Their Constituent Single Fibres. Textile Res. J. 62 (11): 677-685.

[71] Simončič B, Černe L, Tomšič B, Orel B(2008) Surface properties of cellulose modified by imidazolidinone. Cellulose 15:47–58.

[72] Simončič B, Rozman V (2007) Wettability of cotton fabric by aqueous solutions of surfactants with different structures. Colloids and Surfaces A: Physicochem. Eng. Aspects 292 236–245.

[73] Chibowski E (2000) Thin layer wicking – Methods for the determination of acid-base free energies of interaction,. In Mittal KL, editor. Acid-base Interactions: Relevance of Adhesion, Science and technology, Vol.2. Utrech, The Nethelands, VSP BV 2000, pp. 419-437.

[74] Della Volpe C, Siboni S (2001) The evaluation of electrondonor and electron-acceptor properties and their role in the interaction of solid surfaces with water. In: Morra M editor. Water in biomaterials surface science. Chichester: John Wiley & Sons, Ltd. pp. 83–214.

[75] Della Volpe C, Maniglio D, Brugnara M, Siboni S, Morra M (2004) The solid surface free energy calculation I. In defense of the multicoponent approach. J. Colloid Interface Sci. 271: 434–453.

[76] Vesel A (2010) Surf. Modification of polystyrene with a highly reactive cold oxygen plasma. Coat. Technol. 205/2: 490-497.

[77] Vesel A, Junkar I, Cvelbar U, Kovac J, Mozetic M (2008) Surfacemodification of polyester by oxygen-and nitrogen-plasma treatment. Surf. Interface Anal. 40/11: 1444-1453.

[78] Borcia G, Anderson CA, Brown NMD (2003) Dielectric barrier discharge for surface treatment: Application to selected polymers in film and fibre form. Plasma Sources Sci T. 12/3: 335–344.

[79] Borcia G, Anderson CA, Brown NMD (2004) The surface oxidation of selected polymers using an atmospheric pressure air dielectric barrier discharge. Part II. Appl. Surf. Sci. 225: 186-197.

[80] Chan CM, Ko TM, Hiraoka H (1996) Polymer surface modification by plasmas and photons. Surf. Sci. Rep. 24/1-2: 1-54.

[81] Walthera F, Heckla WM, Stark RW (2008) Evaluation of nanoscale roughness measurements on a plasma treated SU-8 polymer surface by atomic force microscopy. Appl. Surf. Sci. 254/22: 7290–7295.

[82] Vourdas N, Kontziampasis D, Kokkoris G, Constantoudis V, Goodyear A, Tserepi A, Cooke M, Gogolides E (2010) Plasma directed assembly and organization: bottom-up nanopatterning using top-down technolog. Nanotechnology. 21: 085302-1 - 085302-8.

[83] Vourdas N, Tserepi A, Gogolides E (2007) Nanotextured super-hydrophobic transparent poly(methyl methacrylate) surfaces using high-density plasma processing. Nanotechnology. 18:125304-1 - 125304-7.

[84] Junkar I, Cvelbar U, Vesel A, Hauptman N, Mozetic M (2009) The Role of Crystallinity on Polymer Interaction with Oxygen Plasma. Plasma Processes Polym. 6/10: 667-675.

[85] Larrieu J, Held B, Martinez H, Tison Y (2005) Ageing of atactic and isotactic polystyrene thin films treated by oxygen DC pulsed plasma. Surf. Coat. Technol. 200: 2310- 2316.

[86] Morent R, De Geyter N, Leys C, Gengembre L, Payen E (2007) Study of the ageing behaviour of polymer films treated with a dielectric barrier discharge in air, helium and argon at medium pressure. Surf. Coat. Technol. 201: 7847- 7854.

[87] Yun YI, Kim KS, Uhm SJ, Khatua BB, Cho K, Kim JK, Park CE (2004) Aging behavior of oxygen plasma-treated polypropylene with different crystallinities. J. Adhesion Sci. Technol. 18/11: 1279–1291.

Chapter 11

Cellulose Functionalysed with Grafted Oligopeptides

Justyna Fraczyk, Beata Kolesinska, Inga Relich and Zbigniew J. Kaminski

Additional information is available at the end of the chapter

1. Introduction

Advantages offered by immobilization of any component of the reacting system are rewarding all additional efforts and the cost of the support. The majority of the methods reported have been based on the principles of solid phase organic synthesis (SPOS) in which the substrate is attached to the polymer support and excesses of reactants and reagents used to drive each synthetic step to completion. Then simple filtration affords a polymer bound product. While this approach is undoubtedly effective, there are a number of drawbacks which include the requirement for additional chemical steps to attach starting material, to develop synthetic methodology for the solid phase and to cleave products. More recently, solution phase methods, which circumvent these difficulties, have been introduced as alternatives to SPOS. These allow the use of excess of reagents followed by sequestrating either the product or excess reagents and byproducts from the reaction mixture using an insoluble functionalized polymer. Isolation and purification can then be achieved by simple filtration and evaporation.

Usually polystyrene and PEG based resins are commonly used as matrixes in SPOS, but nowadays there is observed also increased application of various beaded cellulose supports [1]. These show different solvent swelling profiles relative to those exhibited by the standard organic polymers and, being biomolecules, are biodegradable. Cellulose framework is attracting growing attention due to favorable biophysical properties, biocompatibility, low immunogenicity, relatively high resistance to temperature, inertness under broad range of reaction conditions and solvents and many other unique properties. Moreover, native cellulose microfribrils are abundant within slightly diversified properties dependent on the origin within relatively low cost. These properties make cellulose very useful for biochemical and biological investigations of interactions in aqueous as well as organic media.

Solid supported reagents were found exceedingly useful in all syntheses involving excessive amounts of substrates [2]. Inexpensive cellulose is offering high loading potential but concurrently attended by the threat of side reaction of nucleophilic hydroxyl groups. Using relatively inert towards hydroxylic group under ambient conditions triazine coupling reagents it was possible to obtain monofunctional triazine condensing reagents **1** and bifunctional reagent **2** by the treatment of cellulose with 2,4-dichloro-6-methoxy-1,3,5-triazine or cyanuric chloride respectively [3]. An independent approach towards immobilization of cyanuric chloride was confirmed the general utility of this procedure [4]. The loading of the cellulose carrier has been established by determination of Cl and N contents. For the standard laboratory Whatman filter paper, typical anchoring of triazine condensing reagent gave density of loading $0.6 - 1.0*10^{-6}$ mmole/cm^2.

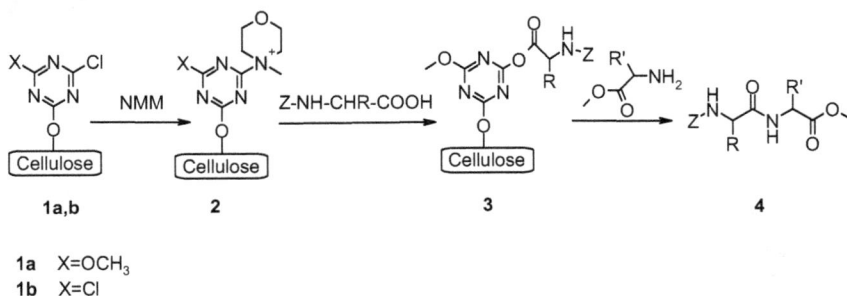

1a   X=OCH$_3$
1b   X=Cl

Scheme 1. Peptide **4** synthesis using monofunctional **1a** and bifunctional **1b** triazine condensing reagents reagents immobilized on cellulose.

An expedient matrix for the preparation of indexed library of amides and oligopeptides has been obtained by the demarcation of the surface of the cellulose plates chess-wise by the thin lines imprinted by polysilane, allocated separated, squared area for parallel synthesis of each individual compound (flat reactors) [5]. Application of carboxylic components into compartments of the matrix afforded "superactive" [6] triazine esters **3** linked to the support. Applying of amino components afforded the indexed library of amides and oligopeptides. The extraction of the final products from the solid support gave chromatographically homogeneous amides and oligopeptides in 60-99% yield. Chromatograms of the crude extracts from the diagonal fields of the part of the 8x12 library of amides and dipeptides are presented on Figure 1. When the size of matrix compartment corresponds to the size of typical ELISA plate, the amount of product recovered by extraction from the single "square flat reactor" was sufficient for elucidation of the structure of product by ES-MS or FAB-MS, for determination of their purity by HPLC, and even for studies involving ^{1}H-NMR.

Increasing the dimensions of the matrix field or application of powdered cellulose enabled "bulk" (1-10 mmole) synthesis of amides and oligopeptides. The further modification of this synthetic procedure as well as the "shape" of cellulose support, opened possibility to design of tailor made system of immobilized triazine coupling reagents the best suited to the given synthetic goal.

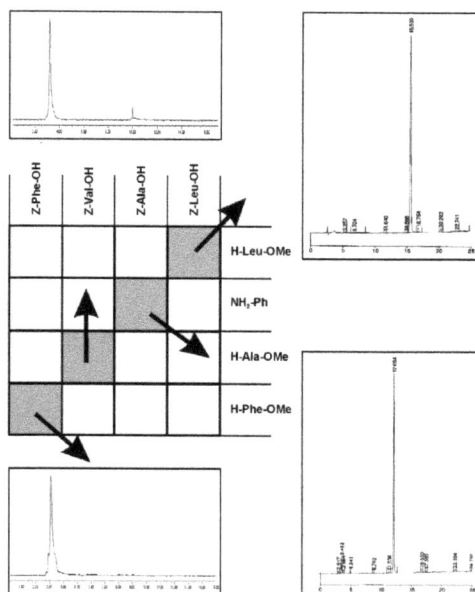

Figure 1. Chromatograms of selected dipeptides or amides obtained by using triazine coupling reagents immobilized on cellulose matrix divided chess-wise into separate "square flat reactors" by separation lines imprinted with polysiloxane.

In the more advanced approach, chiral coupling reagents immobilized on the cellulose were prepared and then used for enantioselective activation of racemic substrates [7]. Traceless enantiodifferentiating reagents [8] were obtained by using the cellulose membrane loaded with 2,4-dichloro-6-methoxy-1,3,5-triazine.

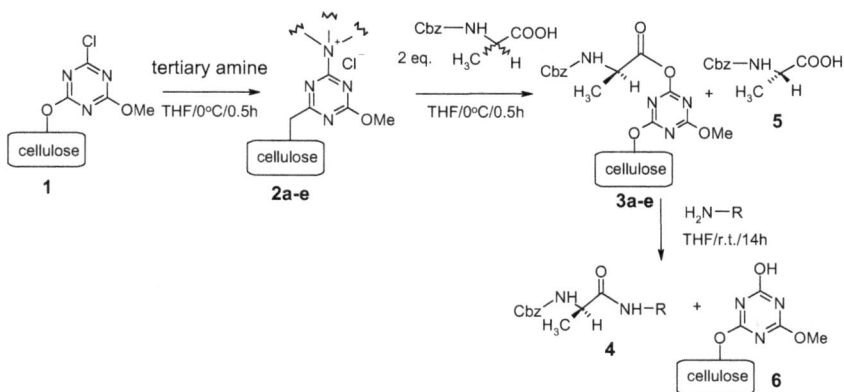

Scheme 2. Tracelss chiral coupling reagents **2a-e** prepared on cellulose.

Chiral quaternary *N*-triazinylammonium derivatives **2a-e** immobilized on the membrane were obtained *in situ* by treatment of **1** with appropriate tertiary amines (*N*-methylmorpholine, column 1; strychnine, column 2, brucine column 3; quinine column 4; and sparteine, column 5). Chirality of cellulose support (column 1) was found sufficient for enantiospecific activation of L enantiomer of racemic Z-Ala-OH with L/D ratio exceeding 90/10.

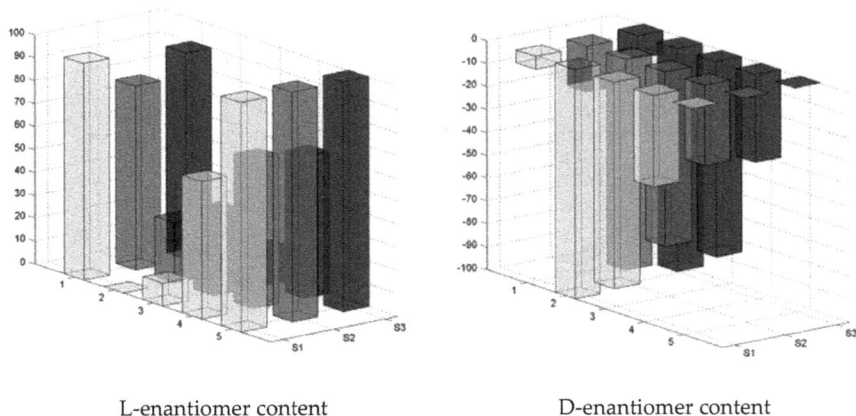

L-enantiomer content   D-enantiomer content

Figure 2. Enantiomeric composition of the products of enantioselective activation of *rac*-Z-Ala-OH and coupling with L-Phe-OMe (S1); D-Phe-OMe (S2), and H-Gly-OMe (S3).

Further structure modification of the immobilized triazine **1** proceeded directly on the membrane using chiral tertiary amines yielding spatially addressed five sub-libraries of enantiodifferentiating condensing reagents (Figure 2, 1-5). In all cases enantiodifferentiating activation of *rac*-Z-Ala-OH afforded triazine "superactive" ester **3a-e** with different enantiomeric composition. It has been found that the effect of chiral amine used as additional chiral selector predominate an effect of cellulose. Enantiomerically enriched esters **3a-e** in reaction with L-Phe-OMe (S1); D-Phe-OMe (S2), and H-Gly-OMe (S3) gave a library of alanine dipeptides of divergent configuration and enantiomeric purity (not linked to the support) and side-products (still immobilized on the cellulose membrane). The method opened an access to L and D alanine derivatives directly from racemic substrates. The best results (ee 92-99%) in the synthesis of L-alanine peptides were obtained in condensations mediated by *N*-methylmorpholine (column 1) or sparteine (column 5) when matching effects of cellulose and chiral selector were cooperated. The best results in the synthesis of D-alanine peptides (ee 91-98%) were obtained in condensations mediated by strychnine (column 2).

The disadvantage of procedure described above is caused by limited stability of *N*-triazinylammonium chlorides **2** prepared on cellulose. Stable immobilized triazine coupling reagents were obtained in reactions of cellulose with *N*-methylmorpholinium ptoluenesulfonates in the presence of sodium bicarbonate or DIPEA. Activation of carboxylic components proceeded under conditions similar to the standard synthesis in solution yielding "superactive" esters of N-protected amino acids anchored to the support *via* triazine ring (see Scheme 3).

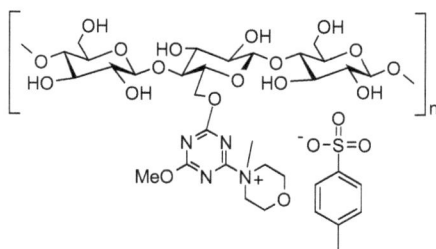

loading: 9.2 *10^{-6} mol/cm^2

Scheme 3. Stable triazine coupling reagents immobilized on cellulose plate.

A synthetic value of triazine reagents immobilized on cellulose was confirmed by dipeptide synthesis. The reagents were found efficient in the synthesis of Z-, Boc, or Fmoc protected chromatographically homogenous dipeptides in 72-91%. Moreover, experiments involving activation of sterically demanding 2-aminoisobutyric acid (Aib) confirmed that an access to the reactive centers of immobilized reagents remains principally unrestricted, although slightly lower yield and purity of respective peptides were noticed in this case [9].

The other modification of cellulosic fibers with tri-functional triazines was applied as control release system. The compounds employed were immobilized on cellulose substituted with monochlorotriazinyl (MCT) anchor group for fixation of an active substance and tuning the reactivity to facilitate release control. While the compounds were completely stable under dry conditions, the active substances were released simply by surrounding humidity. The reagents offered intriguing perspectives for the preparation of modified cellulosic material for single-use application in fields such as healthcare, cosmetics, or personal hygiene [10].

Cellulose was found also useful support for efficient control of selectivity of chemical reactions. In the classic procedure for the nitration of phenols, use of nitric and sulfuric acid mixtures results in the formation of *ortho* and *para* products with a ratio of about 2:1. Nitration of phenols and naphthols in the presence of biodegradable cellulose-supported Ni(NO$_3$)$_2$×6H$_2$O/2,4,6-trichloro-1,3,5-triazine system proceeded in acetonitrile at room temperature regioselectively. *Ortho*- nitrated phenols were obtained within a short reaction time with good yields. The reaction conditions were mild, and the employed cellulose could be recovered several times for further use [11]. The suggested mechanism proposes that cellulose acts as a template by forming hydrogen bonds between OH groups, phenol, and nitrate anions. This complex would transform substrate into the *ortho*-substituted intermediate followed by regioselective rearrangement to *o*-nitro phenol.

2. Cellulose acylated (grafted) with amino acids or peptides

Designing of new materials based on renewable natural resources is one of the most important scientific and technological challenges. The aim of these efforts is to open an access to materials which will have to replace toxic or non-biodegradable materials derived from fossil resources, while offering similar mechanical, thermal, or optical properties. In contrast to polymer membranes, cellulose shows high thermostability up to temperatures of about 180 °C, making it possible to use cellulose for reactions at elevated temperatures [12]. To date filter papers have been mostly used as the solid support.

The classic immobilization procedure involved the use of cyanuric chloride [13] as linker for anchoring broad range of amino acids and peptides on cellulose. Lenfeld and coworkers [14] immobilized 3,5-diiodo-tyrosine (DIT) on cellulose beads activated by the reaction with 2,4,6-trichloro-1,3,5-triazine and used prepared materials as sorbents in affinity chromatography of proteases. Also glutathione-bound cellulose for use in chromatography was prepared with cyanuric chloride as linking agent [15].

The library of p-nitrophenyl esters of oligopeptides anchored with N-terminal amino-acids via triazine linkage to the cellulose were synthesized step by step and after digestion with tissue homogenate were used for colorimetric differentiation of hydrolytic activity of primary subcutaneously growing tumor of Lewis lung carcinoma (LLC) bearing mice, lung metastatic colonies of LLC, blood serum of LLC bearing mice, and appropriate tissue homogenate of the healthy mice [16].

Cellulose is a polysaccharide containing free hydroxyl groups. In the first report cellulose free OH groups were esterified with amino acids activated previously by the transformation into appropriate acid halide or anhydride, in the presence of a catalyst, such as $Mg(ClO_4)_2$, H_2SO_4, H_3PO_4, or $ZnCl_2$ [17]. Recently, the more convenient procedure involved the coupling method of Fmoc protected amino acids such as Fmoc-β-Ala-OH or Fmoc-Gly-OH [18] by using activating reagents such as N,N'-diisopropylcarbodiimide (DIC), 1,1'-carbonyldiimidazole (CDI) in presence of a base, e.g. N-methyl-imidazole (NMI) [19]. There are also recommendations suggesting the use of 1,1'-carbonyl-di-(1,2,4-triazole) (CDT) instead of CDI in order to reduce the risk of the deprotection of Fmoc-amino acids during the coupling reaction [20]. Since the early reports by Frank, cellulose has found widespread application as a support in the synthesis of peptides and oligonucleotides. This involved the use of cellulose in the form of sheets, membranes, disks [21] or cotton thread used as supporting material [22]. Despite these precedents, alternative cellulose supports, notably beads which can offer considerably higher loading levels than that obtained with planar supports [23]. Beaded cellulose can be easily prepared by the coagulation-regeneration technique involving the addition of a solution of a soluble cellulose derivative, commonly the xanthate [24,25] or acetate [26], to a rapidly stirred, inert, immiscible solvent. The beads, thus formed, are precipitated either by a sol-gel process or by a reduction in reaction temperature. Chemical regeneration of the hydroxyl groups and sieving produces the active beads with the desired size distribution.

Since the hydroxyl groups are moderately reactive, the process of functionalization of cellulose is often preceding with more complex modification procedures. A reactive intermediate containing isocyanate groups was prepared by treatment of cellulose with 2,4-tolylene diisocyanate. The reactions of the intermediate with amino acids and their esters gave cellulose derivatives containing amino acid residues. The isocyanate groups reacted with amino acid esters in DMSO at low temp. under nitrogen to give high conversions. The amounts of amino acid esters bound to the cellulose through urea linkage were evaluated as 0.35-1.07 mmol/g. The selective adsorption and chelation of metal ions indicated that celluloses containing lysine and cysteine residues adsorbed 0.051 and 0.056 mmol Cu^{2+}/g, respectively [27].

An essential drawback of the ester linkage applied for anchoring peptides is instability towards aqueous media of pH > 7, not uncommon for bio-assay and stripping conditions. Moreover, cellulose and cellulose membranes show only a limited acid stability. This acid sensitivity severely restricts palette of reagents and reaction conditions that can be applied, even for the most stable commercially available cellulose materials. Therefore, besides the direct esterification of cellulose membranes with amino acids, many publications describe the use of more stable ether or amide linkers.

Cellulose undergoes facile alkaline etherification which, given the availability of up to three hydroxyl groups per glucopyranose residue, offers the potential to provide very high loading supports. Several companies already offer already modified cellulose membranes. Specially prepared cellulose membranes with a stably attached aminated spacer of 8 to 12 PEG units (PEG300-500) are available, which in contrast to common cellulose membranes is stable under strong acidic and basic conditions.

The materials, on which polypeptides were immobilized on different shaped cellulose products *via* chemically stable ether linkage have antimicrobial or anticancer activities. These were used as wound dressings, sutures, artificial blood vessel, catheters, dialysis membranes, clothing, and stents. Thus, material, on which beetle defensin analogue Arg-Leu-Leu-Leu-Arg-Ile-Gly-Arg-Arg was immobilized on cotton fabric showed high antimicrobial activity even after repeated washing and autoclave sterilization [28]. Also other nonapeptide Arg-Leu-Tyr-Leu-Arg-Ile-Gly-Arg-Arg immobilized to amino-functionalized cotton fibers by a modification of the SPOT synthesis technique was active against *S. aureus,* methicillin-resistant *S. aureus* and mouse myeloma cells and human leukemia cells. The assays revealed that these fibers maintained inhibition activity against bacteria and cancer cells after washing and sterilization by autoclaving [29].

Cellulose with intrinsic osteoinductive property useful for the preparation of the bone substitutes was obtained by immobilization of peptides containing Arg-Gly-Asp (RGD) fragment [30]. Biomaterials from bacterial-derived cellulose modified with cell adhesion peptide became a promising material as a replacement for blood vessels in vascular surgery [31].

Application of cellulose as a support for synthesis of complex template-assembled synthetic proteins (TASP) by orthogonal assembly of small libraries of purified peptide building blocks has been reviewed [32]. In most cases the linear template precursor was prepared by standard solid phase peptide synthesis (SPPS) on synthetic resin with orthogonal protecting groups followed by head-to-tail cyclisation of the linear precursor peptide and anchoring the template structure on cellulose. The strategy involving cleavable linker allowed control of the progress of synthesis on polystyrene resin. Final assembly of peptides prepared under standard SPPS conditions proceeded by successive cleavage of orthogonal protecting groups followed by coupling of predefined peptides.

3. Proteins immobilization on cellulose

Cotton is an excellent material for immobilized enzyme active functional textiles because, like the surface of soluble proteins, it is hydrophilic and typically non-denaturing. Many methods are now available for coupling enzymes and other biologically active compounds to solid supports [33]. Several involve the preliminary preparation of carboxymethyl or *p*-amino-benzyl ether derivative of a general support such as cellulose. A simple process involves the use trichloro-1,3,5-triazine [34,35] or chlorotriazine derivatives with solubilizing groups such as methoxycarbonyl or methylcarbamoyl groups which make them very convenient reagents in the coupling with a cellulose carrier [36].

There are also known other proficient approaches to the covalent attachment of enzymes to cotton cellulose. Lysozyme was immobilized on glycine-bound cotton through a carbodiimide reaction. The attachment to cotton fibers was made through a single glycine and a glycine dipeptide esterified to cotton cellulose. Higher levels of lysozyme incorporation were evident in the diglycine-linked cotton cellulose samples. The antibacterial activity of the lysozyme-conjugated cotton cellulose against *B. subtilis* was assessed. Inhibition of *B. subtilis* growth was observed to be optimal within a range of 0.3 to 0.14 mM of lysozyme. This approach has also been applied to organophopsphorous hydrolase and human neutrophil elastase. Immobilizing the chromogenic peptide substrate of human neutrophil elastase on cellulose and studying its interaction with the elastase enzyme provided colorimetric response of human neutrophil elastase [37].

Invertase was immobilized onto the cellulose membrane activated photochemicaly using 1-fluoro-2-nitro-4-azidobenzene as a photolinker and used in a flow through reactor system for conversion of sucrose to glucose and fructose [38].

Over the years, several cellulose affinity ligands have been constructed based on application of noncatalytic domain of glycosidic hydrolase (CBD). This cellulose specific anchor was originally identified in *Trichoderma reesei* and *Cellulomonas fimi*. CBDs is binding on insoluble cellulose through high-affinity noncovalent interactions [39] and it is enabling the further fusion of an antibody-binding domain (i.e., protein A, protein G, protein L). Cellulose-binding domains (CBD) are ideal immobilization domains for affinity ligands because they fold independently and do not interfere with their fusion partner. [40] Coupling to cellulose matrices orients the fusion partner away from the solid support [41] reducing steric hindrance; and their high-affinity binding to cellulose is considered nearly irreversible. [42] At present, many CBD-tagged affinity ligands are purified before attachment to their solid support matrix. [43] For large-scale applications, it could be beneficial to directly immobilize the affinity ligand at the source of production, thus avoiding the cost and time required for purification.

Horseradish peroxidase (HRP) was immobilized to cellulose with cellulose-binding domain (CBD) as a mediator, using a ligand selected from a phage-displayed random peptide library. A 15-mer random peptide library was panned on cellulose-coated plates covered with CBD in order to find a peptide that binds to CBD in its bound form. The sequence LHS, which was found to be an efficient binder of CBD, was fused to a synthetic gene of HRP as an affinity tag. The tagged enzyme (tHRP) was then immobilized on microcrystalline cellulose coated with CBD, thereby demonstrating the indirect immobilization of a protein to cellulose *via* three amino acids selected by phage display library and CBD [44].

As a model system, it has been developed a fusion protein, which consisted of antibody-binding proteins L and G fused to a cellulose-binding domain (LG-CBD) tethered directly onto cellulose. Direct immobilization of affinity purification ligands, such as LG-CBD, onto inexpensive support matrices such as cellulose is an effective method for the generation of functional, single-use antibody purification system. This straightforward preparation of purification reagents make antibody purification from genetically modified crop plants feasible and address one of the major bottlenecks facing commercialization of plant-derived pharmaceuticals [45].

In several cases it could be beneficial to directly immobilize the affinity ligand at the source of production, thus avoiding the cost and time required for purification. A potential use of cellulose-supported affinity ligands for purification of other bioproducts from homogenates from genetically modified plants expressing recombinant proteins is under intensive studies. To examine the potential of immobilizing affinity purification ligands onto cellulose matrices in a single step, the yeast *P. pastoris* were engineered to express and secrete a chimeric protein consisting of antibody-binding proteins L and G[45] fused to a cellulose-binding domain. A similar fusion was recently reported for cell capture in hollow-fiber bioreactors. There are reports on the direct immobilization of chimeric LG-CBD proteins onto cellulosic resins for antibody purification. Both protein L and protein G domains retained dual functionality demonstrated by the specific binding and purification of scFv and IgG antibodies from complex feed stocks of yeast supernatants and tobacco plant homogenates. This is a step towards the rapid generation of inexpensive affinity purification reagents and systems, to reduce the costs associated with downstream processing of pharmaceutical products, including antibodies, from complex production systems such as genetically modified crop plants.

Copolymers having polypeptide side chains grafted on cellulose main chain were used for adhesion of fibroblasts. The factor likely to play a key role in determining the binding ability was the balance between the hydrophilicity and hydrophobicity of the main- and side-chain components [46].

4. Protein sensors

Current research in the field of pathogen detection in food matrixes is aimed at creating fast and reliable detection platforms. Antibody engineering has allowed for the rapid generation of binding agents against virtually any antigen of interest, predominantly for therapeutic applications, development of diagnostic reagents and biosensors. By using engineered antibodies a pentavalent bispecific antibody were prepared by pentamerizing five single-domain antibodies and five cellulose-binding modules. This molecule was dually functional as it bound to cellulose-based filters as well as *S. aureus* cells. When impregnated in cellulose filters, the bispecific pentamer recognized *S. aureus* cells in a flow-through detection assay. The ability of pentamerized CBMs to bind cellulose may form the basis of an immobilization platform for multivalent display of high avidity binding reagents on cellulosic filters for sensing of pathogens, biomarkers and environmental pollutants [47]. Another approach for designing protein sensor used ultrathin films of cellulose modified on surface with small engineered peptides HWRGWV or HWRGWVA as substrate for protein detection. Primary tests run with peptide HWRGWV confirmed that there was an abundant amount of protein absorbed onto the surface, particularly with lower concentration and the sensitivity of a peptide greatly affects the ability to adsorb analytes onto the surface, demonstrating that cellulose substrates can be used to immobilize peptides which can further be used to selectively bind biomolecules [48].

The sensor for human neutrophil elastase (HNE), an enzyme engaged in chronic wounds healing was prepared based on colorimetric determination of enzyme activity. For colorimetric detection of human neutrophil elastase chromogenic peptide substrate Succinyl-Ala-Ala-Pro-Ala-pNA and its analog Succinyl-Ala-Ala-Pro-Val-pNA were attached to derivatized cellulose. Cellulose was pre-treated with 3-aminopropyltriethoxysilane to form the amino-propyloxy ether of cellulose, then reacted with the HNE chromogenic para-nitroanilide peptide substrates to form a covalently linked conjugate of cellulose (Cell-AP-suc-Ala-Ala-Pro-Ala-pNA or Cell-AP-suc-Ala-Ala-Pro-Val-pNA) through amide bond between the Cell-AP amine and the succinyl carboxylate of the substrate. The colorimetric response of the cellulose-bound chromophore was assessed by monitoring release of p-nitroaniline from the derivatized cellulose probe to determine human neutrophil elastase levels from 5.0×10^{-3} to 6.0 units per mL [49].

5. Epitope mapping - SPOT methodology

The SPOT synthesis of peptides, developed by Ronald Frank [50], has become one of the most frequently used methods for synthesis and screening of peptides on arrays. The method is a very useful tool for screening solid-phase and solution-phase assays with the size of arrays changeable from a few peptides up to approximately 8000 peptides [51]. Several hundred papers regarding modification and application of the SPOT method have been published [52].

The method was initiated as an uncomplicated technique for the positionally addressable, parallel chemical synthesis on a membrane support. SPOT synthesis of peptides on cellulose paper is a special type of solid phase peptide synthesis (SPPS) with each spot considered as a separate reaction vessel. The general strategy for parallel peptide assembly on a cellulose membrane is shown in Figure 3.

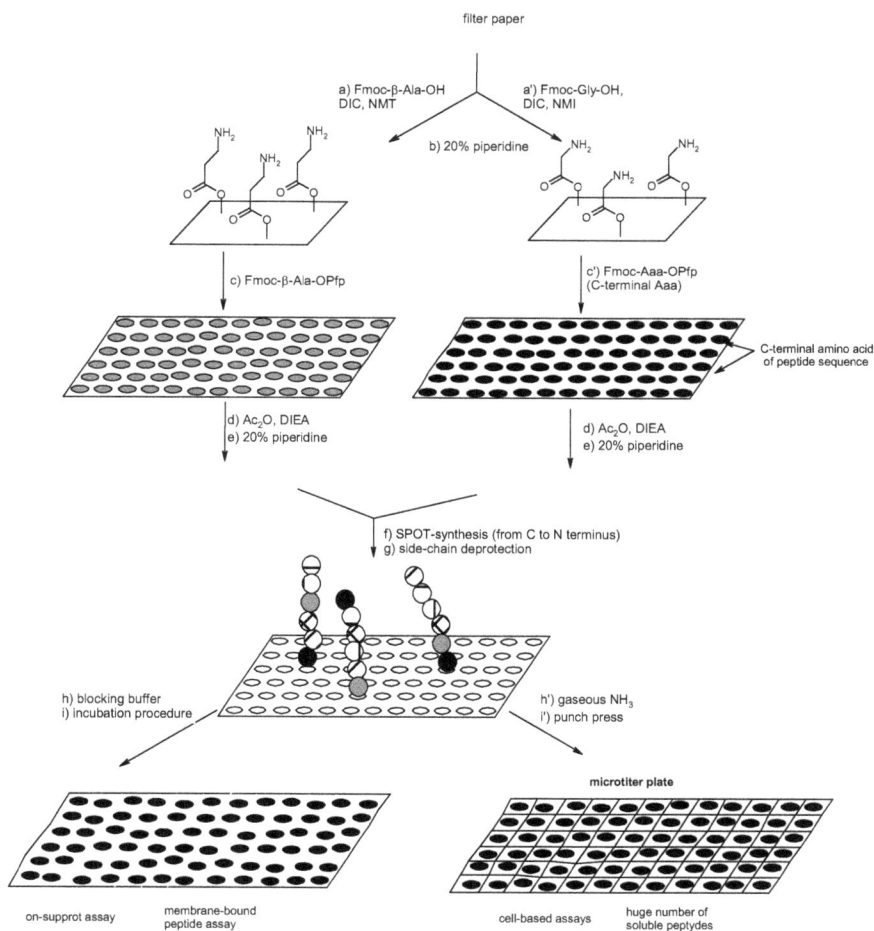

Figure 3. SPOT technology procedure.

Plain cellulose membranes (filter paper, chromatography paper) are commonly used as a support in the SPOT synthesis. These are porous, hydrophilic, flexible and stable in the organic solvents used for peptide synthesis. Cellulose membranes are relatively inexpensive material, which makes them very useful for biochemical and biological studies in aqueous and organic media. However, since cellulose is not stable against harsh chemical conditions, the SPOT synthesis was developed for the milder type of the two major SPPS strategies based on the Fmoc protection of amine function of the main peptide chain [53] and orthogonal protecting groups used for protection of side chains. [54].

5.1. Membrane modifications

Cellulose membranes are still the most widely used supports for SPOT peptide synthesis. The esterification of hydroxy functions of the cellulose with an Fmoc amino acid is a convenient method to introduce a spacer molecule and, after Fmoc deprotection, a free amino function for the SPPS of peptide arrays. The stability of the cellulose to organic solvents and bases allows the synthesis of peptides by utilizing the standard Fmoc methodology. Furthermore, the hydrophilic nature of cellulose offers a high compatibility with a wide variety of biological assay systems. On the other hand, however, cellulose shows only a limited acid stability. This acid sensitivity is severely restricting side chain deprotection conditions and stimulated the search of more convenient supports. Increasing resistance of peptide-cellulose membrane linkage against various types of reagents has been achieved through the development amino-functionalized ether type membranes. Ether type membranes provide stable membrane-bonding of peptides or other compound through the chemical stability of the ether bound. The first example of this type membrane was a cellulose-aminopropyl ether membrane (CAPE membrane) prepared by the treatment of cellulose filter paper with N-protected 2,3-epoxypropylamine [55].

The use of epibromohydrin as an activating reagent allowed introducing reactive bromine attached to the cellulose *via* an ether bond. The bromine moiety is able to react with different diamines [56] such as DAP (1,3-diaminopropane) [57] or TOTD (4,7,10-trioxa-1,13-tridecanediamine) and aminated polyethylene-3 (PEG-3) [58]. DAP modified cellulose membranes are known as N-CAPE membranes [59], while membranes modified with TOTD as a trioxa or TOTD membranes [60]. An amino type linker functionalized planar cellulose support [61] has been obtained by activation of cellulose with tosyl chloride and subsequent reaction with broad variety of diamines.

An additional advantage of CAPE membranes is an excellent signal-to noise ratio during on-support assay because of the very low background signal of this membrane [62]. Due to these properties they were applied in biological studies [63]. Table 1 has shown the characteristic of selected examples of amino-functionalized cellulose membranes.

Another approach to improve SPOT technology involves the use of linker strategies to enable cleavage of peptides from the support. An interesting linker was proposed by Frank. It is known as a Carboxy-Frank-Linker [64]. This linker allows peptide release from the solid support in aqueous solution (pH 7–8). Other linker types used in SPOT technology nowadays are the p-hydroxymethylbenzoic acid (HMB) linker, the Rink-amide linker [65], photolabile linker [66], the Wang linker [67], thioether moieties [68] or 4-hydroxymethyl-phenoxy acetic acid (HMPA) and 4-(4-hydroxymethyl-3-methoxyphenoxy)-butyric acid (HMPB) linkers [69]. An attractive approach based on the use of the C-terminal amino acid as a linker moiety was reported by Ay and co-workers [70]. Proposed solution was applied to sorting peptides according to their C termini using modified membranes with the corresponding C-terminal amino acids anchored either spot- or surface-wise.

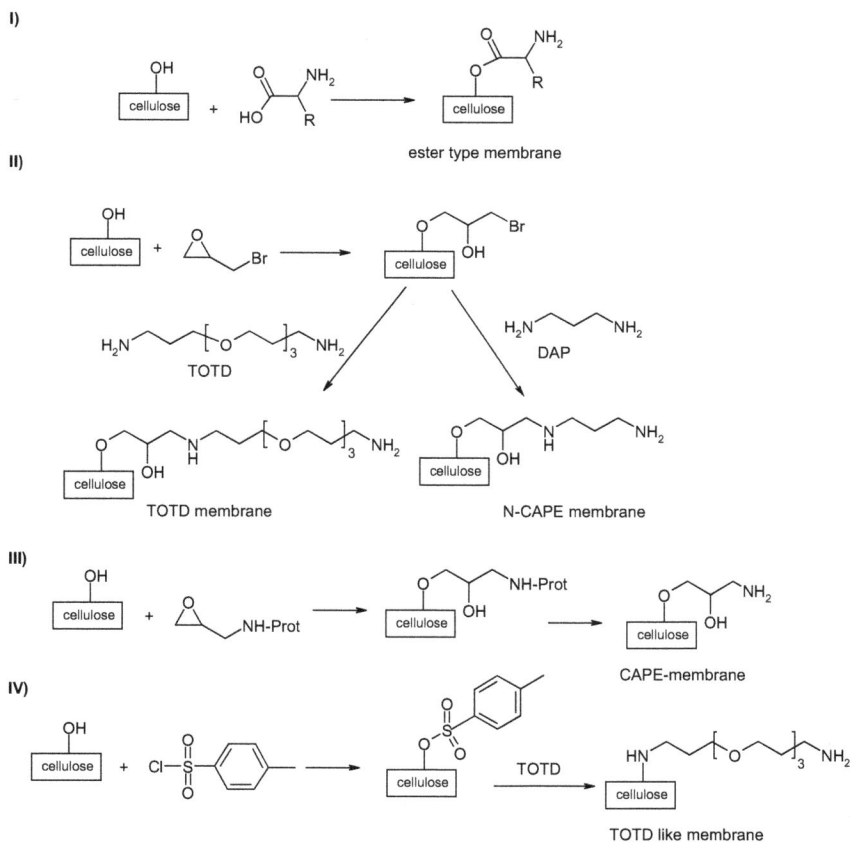

Scheme 4. Different methods of amino functionalization of cellulose membranes: I) preparation of ester type membrane with amino acid; II) functionalization with epibromohydrin and subsequent reaction with TODT or DAP providing TODT or N-CAPE membranes; III) treatment with N-protected 2,3-epoxypropylamine giving CAPE membrane; IV) treatment with tosyl chloride and subsequent reaction with TODT giving a TODT like membrane.

Cellulose type	Cellulose membrane type	Capacity [μmol/cm^2]
Whatman CHr1	Ester type: β-alanine	0.4-0.6
	Amine type: TsCl, diamino-PEG-3	4.0-10.0
	Amine type: TsCl, diamino-PEG-3 + linker	0.45-2.6
Whatman 50	Ester type: β-alanine	0.2-0.4
	Ester type: glycine	0.8-1.9
	Ester type: different amino acids	0.2-1.7
	Ether type: CAPE (amino-epoxy)	0.05-0.20
	Ether type: N-CAPE, trioxa	0.2-1.2
Whatman 540	Ester type: β-alanine	0.2-0.6
AIMS	Amine type: amino-PEG	0.4-0.6 (2.0-5.0)

Table 1. Characterization of amino-modified membranes.

In cases of classical SPOT technology in which the peptide is coupled *via* an ester bond using ß-alanine of glycine spacer, peptide can be released from the cellulose by hydrolysis at pH>9. Numerous reagents were found suitable for this goal, e.g. aqueous solutions of ammonia, sodium hydroxide, trialkylamines or lithium carbonate (see Table 2) [71]. A broadly used method for releasing of soluble peptide amides is based on the treatment of membranes with ammonia vapor [72]. When the cleavage is carried out with nucleophils in an anhydrous environment, the substitution of the ester bond leads to amides, hydrazides and other derivatives.

Linker	Cleavage conditions	C-termini derivatives
Glycine (differ amino acids)	gaseous ammonia	amide
	hydrazine	hydrazide
	hydroxyl amine	hydroxyl amide
	aq. NaOH	free carboxylic group
	aq. triethylamine	free carboxy group
	primary alkyl/aryl amine	alkyl/aryl amide
Boc-Imidazol linker	TFA + aq. buffer	free carboxy group
Allyl linker	Palladium (0)-catalyst	free carboxy group
Boc-Lys-Pro	TFA + aq. buffer	diketopiperazine
HMB linker	gaseous ammonia	amide
Photo-labile linker	UV irradiation at 365 nm	amide
Rink-amide linker	TFA	amide
Thioether	gaseous ammonia	amide
(thiol + coupled by amino acid)	NaOH/H$_2$O/methanol	free carboxy group
haloalkyl esters	NaOH/H$_2$O/acetonitrile	free carboxy group
Wang linker	TFA vapour	free carboxy group

Table 2. Typical linker types bounded to cellulose supports and cleavage methods used for releasing the peptides from the membrane.

Different type of anchoring of the peptide chain to cellulose matrix was proposed by Kaminski [73] and co-workers. 1-Acyl-3,5-dimethyl-1,3,5-triazin-2,4,6(1H,3H,5H)-trion derivatives serve both as a spacer and linker. This isocyanuric linker has been introduced by thermal isomerization [74] of 2-acyloxy-4,6-dimethoxy-1,3,5-triazines immobilized on the cellulose support or isomerization catalysed by the presence of acids. Synthetic procedure leading to peptides anchored to cellulose by 1-acyl-3,5-dimethyl-1,3,5-triazin-2,4,6(1H,3H,5H)-trion (iso-MT) is shown in Scheme 5. In the first step chloro-triazine **a** immobilized on cellulose was treated with N-methylmorpholine yielding N-triazinylammonium chloride **b**. Then compound **b** activated carboxylic function of Fmoc-protected amino acid to superactive ester **c** [75], which finally in refluxing toluene rearranges to stable isocyanuric derivative **d**.

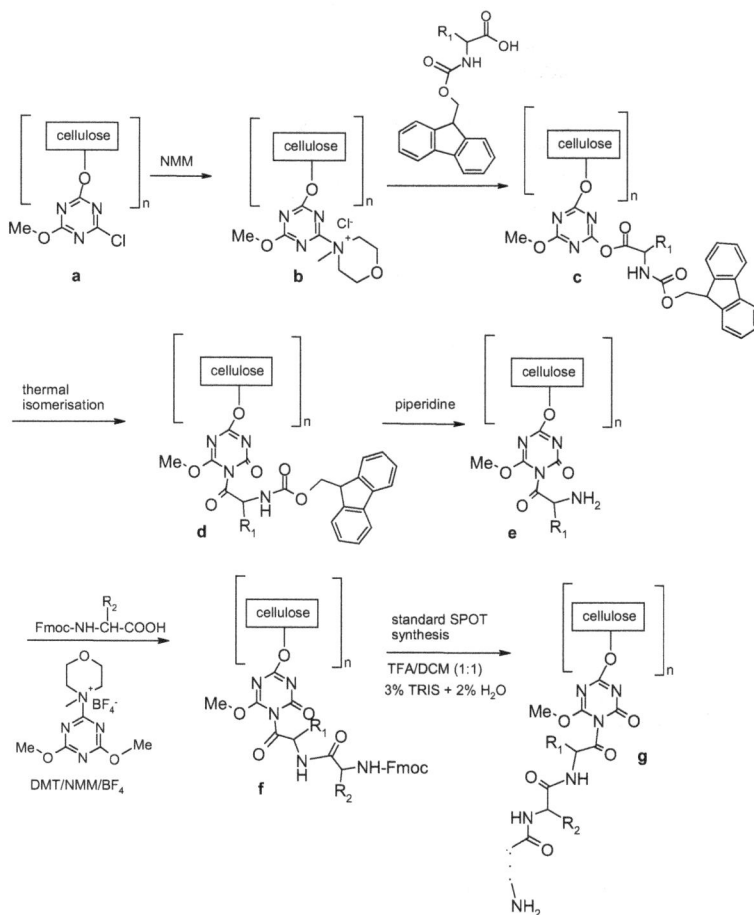

Scheme 5. Synthesis of peptides with free N-termini anchored by iso-MT linker.

Epitope	Peptide sequence	Reaction	Specificity
UB-33	H2N-Cys-His-His-Leu-Asp-Lys-Ser-Ile-Lys-Glu-Asp-Val-Gln-Phe-Ala-Asp-Ser-Arg-Ile-COO-cellulose	-	
	H2N-Cys-His-His-Leu-Asp-Lys-Ser-Ile-Lys-Glu-Asp-Val-Gln-Phe-Ala-Asp-Ser-Arg-Ile-β-Ala-COO-cellulose	-	0%
	H2N-Cys-His-His-Leu-Asp-Lys-Ser-Ile-Lys-Glu-Asp-Val-Gln-Phe-Ala-Asp-Ser-Arg-Ile-β-Ala-β-Ala-β-Ala-COO-cellulose	-	
CSF114	H2N-Thr-Pro-Arg-Val-Glu-Arg-Asn(Glc)-Gly-His-Ser-Val-Phe-Leu-Ala-Pro-Tyr-Gly-Trp-Met-Val-Lys-COO-cellulose	+, m	
	H2N-Thr-Pro-Arg-Val-Glu-Arg-Asn(Glc(OAac)4)-Gly-His-Ser-Val-Phe-Leu-Ala-Pro-Tyr-Gly-Trp-Met-Val-Lys-COO-cellulose	+/-	
	H2N-Thr-Pro-Arg-Val-Glu-Arg-Asp-Gly-His-Ser-Val-Phe-Leu-Ala-Pro-Tyr-Gly-Trp-Met-Val-Lys-COO-cellulose	+, m	90%
	H2N-Thr-Pro-Arg-Val-Glu-Arg-Asn(Glc)-Gly-His-Ser-Val-Phe-Leu-Ala-Pro-Tyr-Gly-Trp-Met-Val-Lys- β-Ala-β-Ala-β-Ala-COO-cellulose	+, m	

UB-33	H2N-Cys-His-His-Leu-Asp-Lys-Ser-Ile-Lys-Glu-Asp-Val-Gln-Phe-Ala-Asp-Ser-Arg-Ile-β-Ala-iso-MT-cellulose	+, m	
	H2N-Cys-His-His-Leu-Asp-Lys-Ser-Ile-Lys-Glu-Asp-Val-Gln-Phe-Ala-Asp-Ser-Arg-Ile-iso-MT-cellulose	+, s	100%
	H2N-Ser-Ile-Lys-Glu-Asp-Val-Gln-Phe-β-Ala-iso-MT-cellulose	+, s	
F-8	H2N-Ser-Ile-Lys-Glu-Asp-Val-Gln-Phe-iso-MT-cellulose	+, m	
CSF114	H2N-Thr-Pro-Arg-Val-Glu-Arg-Asn(Glc)-Gly-His-Ser-Val-Phe-Leu-Ala-Pro-Tyr-Gly-Trp-Met-Val-Lys-iso-MT-cellulose	+, m	
	H2N-Thr-Pro-Arg-Val-Glu-Arg-Asn(GlcAc4)-Gly-His-Ser-Val-Phe-Leu-Ala-Pro-Tyr-Gly-Trp-Met-Val-Lys-iso-MT-cellulose	+/-	70%
	H2N-Thr-Pro-Arg-Val-Glu-Arg-Asp-Gly-His-Ser-Val-Phe-Leu-Ala-Pro-Tyr-Gly-Trp-Met-Val-Lys-iso-MT-cellulose	+/-	

Table 3. Interaction of peptidic epitopes with free N-termini anchored on cellulose with antibodies.

Further stages of the synthesis included the standard SPPS conditions: deprotections of Fmoc group and subsequent condensation with Fmoc/tBu-protected amino acids by using DMT/NMM/BF4 as a coupling reagent [76]. The data summarized in Table 3 shown that in the several cases for the same antigen the strengths of reaction with antibody depends on the anchoring method. Moreover, for isocyanuric linker interactions with antibodies were found more selective [77].

5.2. Cellulose membrane-bound peptides with free C-termini

Unfortunately, cellulose is not suitable for classic SPOT peptides synthesis with free C-termini, due to engagement of C-terminal fragment of peptide for fixation to support. One potential solution to this problem is to synthesize peptides in a nontraditional manner (that is, from the N- towards the C-termini) using amino acid ethyl esters [78]. One major drawback with this approach is the increased risks of epimeriation [79] at all coupling stages due to repeated solid support-bound carboxyl activation. There have been done numerous efforts to develop effective ISPPS strategies (inverse solid-phase peptide synthesis). One of the first reports on ISPPS described the use of amino acid hydrazides [80]. More recently, amino acid 9-fluorenylmethyl (Fm) esters [81], and amino acid allyl esters [82] have been used for ISPPS. However, few if any of these amino acid derivatives are currently commercially available. The Fm ester approach looks attractive considering its similarity to standard Fmoc-based C-towards-N SPPS, but Fm esters are not as stable as Fmoc amino acids, and Fm ester-based inverse peptide synthesis apparently suffers from this limitation. The Fm ester approach also suffers from significant racemization during coupling reactions. The allyl ester-based approach is practicable and appears currently to be the method most competitive with the t-butyl ester-based ISPPS method described below. However, allyl esters are also not readily available commercially, and moreover, their deprotection requires the use of 20 mol% of Pd(PPh$_3$)$_4$, an expensive, heavy metal-based reagent. These strategies for ISPPS, therefore, appear not to be ideal, especially since suitable amino acid substrates are not easy available.

A method for solid-phase peptide synthesis on cellulose in the N- to C-direction that delivers good coupling yields and a relatively low degree of epimerization was reported by Hallberg [83] and co-workers. The optimized method involves the coupling, without preactivation, of the solid support-bound C-terminal amino acid with excess amounts of amino acid tri-tert-butoxysilyl (Sil) esters, using HATU or TBTU as coupling reagent and 2,4,6-trimethylpyridine (TMP, collidine) as a base. For the amino acids investigated, the degree of epimerization was typically 5%, except for Ser(t-Bu) which was more easily epimerized (ca. 20%). Efficiency of proposed methodology was confirmed on the synthesis of five tripeptides: Asp-Leu-Glu, Leu-Ala-Phe, Glu-Asp-Val, Asp-Ser-Ile, and Asp-D-Glu-Leu. The study used different combinations of HATU and TBTU as activating agents, N, N-diisopropylethylamine (DIEA) and TMP as bases, DMF and dichloromethane as solvents, and cupric chloride as an epimerization suppressant. Experiments indicated that the observed suppressing effect of cupric chloride on epimerization in the present system merely seemed to be a result of a base-induced cleavage of the oxazolone system, the key intermediate in the epimerization process. Proposed methodology can provide an attractive alternative for the solid-phase synthesis of short (six residues or less) C-terminally modified peptides, e.g., in library format. On the other hand amino acid silyl esters are difficult to prepare, unstable to store, and unstable under peptide coupling conditions.

The alternative strategy for ISPPS based on amino acid *t*-butyl esters was proposed by Gutheil [84] and co-workers . Favorable features of this approach are that amino acid *t*-butyl esters are stable, a large selection of them are commercially available, and the synthesis of commercially unavailable monomers is relatively straightforward. The *t*-butyl ester strategy also has the benefit that this approach is exactly the inverse of the well-developed Boc strategy for normal C-to-N peptide synthesis, and the extensive knowledge of side chain protection strategies and other chemical details can therefore be transferred from Boc chemistry to *t*-butyl ester chemistry. The effectiveness of the proposed solution has been demonstrated in the synthesis of tripeptides: Tyr-Ala-Phe, Tyr-Gly-Orn, Tyr-Ala-Val, Asn-D-Val-Leu, Asn-Leu-Glu, Gly-Ile-Thr, Phe-Ala-Gly. The consecutive incorporation of amino acids was performed in the presence of HATU as a coupling reagents using an excess of AA-OtBu*HCl. The observed racemization of individual amino acids was <2%.

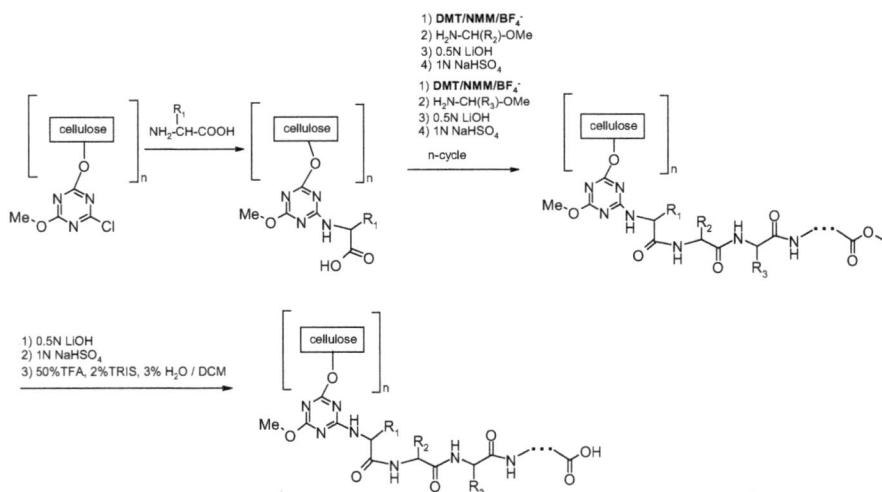

Scheme 6. Synthesis of peptidic epitope: SIKEDVQF and CHHLDKSIKEDVQFADSRI on the cellulose plate from *N*- to *C*-terminus using DMT/NMM/BF₄ as a coupling reagent.

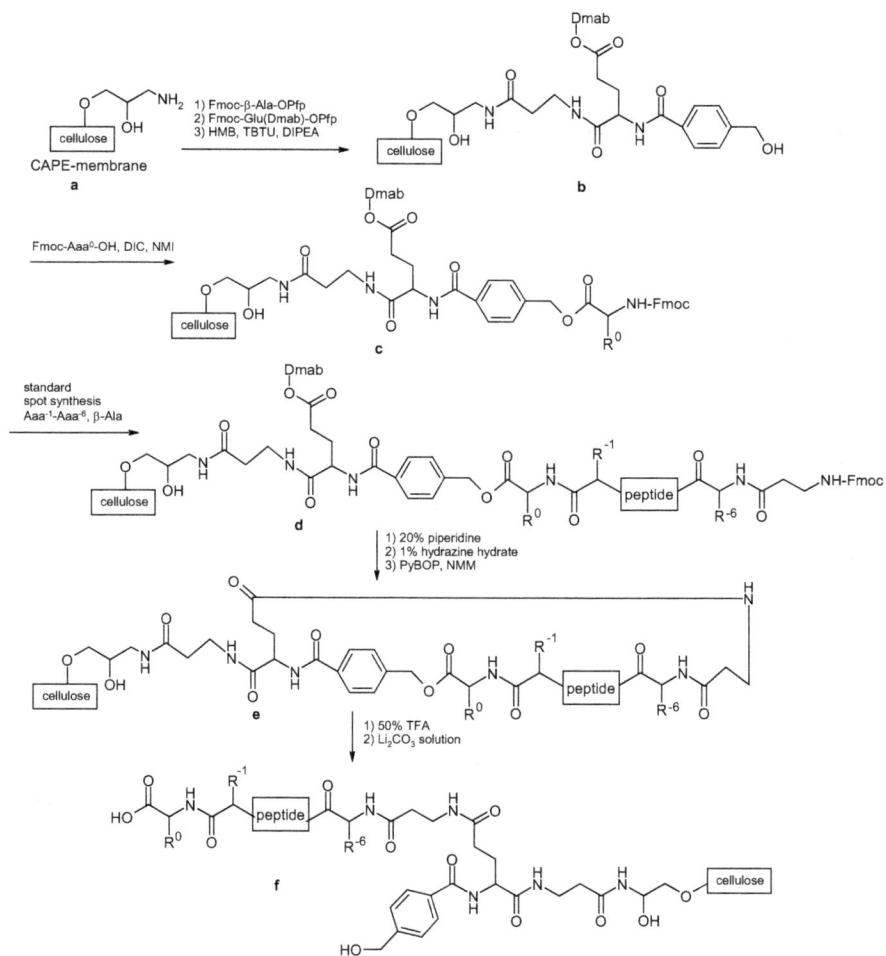

Scheme 7. Synthesis of inverted peptides on cellulose membranes.

Another approach to the synthesis of peptides attached to the cellulose matrix within N-terminus and presenting free C-terminus [85] was based on the utilization of 1,3,5-triazine derivative as an anchoring group. The peptides anchored via N-terminal moiety to the cellulose plate, were synthesized in accord to step-by-step methodology by means of 4-(4,6-dimethoxy-1,3,5-triazin-2-yl)-4-methylmorpholinium tetrafluoroborate (DMT/NMM/BF4) as a coupling [86]. 2-Chloro-1,3,5-triazine fragment, used as an anchoring group, was introduced by the treatment of cellulose with 2,4-dichloro-6-methoxy-1,3,5-triazine (DCMT) [87]. The first amino acid was attached to the triazine ring by the nucleophilic substitution reaction involving amine group. The oligopeptide chain was elongated in accord to step-by-step methodology in the sequence of standard reactions involving: activation of carboxylic function, coupling with the ester of appropriate amino acid, washing, capping, hydrolysis of ester moiety, and washing (Scheme 6).

An amount of natural, all-L diastereomer was sufficiently abundant for selective reaction with sera of patients with medically confirmed atherosclerosis even in the case of long epitope.

A more sophisticated approaches are based on inversion of the peptide chain following conventional ($C \rightarrow N$) synthesis and then modification of the C-terminus. Examples of inverting solid support-bound peptides [88] and methods for the generation of liberated C-terminally modified peptides [89] via a cyclization/cleavage protocols are known. The first example of application of the synthesis of inverting cellulose support-bound peptides according SPOT-methodology with free C-termini via prepared by successive cyclization and re-linearization was described by Hoffmüller and Volkmer-Engert [90].

Aminopropyl ether cellulose (CAPE-membrane) (a) was used as the matrix. β-Alanine serves both as a spacer and to residue directly engaged in the rearrangement. Dmab-glutamic acid was coupled as a bivalent linker followed by introducing hydroxymethylbenzoic acid (HMB) as a base-labile cleavable site ($\rightarrow$ b). The intended C-terminal amino acid was coupled through an ester bond ($\rightarrow$ c). The Fmoc and Dmab protecting groups on the N-terminus and the side chains of the glutamic acids were cleaved off and then construct cyclized ($\rightarrow$ e). Removal of the side chain protecting groups followed by hydrolysis of the ester bond linearizes the construct and generated free C terminus ($\rightarrow$ f). Even if presented above method allowed successful synthesis of inverted peptides arrays, the obtained yields were low and procedure were found troublesome and time-consuming.

Therefore Volkmer-Engert [91] and co-workers developed a more robust and efficient protocol for the preparation of cellulose membrane-bound inverted peptide arrays that could be used for widespread mapping different epitopes anchored on solid support and presenting C-termini peptides.

Synthesis of inverted peptides was performed on a cellulose membrane carrying a stable N-functionalized anchor (N-modified cellulose-amino-hydroxypropyl ether membrane - N-CAPE), which retained the inverted peptides (i). The inverted and N terminally fixed peptides (**i**) display a free *C* terminus resulting from reversal of the peptide orientation by successive thioether-cyclization/ester cleavage transformations. Key intermediates in the synthesis are the 3-brompropyl esters of Fmoc-amino acid (Fmoc-Aaa-OPBr) (**d**), the membrane-bound mercaptopropionyl cysteine adduct (**c**), the matrix-bound amino acid ester derivative (**e**), and the cyclic peptide (**h**). Critical reaction steps are the formation of both the cleavable ester bond and the cyclic peptide.

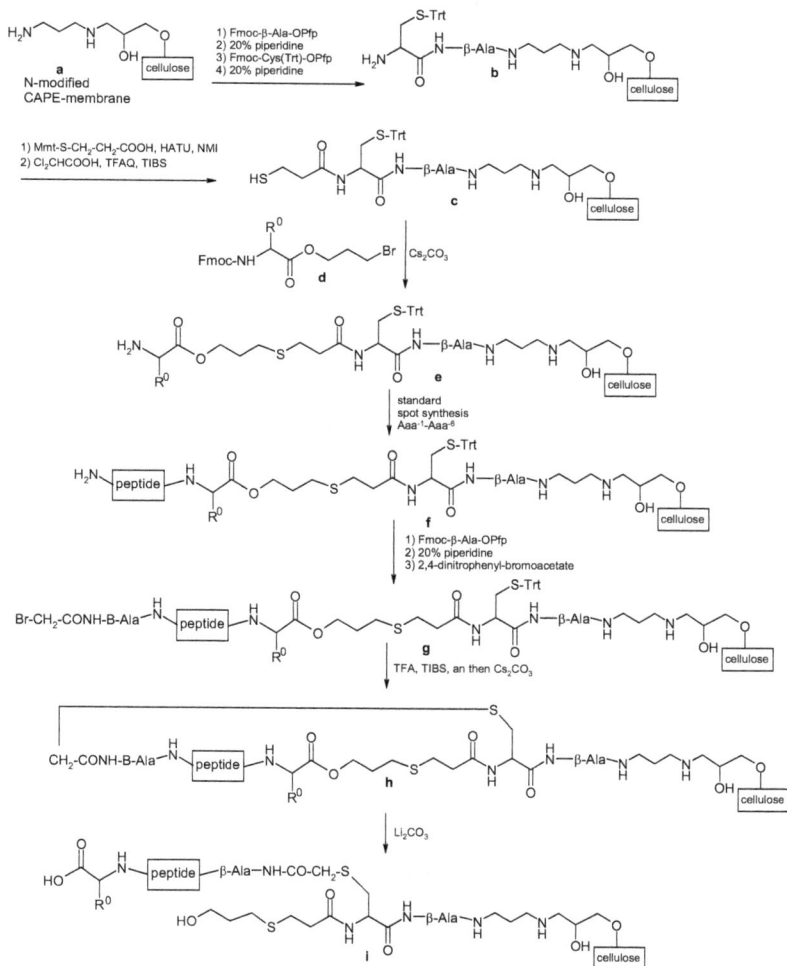

Scheme 8. Synthesis of inverted peptides on cellulose membranes allows further modification of side chains (phosphorylation).

5.3. Application of SPOT technology

Cellulose was found to be the support of choice in the SPOT synthesis. The main area of application of SPOT technology is for epitope mapping:

1. Physiology - Antibodies can identify the structural fragments which allow molecules to interact between themselves or with their specific receptor. They can also be useful in understanding the structure-function relationships.
2. Pathology - Understanding the mechanism by which an immune-mediated pathology develops by a precise identification of both B- and T-cell epitopes on the antigen. Antibodies can, therefore, be useful in analyzing the specificity of antibodies spontaneously formed in a number of diseases in which an immune response is an important parameter. Mapping of epitopes is also essential when one wishes to unravel the mechanisms by which immune tolerance is established and/or broken.
3. Preclinical evaluation of drugs or blood product derivatives - Most drugs act as haptens, that is to say that they are too small for being immunogenic. However, after combination with plasma or tissue proteins, they can become immunogenic.
4. Vaccinations - The identification of both B- and T-cell epitopes on a micro-organism or bacterial derived products such as toxins or enzymes may have a crucial influence on the design of vaccines. This includes not only an increase efficiency of vaccines, but potentially the design of vaccines that could stimulate humoral or the cellular immune response.
5. Diagnosis and subtyping of micro-organisms - Antibodies of defined specificity are currently used to distinguish between micro-organisms that belong to the same strain or to render diagnostic test more specific. The identification of shared antigenic determinants between proteins pertaining to different families can also has an important impact on the understanding of cross-reactions.
6. Mechanism of drug action - an emerging field of interest concerning the use of antibodies to study the mechanism of action of drugs [92].

Today, experiments to identify and characterize linear antibody epitopes using peptide scans, amino acids scans, substitutional analyses, truncation libraries, deletion libraries, cyclization scans, all types of combinatorial libraries and randomly generated libraries of single peptides are standard techniques widely applied even in non-specialized laboratories [93].

The synthesis of non-peptidic compounds or peptides with non-peptidic elements has been carried out on cellulose as well as polypropylene membranes. Using the SPOT technique, one of the most frequently synthesized non-peptidic compounds are a peptoids [94]. These compounds are synthesized pure or as hybrids with peptides, so-called peptomers [95]. Zimmermann et al. [96] investigated the possibility of replacing natural amino acids by peptoidic elements. Screening of an array of 8000 hexapeptoids and peptomers was carried out by Heine et al [97]. Hoffmann et al. [98] described the transformation of a biologically active peptide into peptoid analogues while retaining biological activity. Another application of the SPOT method is the synthesis of chimeric oligomers of peptide nucleic acids [99]. Weiler et al. [100] described the synthesis of a PNA oligomer library, with coupling yields of >97%. The synthesis of small organic compounds is another broad field for the application of SPOT synthesis [101].

6. Supramolecular structures formed by self-organization of N-lipidated peptides anchored to cellulose

Cellulose is a polysaccharide with two different types of hydroxyl groups i.e. primary and secondary. The primary hydroxyl groups are significantly more reactive then the secondary. Since the chains of polyanhydroglucose interacts with each other in the precisely defined way these functional groups are positioned within the reasonably regular fashion on he surface of cellulose. In the crystalline region of cellulose [102] the every second primary hydroxyl groups are exposed and accessible for interaction with reagents making after the transformation relatively regular pattern of anchored molecules separated by the distance of one anhydroglucose residue. Due to this advantageous feature of the cellulose the space available in between molecules anchored on the cellulose surface is sufficient for docking another molecules. Based on this assumption Kaminski and co-workers proposed entirely new approach for designing artificial receptors. According to the proposed concept, appropriate structure of molecules anchored on cellulose creates precisely defined and functionalized space for trapping ligands as presented on Figure 4.

The relatively weak bonding forces and conformational flexibility of both partners make docking of ligands to receptors difficult to study, to categorize by any kind of empirical rules, or to predict based on molecular modeling. Even in the case of interactions between relatively simple molecules, the possible bonding and repulsive forces of mutual host-guest interactions are multifaceted, very numerous, and difficult in terms of molecular modeling [103]. For the more advanced models involving flexible ligands and complex flexible receptor structures the rational construction plan of the host structure still exceeds our capabilities [104]. Thus, design of the molecular trap was done intuitively by mimicking structural features occurring in natural receptors, synthesis of the library of them by methods of combinatorial chemistry and selection of the most efficient representatives.

Strong, yet reversible binding force for the most of potential guest molecules were achieved by introducing into binding pockets most of the structural attributes responsible for weak intermolecular interactions [105]. These include hydrogen-bond donors and acceptors, lipophilic and hydrophilic fragments supplemented with π-donors and π-acceptors as depicted on Figure 5.

All these elements were allocated inside the linear structure forming the matrix of podands in such a way as to separate the flexible N-lipopeptide fragment from the solid support by relatively rigid, aromatic rings. Thus, a bonding "pocket" was composed from the tethered fragments of "walls" constructed from aromatic rings, expanded with a diversity of interactions offered by flexible peptide fragments, and finally closed with a "zipper" of hydrophobic chains of lipidic fragments. Due to the conformational flexibility of interacting partners, the relative direction of the functional groups of a ligand as well as that of the binding pockets could be readjusted to the most energetically favored orientation of both counterparts [106]. Thus structures immobilized on cellulose support *via* triazine linker created the mosaic of binding holes, mimicking the behavior of receptor formed from the neighboring, identical lipidated peptides.

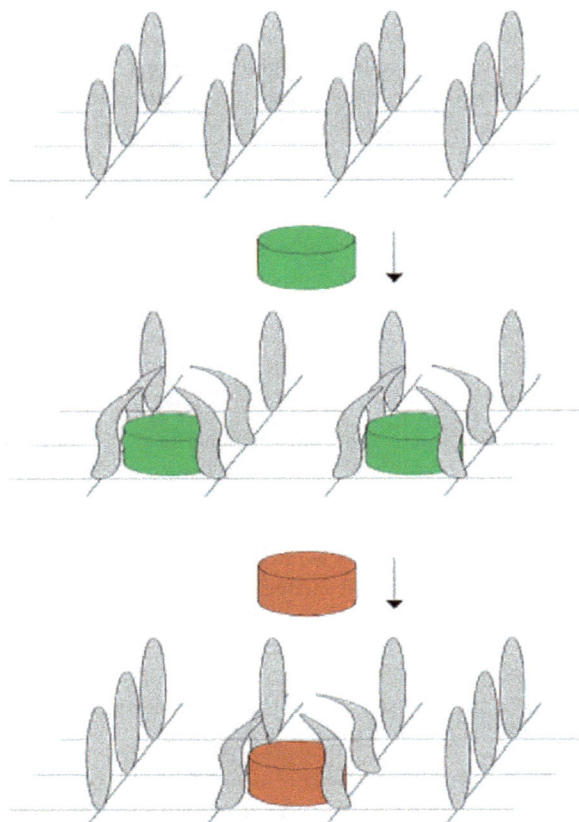

Figure 4. "Molecular traps" formed by podands regularly positioned on the support.

In the absence of some elements (Figure 6. **1-4**) binding process was substantially deteriorated compared to the binding ability of complete receptor structure (Figure 6, **5**).

Thus, the fully serviceable monolayer immobilized on cellulose was prepared in the stepwise process involving functionalization of cellulose with 1,3,5-triazine derivative followed by reaction with *m*-fenylenediamine, attachment of *N*-Fmoc amino acids, deprotection of *N*-terminus and completing the synthetic procedure by binding of carboxylic acid (Scheme 8). In the synthesis of the peptide fragment, DMT/NMM/BF4- was used as a coupling reagent. For the final acylation of immobilized tripeptides with carboxylic acid more lipophilic DMT/NMM/TosO- was found more suitable as a coupling reagent [107].

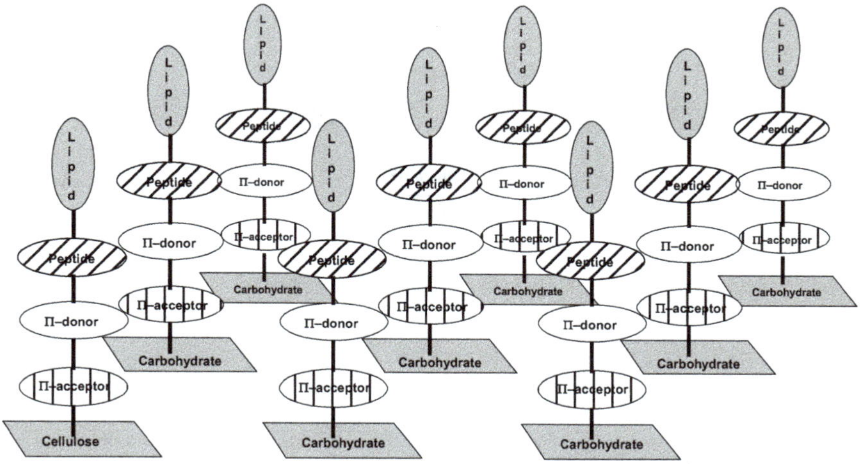

Figure 5. The concept of binding pockets with most of the structural attributes responsible for weak intermolecular interactions.

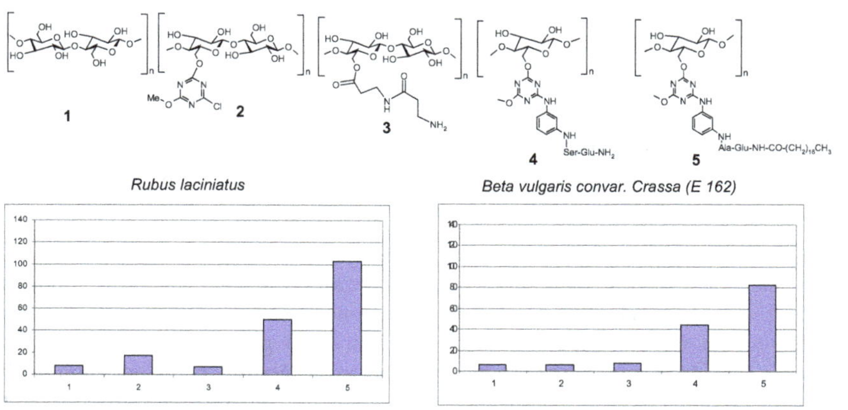

Figure 6. Influence of the structure of podands on binding of antocyane dyes from *Rubus laciniatus* and *Beta vulgaris* extracts.

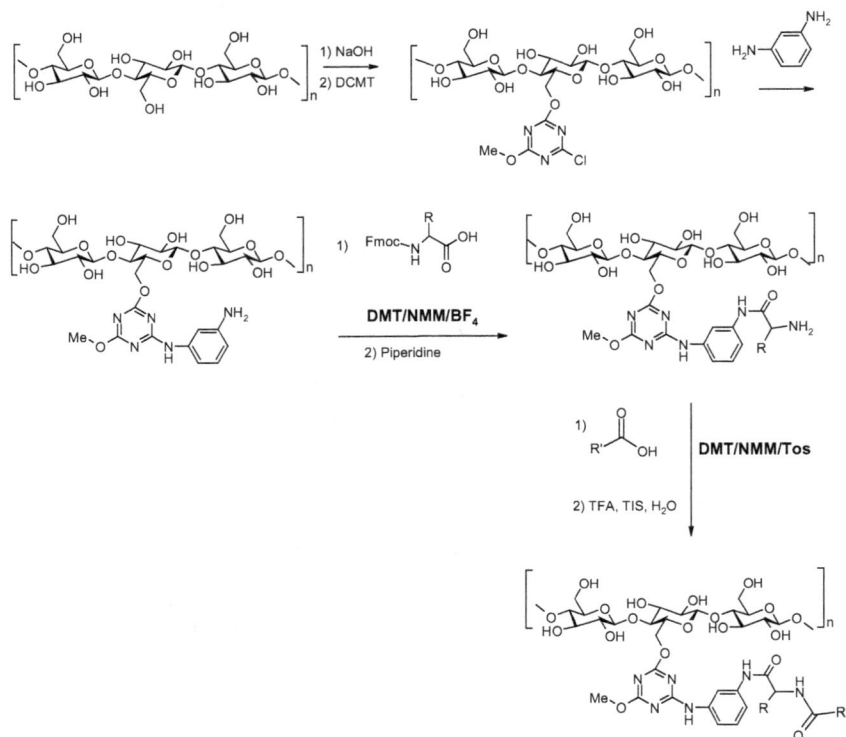

Scheme 9. Synthesis of *N*-lipidated peptides (amino acids) immobilized on cellulose *via* aromatic linker.

Loading of cellulose support was calculated on the basis of N and Cl content determined by elemental analysis 9-10 μmol/cm^2 with the anticipated ratio of molecular fragment triazine/*m*-phenylenediamine/amino acid/carboxylic acid.

The studies of water permeability through the monolayer achieved with 28-element library of *N*-acylated aminoacid prepared from Ala, Phe, Ser and Arg, and cinnamic, 10-undecenic, elaidic, oleic, erucic, palmitic, and ricinic acids confirmed that penetration of water is possible only in the presence of hydrophlilic functional groups incorporated into the monolayer. In their absence the podands were allocated sufficiently dense on cellulose fibers to inhibit penetration of water. This means that lipidic "zip" separate the interior of binding pocket, but the lipidic barrier remains still penetrable at least to the small, highly polar molecules of water.

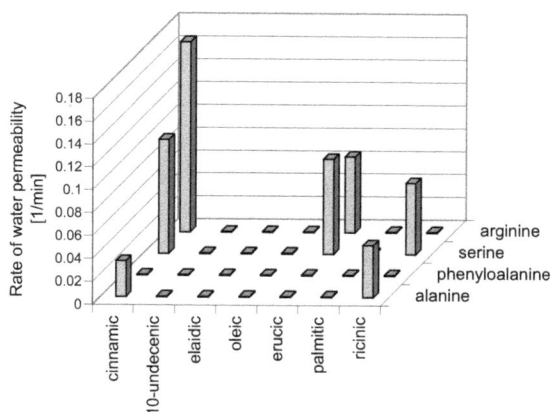

Figure 7. Dependence of water permeability through *N*-lipidated amino acids layer.

Process of binding triphenylmethane dyes with an array of *N*-lipidated dipeptides peptides immobilized on cellulose according to the manner described above was not dependent on initial concentration of ligand reaching equilibrium within 20-30 min. The selectivity and rate of binding depends on the structure of the peptide fragment as well as *N*-lipidic moiety. Even tiny structural changes in guest molecules were detected by monitoring the alteration of the binding pattern [108]. Measurements of fluorescence of fluorescein docked inside the receptor pocket revealed difference in λ_{max}, curvature and intensity of fluorescence depended on the structure of the peptide motif and lipidic fragment of binding pocket. This strongly suggests an alternation of charge distribution inside the receptor pocket [109].

Binding of colorless ligand was monitored by replacement of the reporter dye due to competitiveness of process (Scheme 10).

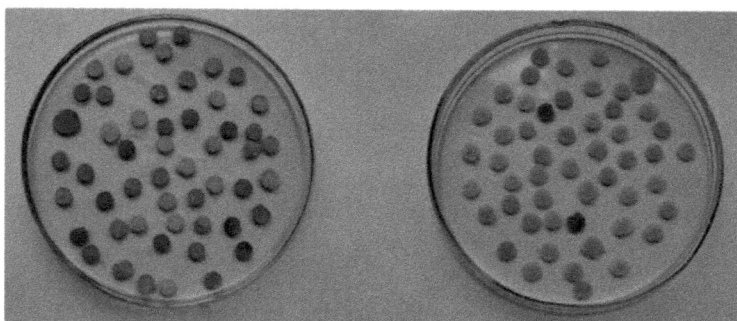

Figure 8. The library of *N*-lipidated dipeptides immobilized on cellulose disks treated with 10 mM/L solution of bromochlorophenol blue (1) (left); the same disks after subsequent treatment with 20 mM/L solution of S-(þ)-naproxen (right).

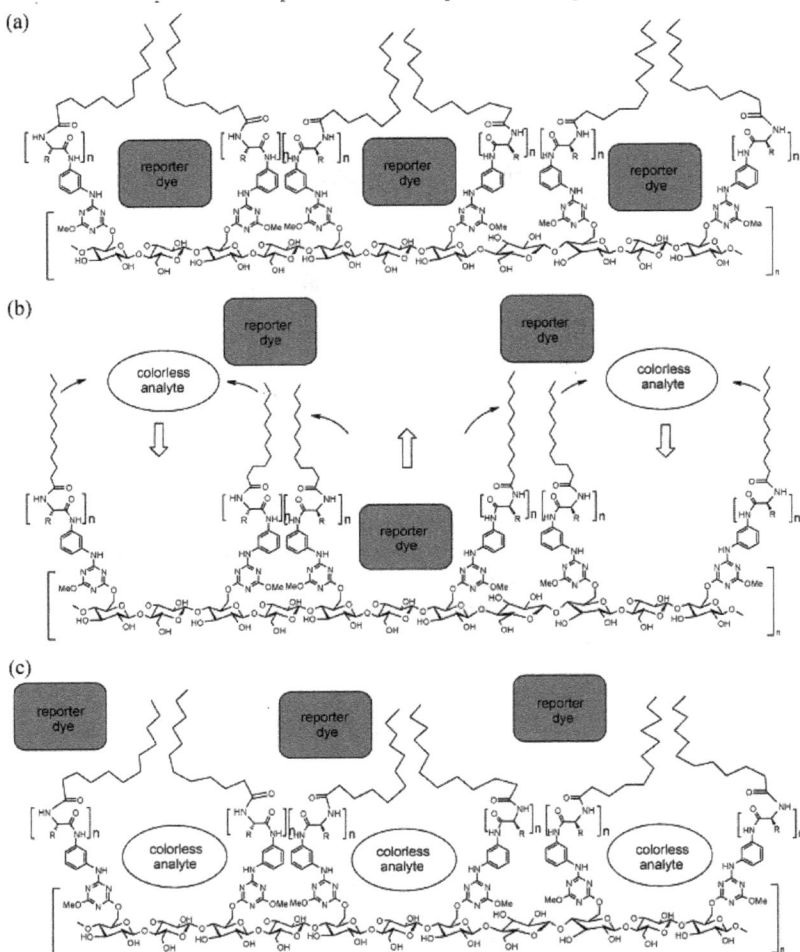

Scheme 10. Proposed mechanism of competitive binding colorless analyte and reporter dye.

Docking colorless N-phenylpiperazines pairs of analogues with or without a fluorine atom in the phenyl ring [110] revealed that using an array of artificial receptors it is possible to verify the presence of such ligand modification.

Analysis of the binding pattern of N-phenylpiperazine derivatives showed two characteristic binding patterns dependent on the structure of amino acid residues interacting with ligands. For most amino acid residues weaker binding of fluorinated analogues and stronger binding of native phenyl substituted analogues was observed with an exception of the receptors bearing tryptophane residue inside the binding pocket [111].

An array of *N*-lipidated peptides immobilized on cellulose was also used in studies of tissue homogenates for early diagnosing thyroid gland cancer [112], which is the most common malignancy of the endocrine system. There were found different binding patterns of healthy and cancer tissue for the various types of cancer.

By incorporation into the peptide fragment of receptor amino acid residues characteristic of catalytic triade of the hydrolytic enzymes the binding pockets demonstrated catalytic activity [113]. These were able to catalyse hydrolysis of esters bond [114]. All members of the library of 36 structures formed by permutations of Ser, Glu, His acylated with 6 long chain carboxylic acids were active as esterase and effectively catalyzed hydrolysis of p-nitrophenyl ester of Z-L-Leu-L-Leu-OH at pH 7-7.5 and temp not exceeding 20°C. The postulated mechanism of catalytic activity of *N*-lipidated oligopeptides is presented on Figure 9.

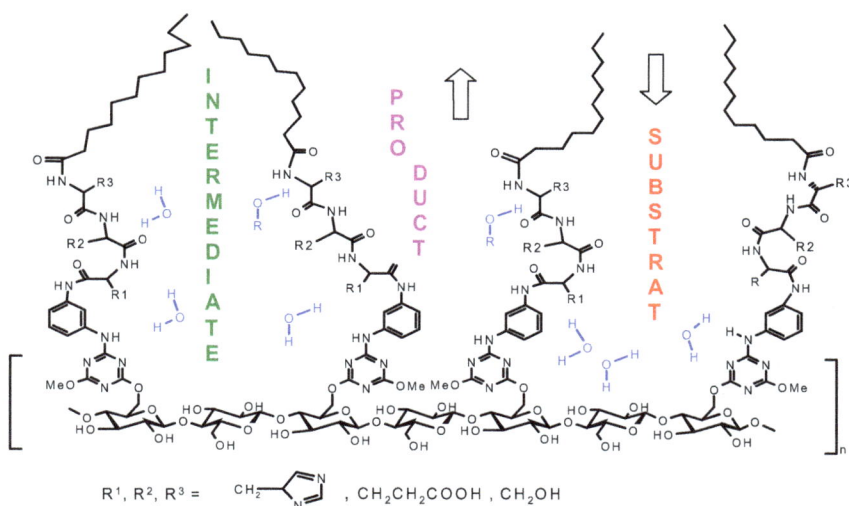

Figure 9. The postulated mechanism of catalytic activity of *N*-lipidated tripeptides of catalytic triade.

In this case the progress of hydrolysis was so fast that under the conditions of the experiment it was difficult to identify the most catalytically active structure. The extinction at 405 nm increased from the initial value of 0.300 for the substrate to more than 0.800 before the first cycle of measurements was completed. The rate of reaction diversified enough for identification of the most active catalytic structures was afforded by using significantly less reactive, sterically hindered substrate Z-Aib-Aib-ONp.

As the final effect of catalytic activity is the transformation of relatively non-polar organic molecule into ionic species, one can expect application of this phenomena for the construction of sensors [115]. There are also many other interesting area of application of catalytically active *N*-lipidated peptides immobilized on cellulose. Most expected are effects stereochemical results of such transformations, expecting open access to essentially unlimited access to configurational arrangement of stereogenic centers of polypeptide fragment [116].

Author details

Justyna Fraczyk, Beata Kolesinska, Inga Relich and Zbigniew J. Kaminski*
Institute of Organic Chemistry, Lodz University of Technology, Zeromskiego, Lodz, Poland

7. References

[1] Chesney, A.; Steel, P.G.; Stonehouse, D.F. High Loading Cellulose Based Poly(alkenyl) Resins for Resin Capture Applications in Halogenation Reactions, *J. Comb. Chem.* 2000, 2, 434-437.

[2] Boguszewski, P.A.; MacDonald, A.A.; Mendonca, A.J.; Warner, F.P. Reactivity, selectivity and compatibility: A practical review of polymer supported coupling reagents. *Abstracts of Papers, 232nd ACS National Meeting, San Francisco, CA, United States*, 2006, Sept. 10-14, (2006), Publ. A. C. S. Washington, D. C. Conference; Meeting Abstract; Computer Optical Disk

[3] Kaminski, Z,J.*; Kolesinska; B,; Cierpucha, M. Preparation of Indexed Library of Amides and Oligopeptides by Means of Triazine Condensing Reagent Immobilized on Cellulose, *Peptides 2000. Proceedings of the 26th European Peptide Symposium*, Ed. Martinez, J.; Fehrentz, J-A, EDK-Paris, 2001, 965-966.

[4] Benes, M. J.; Adamkova, K.; Turkova, J. Activation of beaded cellulose with 2,4,6-trichlorotriazine, *J. Bioact. Compat. Polym.* 1991, 6(4), 406-13.

[5] Kaminski, Z,J.; Kolesinska; B,; Cierpucha, M. Preparation of Indexed Library of Amides and Oligopeptides by Means of Triazine Condensing Reagent Immobilized on Cellulose, *Peptides 2000. Proceedings of the 26th European Peptide Symposium*, Ed. Martinez, J.; Fehrentz, J-A, EDK-Paris, 2001, 965-966.

[6] Kaminski Z.J. "Concept of superactive esters. Could the peptide synthesis be improved by inventing superactive esters" *Int. J. Peptide Protein Res.* 1994, 43, 312-319.

[7] Kolesinska, B.; Kaminski, Z.J. Synthesis of Spatially Addressed Library of Alanine Dipeptides from rac-Z-Ala-OH by Means of Sub-Library of Chiral Triazine Condensing Reagents Immobilized on Cellulose. *Peptides 2002. Proceedings of the 27th European Peptide Symposium*, Ed. Benedetti, E.; Pedone, C, Edizioni Ziino, Napoli, 2002, 172-173.

[8] Kolesinska, B.; Kaminski, Z,J. Design, synthesis and application of enantioselective coupling reagent with a traceless chiral auxiliary. *Org. Lett.* 2009, 11, 765-768.

[9] Kaminski, Z.J.; Kolesinska, B.; Kolesinska, J,; Wasikowska, K.K.; Piatkowska, N. *N*-Triazinylammonium salts immobilized on solid support as coupling reagents. *Peptides 2006. Proceedings of the 29th European Peptide Symposium*, Gdańsk 2006, K. Rolka, P. Rekowski and J. Silberring , Eds.; Kenes International, 2007, 590-591.

* Corresponding Author

[10] Rosenau, T.; Renfrew, A.H.M.; Adelwoehrer, C.; Potthast, A.; Kosma, P. Cellulosics modified with slow-release reagents. Part I. Synthesis of triazine-anchored reagents for slow release of active substances from cellulosic materials. Polymer 2005, 46, 1453–1458.

[11] Nemati, F.; Kiani, H.; Hayeniaz, Y.S. Cellulose-Supported Ni(NO3)2×6H2O/2,4,6-Trichloro-1,3,5-Triazine (Tct) as a Mild, Selective, and Biodegradable System for Nitration of Phenols", Synthetic Commun. 2011, 41, 2985–2992.

[12] Blackwell, H.E. Hitting the SPOT: small-molecule macroarrays advance combinatorial synthesis. Current Opinion in Chemical Biology, 2006, 10, 203-212.

[13] Benes, M.J.; Adamkova, K.; Turkova, J. Activation of beaded cellulose with 2,4,6-trichlorotriazine. J. Bioact. Comp. Polym. 1991, 6, 406-413.

[14] Lenfeld, J.; Benes, M.J.; Kucerova Z., 3,5-Diiodo-L-tyrosine immobilized on bead cellulose. React. Funct. Polym. 1995, 28, 61-68.

[15] Danner, J.; Lenhoff, H.M.; Heagy, W. Glutathione-bound celluloses: preparation with the linking reagent s-triazine trichloride and use in chromatography. J. Solid-Phase Biochem. 1976, 1, 177-188.

[16] Kaminski, Z.J.; Kolesinska, B.; Kinas, R.W.; Wietrzyk, J.; Opolski, A. The library of p-nitrophenyl esters of oligopeptides immobilized on cellulose membrane . Synthesis and degradation by tissue homogenates of Lewis Lung Carcinoma bearing mice. Peptides 2002. Proceedings of the 27th European Peptide Symposium, Ed. Benedetti, E.; Pedone, C, Edizioni Ziino, Napoli, 2002, 782-783.

[17] Gardner, Thomas S. Cellulose esters of amino acids. (Eastman Kodak Co.). (1949), US Patent: 2461152 from 19490208, US Pat. appl. 1945-627549.

[18] Volkmer, R. Synthesis and application of peptide arrays: Quo vadis SPOT technology. ChemBioChem, 2009, 10, 1431-1442.

[19] a) Deswal, R.; Singh, R.; Lynn, A.M.; Frank, R. Identification of immunodominant regions of Brassica juncea glyoxalase I as potential antitumor immunomodulation targets. Peptides, 2005, 26, 395-404. b) Ay, B.; Streitz, M.; Boisguerin, P.; Schlosser, A.; Mahrenholz, C.C.; Schuck, S.D.; Kern, F.; Volkmer, R. Sorting and Pooling Strategy: A novel tool to map a virus proteome for CD8 T-cell epitopes. Biopolymers (Pept. Sci.). 2007, 88, 64-75.

[20] Ay, B.; Volkmer, R.; Boisguerin, P. Synthesis of cleavable peptides with authentic C-termini: An application for fully automated SPOT synthesis. Tetrahedron Lett., 2007, 48, 361-364.

[21] a) Frank, R.; Doring, R. Simultaneous multiple peptide synthesis under continuous flow conditions on cellulose paper discs as segmental solid supports. Tetrahedron 1988, 44, 6031-6040. b) Dittrich, F.; Tegge, W.; Frank, R. "Cut and combine": An easy membrane-supported combinatorial synthesis technique. Bioorg. Med. Chem. Lett. 1998, 8, 2351-2356.

[22] Schwabacher, A. W.; Shen, Y. X.; Johnson, C. W. Fourier transform combinatorial chemistry. J. Am. Chem. Soc. 1999, 121, 8669-8670.

[23] Tegge, W.; Frank, R. Peptide synthesis on Sepharose(TM) beads. J. Pept. Res. 1997, 49, 355-362.

[24] Determann, H.; Rehner, H.; Wieland, T. Cellulose gel beads for chromatography. Makromol. Chem. 1968, 114, 268-274.

[25] Chitumbo, K.; Brown, W. Separation of oligosaccharides on cellulose gels. *J. Polym. Sci.* 1971, *36*, 279-292.

[26] Chen, L. F.; Tsao, G. T. Physical characteristics of porous cellulose beads as supporting material for immobilized enzymes. *Biotech. Bioeng. 1976*, 18, 1507-1516.

[27] Sato, T.; Karatsu, K.; Kitamura, H.; Ohno, Y. Synthesis of cellulose derivatives containing amino acid residues and their adsorption of metal ions, *Sen'i Gakkaishi*, 1983, *39*(12), CAN 100:158390. AN 1984:158390.

[28] Ishibashi, J.; Yamakawa, M.; Iwasaki, T.; Nakamura, M.; Tokino, S.; Ohagi, S. Antimicrobial peptide-immobilized materials, their manufacture, and their use for antimicrobial or anticancer shaped products and for sterilization of liquids. Patent JP 2010184022 A 20100826 (2010), Application: JP 2009-29726, 20090212. Priority: JP 2009-29726 20090212. CAN 153:368472, AN 2010:1065072.

[29] Nakamura, M.; Iwasaki, T.; Tokino, S.; Asaoka, A.; Yamakawa, M.; Ishibashi, J. Development of a Bioactive Fiber with Immobilized Synthetic Peptides Designed from the Active Site of a Beetle Defensin. *Biomacromolecules* 2011, *12*, 1540–1545.

[30] Bartouilh de Taillac, L.; Porte-Durrieu, M.C.; Labrugere, Ch.; Bareille, R. Amedee, J.; Baquey, Ch. Grafting of RGD peptides to cellulose to enhance human osteoprogenitor cells adhesion and proliferation. *Composites Science and Technology* 2004, *64*, 827–837.

[31] Fink, H.; Ahrenstedt, L.; Bodin, A.; Brumer, H.; Gatenholm, P.; Krettek, A.; Risberg, B. Bacterial cellulose modified with xyloglucan bearing the adhesion peptide RGD promotes endothelial cell adhesion and metabolism - a promising modification for vascular grafts. *J. Tissue Eng. Regenerative Med.* 2011, *5*, 454-463.

[32] Haehnel, W. Chemical synthesis of TASP arrays and their application in protein design. Mol. Div. 2004, *8*: 219–229.

[33] Baneyx, F.; Schwartz, D.T. Selection and analysis of solid-binding peptides. *Curr.Opinion Biotechn.* 2007, *18*, 312–317.

[34] Pechan, Z. Preparation and use of insoluble enzymes, *Chemie a Lide* 1971, *2*, 3-6. CAN 75:45122 AN 1971:445122.

[35] Smith, N.L., III; Lenhoff, H.M. Covalent binding of proteins and glucose 6-phosphate dehydrogenase to cellulosic carriers activated with s-triazine trichloride. *Anal. Biochem.* 1974, *61*, 392-415.

[36] Kay, G,; Crook, E.M. "Coupling of Enzymes to Cellulose using Chloro-s-triazines", *Nature* 1967, *216*, 514-515.

[37] Edwards, J.V.; Batiste, S.; Ullah, A.J.; Sethumadhavan, K.; Pierre, S. New uses for immobilized enzymes and substrates on cotton and cellulose fibers. *Abstracts of Papers, 230th ACS National Meeting, Washington, DC, United States*, Aug. 28-Sept. 1, 2005 (2005), American Chemical Society, Conference; Meeting Abstract; Computer Optical Disk.

[38] Utpal, B.; Krishnamoorthy, K.; Pradip, N. A simple method for functionalization of cellulose membrane for covalent immobilization of biomolecules. *J. Membrane Sci.* 2005, *250*, 215–222.

[39] a) Gilkes, N. R.; Warren, R. A.; Miller, R. C., Jr.; Kilburn, D. G. Precise excision of the cellulose binding domains from two *Cellulomonas fimi* cellulases by a homologous protease and the effect on catalysis. *J. Biol. Chem.* 1988, *263*, 10401–10407. b) van Tilbeurgh, H.; Tomme, P.; Claeyssens, M.; Bhikhabhai, R.; Pettersson, G. Limited

proteolysis of the cellubiohydrolase I from Trichoderma reesei: Separation of functional domains. *FEBS Lett.* 1986, *204*, 223–227.

[40] Shoseyov, O.; Shani, Z.; Levy, I. Carbohydrate binding modules: biochemical properties and novel applications. *Microbiol. Mol. Biol. Rev.* 2006, *70*, 283–295.

[41] Craig, S. J.; Shu, A.; Xu, Y.; Foong, F. C.; Nordon, R. Chimeric protein for selective cell attachment onto cellulosic substrates. *Protein Eng. Des. Sel.* 2007, *20*, 235–241.

[42] Tomme, P.; Boraston, A.; McLean, B.; Kormos, J.; Creagh, A. L.; Sturch, K.; Gilkes, N. R.; Haynes, C. A.; Warren, R. A.; Kilburn,D. G. Characterization and affinity applications of cellulose-binding domains. *J. Chromatogr., B* 1998, *715*, 283–96.

[43] Rodriguez, B.; Kavoosi, M.; Koska, J.; Creagh, A. L.; Kilburn, D. G.; Haynes, C. A. Inexpensive and generic affinity purification of recombinant proteins using a family 2a CBM fusion tag. *Biotechnol. Prog.* 2004, *20*, 1479–1489.

[44] Levy, I.; Shoseyov, O. Expression, Refolding and Indirect Immobilization of Horseradish Peroxidase (HRP) to Cellulose via a Phage-selected Peptide and Cellulose-binding Domain (CBD). *J. Peptide Sci.* 2001, *7*, 50–57.

[45] Hussack, G.; Grohs, B.M.; Almquist, K.C.; Mclean, M.D.; Ghosh, R.; Hall, J.C. Purification of Plant-Derived Antibodies through Direct Immobilization of Affinity Ligands on Cellulose *J. Agric. Food Chem.* 2010, *58*, 3451–3459.

[46] Hasegawa, O.; Fukuda, T.; Miyamoto, T.; Akaike, T. Adhesion behavior of fibroblasts on oligopeptide-grafted cellulose derivatives. *Cellulose* 1990, 465-72.

[47] Hussack, G.; Luo, Y.; Veldhuis, L.; Hall, J.C.; Tanha, J.; MacKenzie, R. Multivalent Anchoring and Oriented Display of Single-Domain Antibodies on Cellulose. *Sensors* 2009, *9*, 5351-5367.

[48] Webb, A.L.; Islam, N.; Rojas, O.J.; Davis, S.; Russell, H.F.; Champion, T.D. Protein sensor based on peptide ligands immobilized on thin films of cellulose. Abstracts, 63rd Southeast Regional Meeting of the American Chemical Society, Richmond, VA, United States, 2011, October 26-29. Publ. A.C.S. Washington, D. C.

[49] Edwards, J.V.; Caston-Pierre, S.; Howley, P.; Condon, B.; Arnold, J. A bio-sensor for human neutrophil elastase employs peptide-p-nitroanilide cellulose conjugates. *Sensor Lett.* 2008, *6*, 518-523.

[50] Frank, R. Spot-synthesis: an easy technique for the positionally addressable, parallel chemical synthesis on a membrane support. *Tetrahedron*, 1992, *48*, 9217-9232.

[51] a) Schneider-Mergener, J.; Kramer, A.; Reineke, U. In: Combinatorial Libraries: Synthesis, Screening and Application Potential; Cortese, R. Ed.; Walter de Gruyter: Berlin, 1996, pp. 53-68; b) Heine, N.; Ast, T.; Schneider-Mergener, J.; Reineke, U.; Germeroth, L.; Wenschuh, H. Synthesis and screening of peptoid arrays on cellulose membranes. *Tetrahedron*, 2003, *59*, 9919-9930.

[52] a) Reineke, U.; Volkmer-Engert, R.; Schneider-Mergener, J. Applications of peptide arrays prepared by the SPOT-technology. *Curr. Opin. Biotechnol.*, 2001, *12*, 59-64; b) Reineke, U.; Schneider-Mergener, J.; Schutkowski, M. In BioMEMS and Biomedical Nanotechnology, Micro/Nano Technologies for Genomics and Proteomics; Ozkan, M.; Heller, M.J. Eds.; Springer: New York, 2006, Vol. II, pp. 161-282; c) Winkler, D.F.H. Chemistry of SPOT Synthesis for the Preparation of Peptide Macroarrays on Cellulose Membranes. *Mini-Reviews in Organic Chemistry*, 2011, *8*, 114-120, d) Koch, J. SPOT

Peptide Arrays to Study Biological Interfaces at the Molecular Level. *Mini-Reviews in Organic Chemistry*, 2011, *8*, 111-113; e) Breitling, F.; Löffler, F.; Schirwitz, C.; Cheng, Y.-C.; Märkle, F.; König, K.; Felgenhauer, T.; Dörsam, E.; Bischoff, F.R.; Nesterov-Müller, A. Alternative Setups for Automated Peptide Synthesis. *Mini-Reviews in Organic Chemistry*, 2011, *8*, 121-131; e) Maisch, D.; Schmitz, I.; Brandt, O. CelluSpots Arrays as an Alternative to Peptide Arrays on Membrane Supports. *Mini-Reviews in Organic Chemistry*, 2011, *8*, 132-136; f) Reimer, U.; Reineke, U.; Schutkowski, M. Peptide Arrays for the Analysis of Antibody Epitope Recognition Patterns. *Mini-Reviews in Organic Chemistry*, 2011, *8*, 137-146; g) Thiele, A.; Pösel, S.; Spinka, M.; Zerweck, J.; Reimer, U.; Reineke, U.; Schutkowski, M. Profiling of Enzymatic Activities Using Peptide Arrays. *Mini-Reviews in Organic Chemistry*, 2011, *8*, 147-156; h) Hilpert, K. Identifying Novel Antimicrobial Peptides with Therapeutic Potential Against Multidrug-Resistant Bacteria by Using the SPOT Synthesis. *Mini-Reviews in Organic Chemistry*, 2011, *8*, 157-163; i) Volkmer, R.; Tapia, V. Exploring Protein-Protein Interactions with Synthetic Peptide Arrays. *Mini-Reviews in Organic Chemistry*, 2011, *8*, 164-170; j) Kato, R.; Kaga, C.; Kanie, K.; Kunimatsu, M.; Okochi, M.; Honda, H. Peptide Array-Based Peptide-Cell Interaction Analysis. *Mini-Reviews in Organic Chemistry*, 2011, *8*, 171-177; k) Frank, R. The SPOT-synthesis technique. Synthetic peptide arrays on membrane supports-principles and applications. *J. Immunol. Meth.*, 2002, *267*, 13-26; l) Hilpert, K.; Winkler, D.F.H.; Hancock, R.E.W. Peptide arrays on cellulose support: SPOT synthesis - a time and cost efficient method for synthesis of large numbers of peptides in a parallel and addressable fashion. *Nat. Protoc.*, 2007, *2*, 1333-1349.

[53] Chan, W. C.; White, P.D. Fmoc solid phase peptide synthesis. A practical approach, Oxford University Press: Oxford, 2000.

[54] a) Fields, G. B.; Noble, R. L. Solid phase synthesis utilizing 9-fluorenyl-methoxycarbonyl amino acids. *Int. J. Pept. Prot. Res.*, 1990, *35*, 161-214; b) Zander, N.; Gausepohl, H. In Peptide Arrays on Membrane Supports. Synthesis and Applications; Koch, J.; Mahler, M. Eds.; Springer: Heidelberg, 2002, pp. 23-39.

[55] a) Volkmer-Engert, R.; Hoffmann, B.; Schneider-Mergener, J. Stable attachment of the HMB-linker to continuous cellulose membranes for parallel solid phase spot synthesis. *Tetrahedron Lett.*, 1997, *38*, 1029-1032.

[56] Wenschuh, H.; Volkmer-Engert, R.; Schmidt, M.; Schulz, M.; Schneider-Mergener, J.; Reineke, U. Coherent membrane supports for parallel microsynthesis and screening of bioactive peptides. *Biopolymers (Pept. Sci.)*, 2000, *55*, 188-206.

[57] Licha, K.; Bhargava, S.; Rheinländer, C.; Becker, A.; Schneider-Mergener, J.; Volkmer-Engert, R. Highly parallel nano-synthesis of cleavable peptide-dye conjugates on cellulose membranes. *Tetrahedron Lett.*, 2000, *41*, 1711-1715.

[58] a) Ast, T.; Heine, N.; Germeroth, L.; Schneider-Mergener, J.; Wenschuh, H. Efficient assembly of peptomers on continuous surfaces. *Tetrahedron Lett.*, 1999, *40*, 4317-4318; b) Reineke, U.; Ivascu, C.; Schlief, M.; Landgraf, C.; Gericke, S.; Zahn, G.; Herzel, H.; Volkmer-Engert, R.; Schneider-Mergener, J. Identification of distinct antibody epitopes and mimotopes from a peptide array of 5520 randomly generated sequences. *J. Immunol. Methods*, 2002, *267*, 37-51.

[59] a) Hilpert, K.; Elliott, M.; Jenssen, H.; Kindrachuk, J.; Fjell, C.D.; Körner, J.; Winkler, D.F.H.; Weaver, L.L.; Henklein, P.; Ulrich, A.S.; Chiang, S.H.Y.; Farmer, S.W.; Pante, N.; Volkmer, R.; Hancock, R.E.W. Screening and characterization of surface-tethered cationic peptides for antimicrobial activity. *Chem. Biol.*, 2009, *16*, 58-69; b) Tapia, V.; Ay, B.; Triebus, J.; Wolter, E.; Boisguerin, P.; Volkmer, R. Evaluating the coupling efficiency of phosphorylated amino acids for SPOT synthesis. *J. Pept. Sci.*, 2008, *14*, 1309-1314.

[60] a) Volkmer, R. Synthesis and application of peptide arrays: Quo vadis SPOT technology. *ChemBioChem*, 2009, *10*, 1431-1442; b) Winkler, D.F.H.; Hilpert, K. Synthesis of antimicrobial peptides using the SPOT technique. *Methods Mol. Biol.*, 2010, *618*, 111-124.

[61] a) Akita, S.; Umezawa, N.; Kato, N.; Higuchi, T. Array-based fluorescence assay for serine/threonine kinases using specific chemical reaction. *Bioorg. Med. Chem.*, 2008, *16*, 7788–7794. b) Bowman, M.D.; Jeske, R.C.; Blackwell, H.E. Microwave-accelerated SPOT-synthesis on cellulose supports. *Org. Lett.*, 2004, *6*, 2019-2022; c) Lin, Q.; O'Neil, J.C.; Blackwell, H.E. *Org. Lett.* 2005, *7*, 4455-4458; d) Bowman, M.D.; Jacobsen, M.M.; Pujanauski, B.G.; Blackwell, H.E. *Tetrahedron*, 2006, *6*, 2019-2022.

[62] Landgraf, C.; Panni, S.; Montecchi-Palazzi, L.; Castagnoli, L.; Schneider-Mergener, J.; Volkmer-Engert, R.; Cesareni, G. Protein interaction networks by proteome peptide scanning. *PLoS Biol.*, 2004, *2*, 94-103.

[63] a) Rottensteiner, H.; Kramer, A.; Lorenzen, S.; Stein, K.; Landgraf, C.; Volkmer-Engert, R.; Erdmann, R. *Mol. Biol. Cell*, 2004, *15*, 3406–3417; b) Schell-Steven, A.; Stein, K.; Amoros, M.; Landgraf, C.; Volkmer-Engert, R.; Rottensteiner, H.; Erdmann, R. *Mol. Cell. Biol.*, 2005, *25*, 3007–3018; c) Saveria, T.; Halbach, A.; Erdmann, R.; Volkmer-Engert, R.; Landgraf, C.; Rottensteiner, H.; Parsons, M. *Eukaryotic Cell*, 2007, *6*, 1439–1449; d) Pires, J.R.; Hong, X.; Brockmann, C.; Volkmer-Engert, R.; Schneider-Mergener, J.; Oschkinat, H.; Erdmann, R. *J. Mol. Biol.*, 2003, *326*, 1427–1435; e) Halbach, A.; Lorenzen, S.; Landgraf, C.; Volkmer-Engert, R.; Erdmann, R.; Rottensteiner, H. *J. Biol. Chem.*, 2005, *280*, 21176–21182; f) Fest, S.; Huebener, N.; Weixler, S.; Bleeke, M.; Zeng, Y.; Strandsby, A.; Volkmer-Engert, R.; Landgraf, C.; Gaedicke, G.; Riemer, A.B.; Michalsky, E.; Jaeger, I.S.; Preissner, R.; Forster-Wald, E.; Jensen-Jarolim, E.; Lode, H.N. *Cancer Res.*, 2006, *66*, 10567–10575.

[64] a) Hoffmann, S.; Frank, R. *Tetrahedron Lett.*, 1994, 35, 7763–7766; b) Panke, G.; Frank, R. *Tetrahedron Lett.*, 1998, *39*, 17–18.

[65] a) Scharn, D.; Wenschuh, H.; Reineke, U.; Schneider-Mergener, J.; Germeroth, L. *J. Comb. Chem.* 2000, *2*, 361–369; b) Rau, H.K.; DeJonge, N.; Haehnel, W. *Angew. Chem. Int. Ed.* 2000, *39*, 250–253; c) Haehnel, W. *Mol. Diversity* 2004, *8*, 219–229.

[66] Scharn, D.; Germeroth, L.; Schneider-Mergener, J.; Wenschuh, H. *J. Org. Chem.* 2001, *66*, 507–513.

[67] Bowman, M.D.; Jeske, R.C.; Blackwell, H.E. *Org. Lett.* 2004, *6*, 2019–2022.

[68] Bhargava, S.; Licha, K,; Knaute, T.; Ebert, B.; Becker, A.; Grotzinger, C.; Hessenius, C.; Wiedenmann, B.; Schneider-Mergener, J.; Volkmer-Engert, R. *J. Mol. Recogn.* 2002, *15*, 145–153.

[69] Ay, B.; Landgraf, K.; Streitz, M.; Fuhrmann, S.; Volkmer, R.; Boisguerin, P. *Bioorg. Med. Chem. Lett.* 2008, *18*, 4038–4043.

[70] a) Ay, B.; Streitz, M.; Boisguerin, P.; Schlosser, A.; Mahrenholz, C.C.; Schuck, S.D.; Kern, F.; Volkmer, R. *Biopolymers* 2007, *88*, 64-75; b) Ay, B.; Volkmer, R.; Boisguerin, P. *Tetrahedron Lett.* 2007, *48*, 361–364.

[71] Hilpert, K.; Winkler, D.F.H.; Hancock, R.E.W. *Biotechn. Gen. Eng. Rev.* 2007, *24*, 31-106.

[72] Bray, A.M.; Maeji, N.J.; Jhingran, A.G.; Velerio, R.M. *Tetrahedron Lett.* 1991, *32*, 6163-6166; b) Bray, A.M.; Valerio, R.M.; Maeji, N.J. Tetrahedron Lett. 1993, 34, 4411-4414.

[73] a) Arabski, M.; Konieczna, I.; Sołowiej, D.; Rogon, A.; Kaca, W.; Kolesinska, B.; Kaminski, Z.J. Are anti-H. pylori urease antibodies involved in atherosclerotic diseases? *Clinical Biochemistry,* 2010, *43*, 115–123; b) Kaca, W.; Kaminski, Z.J.; Kolesinska, B.; Kwinkowski, M.; Arabski, M.; Konieczna, I. Peptides mimicking urease, methods of manufacturing, application in diagnostic tests and the way of performance the test. Patent Applications PCT/PL2009/000106, WO/2010/071462 from 16 12 2009.

[74] Kaminski, Z.J.; Glowka, M.L.; Olczak, A.; Martynowski, D. *Pol. J. Chem.,* 1996, 70, 1316-1323.

[75] Kaminski, Z.J.; Paneth, P.; Rudzinski, J. *J. Org. Chem.,* 1998, 63, 4248-4255; b) Kaminski, Z.J.; Paneth, P.; O'Leary, M. *J. Org. Chem.,* 1991, 56, 5716-5719.

[76] Kaminski, Z.J.; Kolesinska, B.; Kolesinska, J.; Sabatino, G.; Chelli, M.; Rovero, P.; Błaszczyk, M.; Główka, M. L.; Papini, A. M. „*N*-Triazinylammonium tetrafluoroborates. A New Generation of Efficient Coupling Reagents Useful for Peptide Synthesis" *J. Am. Chem. Soc.,* 2005, *127*, 16912-16920.

[77] Kolesinska, B.; Grabowski, S.; Konieczna, I.; Kaca, W.; Peroni, E.; Papini, A.M.; Rovero, P.; Kaminski, Z.J. „Immunoenzymatic assay with peptide antigens immobilized on cellulose: effect of the linker on antibody recognition". *Peptides 2006. Proceedings of the 29th European Peptide Symposium.* Ed. Rolka, K.; Rekowski, P.; Silberring, J.; Kenes Int., Gdańsk 2006, 616-617 (2007).

[78] a) Letsinger, R. L.; Kornet, M.J. *J Am. Chem. Soc.* 1963, *85*, 3045-3046; b) Letsinger, R. L.; Kornet, M.J.; Mahadevan, V.; Jerina, D. M. ibid. 1964, *86*, 5163-5165; c) Felix. A.M.; Merrifield. R.B. ibid. 1970, *92*, 1385-1391.

[79] Benoiton. N.L.; Lee, Y.C.; Chen, F.M.F. *Int. J Pept Protein Res.,* 1993, *41*, 512-516.

[80] Felix, A.M.; Merrifield, R.B. *J. Am. Chem. Soc.* 1970, *92*, 1385-1391.

[81] Henkel, B.; Zhang, L.S.; Bayer, E. *Liebigs Ann.-Recueil* 1997, 2161-2168.

[82] Thieriet, N.; Guibe, F.; Albericio, F. *Org. Lett.* 2000, *2*, 1815-1817.

[83] Johansson, A.; Akerblom, E.; Ersmark, K.; Lindeberg, G.; Hallberg, A., An Improved Procedure for *N*- to *C*-Directed (Inverse) Solid-Phase Peptide Synthesis. *J. Comb. Chem.* 2000, *2*, 496-507.

[84] a) Gutheil, W.G.; Xu, Q. *N*-to-*C* solid-phase peptide and peptide trifluoromethylketone synthesis using amino acid tert-butyl esters. *Chem. Pharm. Bull.* 2002, *50*, 688-691; b) Sasubilli, R.; Gutheil, W.G. General inverse solid-phase synthesis method for C-terminally modified peptide mimetics. *J Comb. Chem.* 2004, *6*, 911-915.

[85] Kolesinska, B.; Kaminski, Z.J.; Kaca, W.; Grabowski, S. Synthesis and Serological Interactions of H.Pylori Urease Fragment 321-339 *N*-Terminally Immobilized On The Cellulose *Acta Pol. Pharm.,* 2006, *63*, 265-269.

[86] Kaminski, Z.J.; Kolesinska, B.; Kolesinska, J.; Sabatino, G.; Chelli, M.; Rovero, P.; Błaszczyk, M.; Główka, M.L.; Papini, A.M. „*N*-Triazinylammonium tetrafluoroborates.

A New Generation of Efficient Coupling Reagents Useful for Peptide Synthesis *J. Am. Chem. Soc.*, 2005, *127*, 16912-16920.

[87] Kaminski Z.J., Kolesinska B., Cierpucha, M.: Pat. Appl. nr. P- 338931 from 2000, 03, 08.

[88] a) Kania, R.S.; Zuckermann, R.N.; Marlowe, C.K. *J. Am. Chem. Soc.* 1994, *116*, 8835-8836; b) Lebl, M.; Krchnak, V.; Sepetov, N. F.; Nikolaev, V.; Stierrandova, A.; Safar, P.; Seligmann, B.; Stop, P.; Thorpe, P.; Felder, S.; Lake, D.F.; Lam, K.S; Salmon S.E. in *Innovation and Perspectives in Solid Phase Synthesis* (Ed.: R. Epton), Mayflower Worldwide, Oxford, 1994, p. 233.

[89] Davies, M.; Bradley, M. C-Terminally Modified Peptides and Peptide Libraries-Another End to Peptide Synthesis. *Angw. Chem. Int. Ed.*, 1997, *36*. 1097-1099.

[90] Hoffmüller, U.; Russwurm, M.; Kleinjung, F.; Ashurst, J.; Oschkinat, H.; Volkmer-Engert, R.; Koesling, D.; Schneider-Mergener J. Interaction of a PDZ Protein Domain with a Synthetic Library of All Human Protein C Termini. *Angew. Chem. Int. Ed.*, 1999, *38*, 2000-2004.

[91] Boisguerin, P.; Leben, R.; Ay, B.; Radziwill, G.; Moelling, K.; Dong, L.; Volkmer-Engert, R. An Improved Method for the Synthesis of Cellulose Membrane-Bound Peptides with Free C Termini Is Useful for PDZ Domain Binding Studies. *Chem.Biol.*, 2004, *11*, 449–459.

[92] Saint-Remy J.M. Epitope mapping: a new method for biological evaluation and immunotoxicology. *Toxicology*, 1997, *119*, 77-81.

[93] Reineke, U.; Volkmer-Engert, R.; Schneider-Mergener, J. Applications of peptide arrays prepared by the SPOT-technology. *Curr. Opin. Biotechnol.*, 2001, *12*, 59-64.

[94] Zuckermann, R.N.; Kerr, J.M.; Kent, S.B.H.; Moos, W.H. Efficient method for the preparation of peptoids [oligo(n-substituted glycines)] by submonomer solid-phase synthesis. *PNAS*, 1992, *114*, 10646-10647.

[95] Ast, T.; Heine, N.; Germeroth, L.; Schneider-Mergener, J.;Wenschuh, H. Efficient assembly of peptomers on continuous surfaces. *Tetrahedron Lett.*, 1999, *40*, 4317-4318.

[96] Zimmermann, J.; Kühne, R.; Volkmer-Engert, R.; Jarchau, T.; Walter, U.; Oschkinat, H.; Ball, L.J. Design of N-substituted peptomer ligands for EVH1 domains. *J. Biol. Chem.*, 2003, *278*, 36810-36818

[97] Heine, N.; Ast, T.; Schneider-Mergener, J.; Reineke, U.; Germeroth, L.; Wenschuh, H. Synthesis and screening of peptoid arrays on cellulose membranes. *Tetrahedron*, 2003, *59*, 9919-9930.

[98] Hoffmann, B.; Ast, T.; Polakowski, T.; Reineke, U.; Volkmer-Engert, R. Transformation of a biologically active peptide into peptoid analogs while retaining biological activity. *Prot. Pept. Lett.*, 2006, *13*, 829-833.

[99] Gausepohl, H.; Behn, C. A rapid and convenient assay to compare coupling activities of activated compounds. In: *Peptides 1996: Proceedings of the 24th European Peptide Symposium*, eds. R. Ramage and R. Epton, Kingswinford: Mayflower Scientific Ltd. 1998, pp.409-410.

[100] Weiler, J.; Gausepohl, H.; Hauser, N.; Jensen, O.N.; Hoheisel, J.D. Hybridisation based DNA screening on peptide nucleic acid (PNA) oligomer arrays. *Nuc. Acids Res.*, 1997, *25*, 2792-2799.

[101] Blackwell, H.E. Hitting the SPOT: small-molecule macroarrays advance combinatorial synthesis. *Curr. Opin. Chem. Biol.*, 2006, *10*, 203-212.

[102] Baker, A. A.; Helbert, W.; Sugiyama, J.; Miles, M. J. *Biophys. J.* 2000, 79, 1139–1145.

[103] a) Kaemper, A.; Apostolakis, J.; Rarey, M.; Marian, C. M.; Lengauer, T. *J. Chem. Inf. Model.* 2006, *46*, 903–911. b) Zhao, Y.; Sanner, M. F. *Proteins: Struct., Funct., Bioinf.* 2007, 68, 726–737.

[104] Zhou, Z.; Felts, A. K.; Friesner, R. A.; Levy, R. M. *J. Chem. Inf. Model.* 2007, *47*, 1599–1608.

[105] a) Steed, J. W.; Atwood, J. L. *Supramolecular Chemistry* J. Wiley: Chichester, U.K., 2000; pp 19-29. b) Eric, V. *J. Org. Chem.* 2007, *72*, 687–699.

[106] Majchrzak, J.; Fraczyk, J.; Kaminski, Z.J.. "Rate of binding of host molecules to artificial receptors formed by self-organisation of lipidated oligopeptides" *Acta Poloniae Pharm.* 65(6), 2008, 703-708.

[107] Kolesinska, B.; Fraczyk, J.; Papini, A.M.; Kaminski, Z.J. „Sulfonates of N-triazinylammonium salts as highly efficient, inexpensive and environmentally friendly coupling reagents for peptide synthesis in solution".*Chemistry Today (Chimica Oggi)* 2007, *25*, 26-29.

[108] Fraczyk, J.; Kaminski, Z.J. "Designing, synthesis and application of a library of supramolecular structures formed by N-lipidated peptides immobilized on cellulose. Artificial receptors". *J. Comb. Chem.* 2008, 10, 934-940.

[109] Kolesinska, J.; Kolesinska, B.; Wysocki, S.; Kaminski, Z.J. „Nanostructures formed from lipidated oligopeptides immobilized on cellulose as artificial receptors". *Peptides 2006. Proceedings of the 29th European Peptide Symposium, Gdańsk 2006*, Rolka, P. Rekowski and J. Silberring , Eds.; Kenes International, 2007, 774-775.

[110] Fraczyk, J.; Malawska, B.; Kaminski, Z. J. "The application of a library of artificial receptors formed by the self-organization of N-lipidated peptides immobilized on cellulose in studying the effects of the incorporation of a fluorine atom." *J. Comb. Chem.*, 11, 2009, 446-451.

[111] Fraczyk, J.; Kolesinska, B.; Czarnecka, A.; Malawska, B.; Wieckowska, A.; Bajda, M.; Kaminski, Z.J. "The application of a library of artificial receptors formed by the self-organization of N-lipidated peptides immobilized on cellulose for preliminary studies of binding of N-phenylpiperazines". *QSAR & Comb. Sci.* 2009, *28*, 728-736.

[112] Balcerzak, W.; Bednarz, W.; Domosławski, P.; Olewinski, R.; Kolesinska, J.; Kaminski, Z.J.; Dziarkowska, K.; Wieczorek, P. "Preliminary approach towards construction of peptide libraries as potential tools for diagnosis of malignant thyroid tumors" *Endokrynologia Polska (Polish J. Endocrin.)*, 57(4), 2006, 308-313.

[113] Majchrzak, J.; Fraczyk, J.; Kaminski, Z.J.; Kolesinska, B. Design and synthesis of libraries of artificial enzymes, *SMCBS*, 2009, p. P-9.

[114] Fraczyk, J.; Kujawska, N.; Kaminski, Z.J. Artificial esterases formed by self-organization of N-lipidated tripeptides of a catalytic triad immobilized on cellulose. *Proc. 30-th EPS*, 2008, ed. Hilkka Lankinen, Helsinki, Finlandia, p. 598-599.

[115] Fraczyk, J.; Mrozek, A.; Kaminski, Z.J. Structure–activity relationship in binding ligands to library of artificial receptors. The search for biocompatible sensor. *Bioelectrochemistry* 2010, *80*, 2–9.

[116] Fraczyk, J.; Kaminski, Z.J.; Kolesinska, B, in preparation.

Chapter 12

The Development and Application of Cellulose-Based Stationary Phases in Stereoselective Separation of Chiral Pesticides

Jing Qiu, Shouhui Dai, Tingting Chai, Wenwen Yang, Shuming Yang and Hualin Zhao

Additional information is available at the end of the chapter

1. Introduction

In the 1980s, polysaccharide-based chiral stationary phases (CSPs) were identified as versatile and useful chiral sorbents for separation of enantiomers/stereoisomers in high performance liquid chromatography (HPLC). Chiral discrimination abilities of these CSPs can be derived from the highly organized structure of the left-handed 3/2 helical chain conformations [1]. Some chiral cavities with specific configuration can be formed on the CSPs, which provide the suitable site for a particular enantiomer and make it easier to interact with CSPs by hydrogen bonding and π-π interactions. This leads to enantioseparation of chiral compounds by different retention and elution on CSPs between their enantiomers [2]. Okamoto et al. reported that the introduction of various kinds of substituents on the hydroxyl group of polysaccharides can improve their stereoslectivity [3].

Cellulose is an important polysaccharide, it is also a highly crystalline polymer which occurs with various crystal structures. In the 1970s, Hesse and Hagel first synthesized microcrystalline cellulose triacetate (MCTA), and thought its chiral recognition ability might originate from secondary structures creating chiral cavities upon swelling, which can clamp stereoselectively compounds with aromatic residues [4]. In recent years, different cellulose derivatives have been synthesized, coated or covalently bonded on decorative silica gel, and broadly used as CSPs in enantiomeric separation of chiral compounds especially on pesticides and pharmaceuticals. These derivatives exhibit powerful chiral recognition ability towards a wide number of different racemic compounds. More and more commercial cellulose-based CSPs including cellulose acetate, benzoate and phenylcarbamates are being developed and applied in enantioseparation [2,3].

Chiral compounds account for 25% of all agrochemical compounds used commercially and for 26% of the total value of the world agrochemical market [5]. The enantiomers of chiral pesticides possess similar physicochemical properties in a non-chiral environment while they show different activities in biological systems due to enantioselective interactions with enzymes, receptors, and other enantiomeric biological entities. For example, triadimenol is a systemic fungicide and has four stereisomers due to the presence of two chiral centers in its molecule. Of the four, the (1S, 2R)-isomer shows the highest fungicidal activity (up to 1000-fold more active than the other three) [6]. However, most chiral pesticides are produced and formulated as racemic mixture even though the desired biological activity may be derived from only one enantiomer. It is therefore very important to be able to separate enantiomers of chiral pesticides in order to prepare single enantiomers, develop enantiomeric analysis methods and evaluate their bioactivity and environmental fates.

This work focuses mainly on a review of the development of cellulose derivatives for CSPs which are prepared as cellulose-based chiral columns by coating and bonding on supports, and their applications in stereoselective separations of chiral pesticides.

2. The development of cellulose-based CSPs

The cellulose-based CSPs generally are of two types: the coated and the bonded. The coated cellulose-based CSPs consisting of the low-molecular-weight cellulose benzoate or phenyl carbamate showed higher chiral recognition than the covalently bonded CSPs for most racemates. The major reason was considered to be an optimal secondary and supermolecular structure for the chiral recognition mechanism of polysaccharide derivatives under coated conditions [1,3]. However, the coated CSPs can only be used with a limited range of solvents as mobile phases such as alkanes, alcohols, acetonitrile, or aqueous solvents including alcohols or acetonitrile because CSPs may dissolve in 'strong' solvents such as tetrahydrofuran (THF) and chloroform (CHCl₃). Such a dissolution would damage or destroy the CSPs. This limited the application range of the coated CSPs on separation and preparation of chiral compounds, because the solubility of the sample in the mobile phase is very important to increase the amount of racemates loaded on CSPs, especially on a preparative large-scale separation [7].

The bonded CSPs were prepared by covalently bonding cellulose derivates to silica gel. They can be applied to a wider range of resolving conditions than the coated type. The fixation can affect the conformation of cellulose derivates and make it difficult to obtain optimal supermolecular structure. This results in lower chiral recognition ability of the bonded-type CSPs. However, the fixation improves versatility in the solvent selection, and allows the use of some solvents that cannot usually be applied on the coated CSPs as mobile phases or sample dissolving reagents [8].

The commercial cellulose-based CSPs including the coated and the bonded CSPs currently in use are summarized in Table 1. As can be seen, there are only two columns (Chiralpak IB and Chiralpak IC) prepared from cellulose derivatives by bonding out of 13 commercial chiral columns. This means that the coated CSPs include more cellulose derivatives and are more frequently used for the resolution of chiral compounds than the bonded CSPs. Some of these chiral columns can be selectively used in normal-phase HPLC (NP-HPLC), like Chiralcel OD, Chiralcel OA, Chiralcel OB, Chiralcel OC, Chiralcel OF, Chiralcel OG and Chiralcel OJ etc.; some can be used in reversed-phase HPLC (RP-HPLC), like Chiralcel OD-R, Chiralcel OZ-R and Chiralcel OJ-R; and some can be used in both NP-HPLC and RP-HPLC, like Lux Cellulose-1, Lux Cellulose-2 , Lux Cellulose-3, Lux Cellulose-4, Chiralpak IB and Chiralpak IC [9,10]. Some studies have been done to evaluate comparatively the enantioselective and chromatographic properties of Chiralcel OD and Chiralpak IB using a set of 48 compounds that differ in their physical and chemical properties [11]. The uses of these CSPs in different mobile phases mainly depend on their different preparation methods.

No.	Chemical name	Shortened name	Commercial product [9,10]	Type	Chemical structure of cellulose derivative
1	cellulose-*tris*-(3,5-dimethylphenylcarbamate)	CDMPC	Chiralcel OD-H; Chiralcel OD; Chiralcel OD-RH; Chiralcel OD-R; Lux Cellulose-1; Kromasil CelluCoatTM	Coating	
2	cellulose-*tris*-phenylcarbamate	CTPC	Chiralcel OC	Coating	
3	cellulose-*tris*-(4-fluoro-phenylcarbamate)	CFPC	Chiralcel OF	Coating	
4	cellulose-tris(4-chloro-3-methylphenylcarbamate)		Chiralcel OX-H; Lux Cellulose-4	Coating	
5	Cellulose-tris(3-chloro-4-methylphenylcarbamate		Chiralcel OZ-H; Chiralcel OZ-RH; Lux Cellulose-2	Coating	
6	cellulose-*tris*-(4-methylphenylcarbamate)	CMPC	Chiralcel OG	Coating	
7	cellulose-*tris*-(4-methylbenzoate)	CTMB	Chiralcel OJ-H; Chiralcel OJ; Chiralcel OJ-RH; Lux Cellulose-3	Coating	

No.	Chemical name	Shortened name	Commercial product [9,10]	Type	Chemical structure of cellulose derivative
8	cellulose-*tris*-benzoate	CTB	Chiralcel OB-H Chiralcel OB	Coating	
9	cellulose-*tris*-acetate	CTA	Chiralcel OA	Coating	
10	Mricocrystalline cellulose-*tris*-acetate	MCTA	Chiralcel CA-1	Coating	
11	cellulose-*tris*-cinnamate	CTC	Chiralcel OK	Coating	
12	cellulose-*tris*-(3,5-dimethylphenylcarbamate)	Bonded CDMPC	Chiralpak IB	Bonding	
13	cellulose-*tris*-(3,5-dichloro-phenylcarbamate)	Bonded CDCPC	Chiralpak IC	Bonding	

Table 1. The list of commercial cellulose-based CSPs in the present.

2.1. The development of coated cellulose-based CSPs

Various cellulose derivatives were reported as CSPs in recent years, especially on cellulose benzoates and phenylcarbamates because of their higher enantiomeric discrimination ability and wide applications. Okamoto et al, synthesized some cellulose triphenylcarbamate derivatives and absorbed them on silica gel as CSPs, and then compared optical resolution abilities with the characteristics of the substituents on the phenyl rings. The results showed that dimethylphenyl- and dichlorophenylcarbamates substituted at 3,4- or 3,5-positions exhibited better chiral recognition for most reacemates than monosubstituted derivaties. Of the these, cellulose tris-(3,5-dimethylpheyl-carbamate) (CDMPC) offered the highest enantiomeric separability [12]. In another investigation on chiral recognition ability of cellulose phenylcarbamate derivatives, cellulose-tris-(3-fluoro-5-methylphenylcarbamate) was reported to be better than 3,5-difluoro- and 3,5-dimethylphenylcarbamates of cellulose for enantioseparation of ten racemates [13].

The investigations of four regioselectively substituted cellulose derivatives having two different substituents at 2-, 3-, and 6-positions showed better enantioseparations were sometimes obtained on these CSPs, compared to the corresponding homogeneously tris-substituted cellulose derivatives-based CSPs. Cellulose 2,3-(3-chloro-4-methylphenylcarbamate)-6-(3,5- dimethylphenylcarbamate), and 2,3- (3,5-dimethylphenyl-carbamate)-6-(3-chloro-4-methylphenylcarbamate) exhibited the most efficient enantioseparations for tested racemates in four CSPs [14]. The cellulose derivative of benzoylcarbamate also showed a higher chiral discrimination ability compared to those of phenylcarbonate, p-toluenesulfonylcarbamate, and benzoylformate when they used as CSPs on HPLC. This discrimination could be achieved by hydrogen bonding of the racemates' hydrogen atoms with the carbonyl group of the benzoylcarbamates [15].

Chiral recognition abilities of cellulose-methoxyphenylcarbamates were significantly influenced by the position, bulkiness, and number of alkoxy groups introduced on the phenyl group. The 3-position was found to be the best for introducing an alkoxy group, and cellulose-tris-(3-methoxyphenylcarbamates) exhibited much higher recognitions. Additionally, the recognition abilities also increased with the increases of the bulkiness of the 3-alkoxy group [16]. Cellulose-tris- (3-trifluoromethylphenylcarbamate) also exhibited characteristic enantioseparation and were better to resolve some chiral compounds than Chiralcel OD [17].

During the preparation of polymer cellulose-based CSPs by coating on silica gel, chiral additives such as (+)-L-Mandelic acid, (+)-1-phenyl-1,2-ethanediol and (-)-2-phenyl-1-propanol for CSPs of cellulose tribenzoate, and (-)-2-phenyl-1- propanol and (+)-phenylsuccinic for CSPs of cellulose trisphenylcarbamate have a substantial effect on the resolution and efficiency of the CSPs, and can improve chiral recognition ability compared to the original CSPs [18].

Some new supports other than decorative silica gel were also used to prepare the coated CSPs. For example, a new CSP of CDMPC was prepared by coating CDMPC on TiO_2/SiO_2 particles. Its good chiral separation ability and a comparably low column pressure proved that TiO_2/SiO_2 could be used as an alternative to silica gel, and could enlarge the range of base materials when preparing CSP [19].

2.2. The development of bonded cellulose-based CSPs

CDMPC and CDCPC were covalently bonded to decorative silica gel to obtain the bonded chiral columns of Chiralpak IB and Chiralpak IC respectively [9]. CTPC regioselectively bonded at the 6-position to silica gel exhibited a higher chiral recognition than either CTPC regioselectively bonded at the 2- or 3-position or non-regioselectively bonded at the 2-, 3-, and 6-positions [20]. When cellulose derivatives bearing pyridyl and bipyridyl residues were compared in chiral recognition abilities, the results showed that the regioselectively substituted derivatives exhibited higher recognition compared with cellulose derivatives bearing these residues at the 2-, 3- and 6-positions of a glucose ring. This ability was significantly influenced by the coordination of Cu(II) ion to the bipyridyl groups that resulted in the difference of the higher-order structures of cellulose derivatives [21].

CSP with poly[styrene-*b*-cellulose 2,3-bis-(3,5-diphenylcarbamate)] was prepared by the surface-initiated atom transfer radical polymerization (SI-ATRP) of cellulose 2,3-bis-(3,5-dimethylphenylcarbamate)-6-acrylate after the SI-ATRP of styrene on the surface of silicon dioxide supports in pyridine. This CSP showed considerably high column efficiency for the resolution of tested racemates [22].

Laureano Oliveros et al, prepared five mixed 10-undecenoate/benzoates of cellulose and linked them to allyl silica gel by means of a radical reaction. The investigation of chiral recognition ability showed that CSP5 (10-undecenoate/3,5-dichlorobenzoate) has the highest enantioselectivity for most of tested racemates, followed by CSP3 (10-undecenoate/4-methylbenzoate) and CSP4 (10-undecenoate/benzoate). These CSPs showed lower resolution than the coated CSPs although they have higher column efficiency. The reason may be the lack of polar amino groups on the surface of the CSPs. However, when being compared with the coated CSPs, these CSPs can tolerate the use of more polar solvents such as chloroform in the mobile phase [23].

Three cellulose-based CSPs were prepared by reticulation of the same cellulose derivative on three end-capped silica gels with different pore sizes (50Å, 100Å and 4000Å). The comparison of chiral recognition ability among them showed that CSPs with higher pore size exhibited higher selectivity factors, because it can accommodate a larger amount of accessible cellulose derivative on its surface [7].

Four mixed 10-undecenoyl-3,5-dimethylphenylaminocarbonyl derivatives of cellulose with increased proportion of alkenoyl groups were bonded on allylsilica gel. Their comparison showed that CSPB presents the best chiral recognition and can separate the widest range of the tested racemates. The reason may be the higher number of substitution of glucose units. The important decrease in the recognition ability of these CSPs could be attributed to their higher degree of reticulation. More heterogeneous reaction sites of allysilica gel with cellulose derivatives can result in lower degree of reticulation in CSPs and therefore improve their recognition ability [24].

Azido cellulose phenylcarbamate (AzCPC) was synthesized regioselectively and chemically immobilized onto amino-functionalized silica gel to obtain urea-bonded CSPs. Enantioseparation using CHCl₃ on these CSPs showed better separation than traditional hexane/2-propanol in mobile phases for some tested racemates. The pre-coating of AzCPC onto silica gel prior to chemical immobilization could significantly improve immobilization efficiency, and obtained better enantioselectivity [25].

3. The preparation method of cellulose-based CSPs

3.1. The preparation method of coated CSPs

Generally, benzoate and phenylcarbamate derivatives of cellulose were prepared by reaction between cellulose and excess benzoyl chloride or phenyl isocyanate derivatives in dry pyridine (Figure 1). These derivatives are then coated onto macro-porous 3-aminopropylsilica (APS) from a solution by evaporation of the solvent to obtain coated

CSPs. The APS was prepared beforehand by silanizing silica gel with a solution of 3-aminopropyltriethoxysilane. Finally, the CSPs were packed into HPLC columns by the slurry method, to obtain coated chiral columns [18, 26]. For example, CDMPC was synthesized by reaction of microcrystalline cellulose with 3,5-dimethylphenylcarbimide in pyridine; the product was filtered off, washed with methanol and dried at 60° C for 24h. CDMPC was then dissolved in THF and coated on the APS under vacuum to dryness. Finally, the coated CDMPC were packed into a stainless-steel column at 3.7×10^7Pa by the high-pressure slurry method to obtain the corresponding CSP [26].

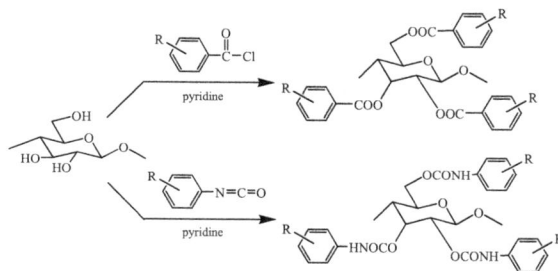

Figure 1. The synthesized routes of cellulose benzoates or phenylcarbamates.

Investigations on the influence of the pore size of silica gel, the coating amount , the coating solvent, and the column temperature on chiral discrimination of CDMPC showed that CSPs prepared with a large-pore silica gel having a small surface area exhibited higher recognition abilities. An increase in the amount of coating of CDMPC on the silica gel can improve the loading capacity of racemates, and a CSP coated with 45% CDMPC by weight can be used for both analytical scale and semi-preparative scale separations. CSPs coated with acetone showed higher enantioselectivity than those coated with THF or a mixture of CH_2Cl_2 and phenol [27].

3.2. The preparation method of covalently bonded CSPs

Generally, cellulose-derived CSPs covalently bonded on silica gel are prepared by using a benzoyl chloride or a phenyl isocyanate to react with cellulose in homogeneous conditions, to obtain the corresponding benzoates or carbamates. However, other methods to prepare this type of CSP have been reported. Ikai et al. summarized various immobilization methods of the polysaccharide derivatives mainly onto silica gel: immobilization using diisocyanate, vinyl groups by polymerization and copolymerization with a vinyl monomer etc. [28,29]. Several methods of synthesis are shown in Figures 2 to 4.

CDMPC can be efficiently immobilized on silica gel as CSPs by copolymerizing with vinyl monomers. The introduction of vinyl groups or the employment of vinyl monomers can readily tune the immobilization efficiency and the chiral recognition of cellulose derivatives [30]. The new method was applied to immobilize CDMPC onto bare silica gel via the intermolecular polycondensation of triethoxysilyl groups, which were introduced onto the glucose unit by the epoxide ring-opening reaction under acidic conditions. The CSPs thus obtained also exhibited high chiral recognition ability for 10 tested racemates and could be used with various eluents that are not compatible with the conventionally coated CSPs [31]. One-pot method was applied to synthesize CDCPC bearing a small amount of 3-(triethoxysilyl) propyl residues, and then immobilized onto silica gel through intermolecular polycondensation. The immobilized CSPs exhibited chiral recognition abilities similar to the corresponding coated CSP and slightly different from the commercial Chiralpak IC [32].

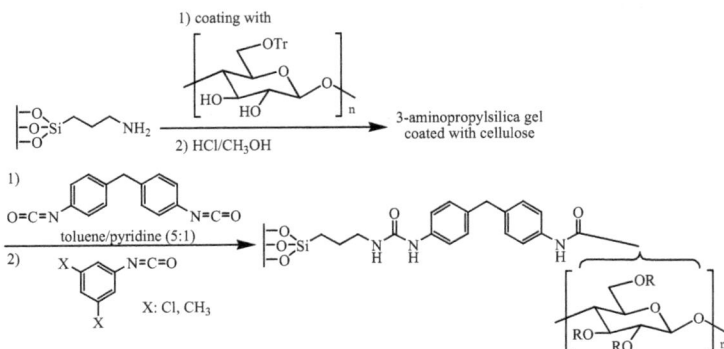

Figure 2. The covalent bonding of 3,5-dichloro- and 3,5-dimethylphenylcarbamate of cellulose onto APS [33].

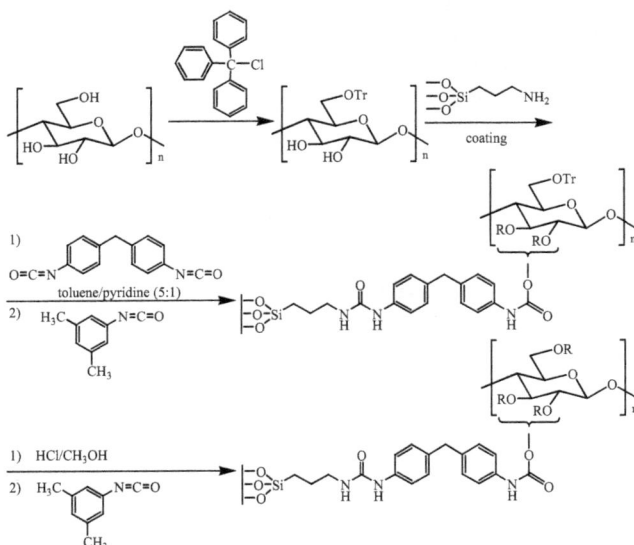

Figure 3. Regioselective covalent bonding of CDMPC to positions 2 and 3 of the glucosidic rings.

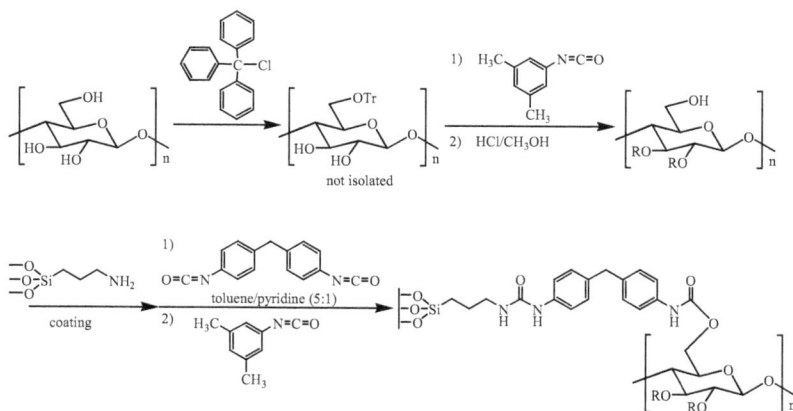

Figure 4. Regioselective covalent bonding of CDMPC to position 6 of the glucosidic rings [34].

Cellulose-(diphenymethyldicarbamate/phenylcarbamate) covalently bonded to APS showed some chiral recognition ability [35]. Cellulose-tris-phenylcarbamate was covalently bonded to silica gel with different spacers. The results showed CSPs prepared with spacer 1(4-(1-(3-(triethoxysilyl)-propyl)urea)-benzyl-4-isocyanatobenzene) exhibited higher resolution ability than spacer TEPI (3-(triethoxysilyl) propyl isocyanate) with the same preparation procedure. The amount of spacer in the synthesis influences the optical resolution ability of CSPs, and a lower amount can produce higher resolution ability [36].

Polar monodisperse amine terminated polymer (2-aminoethyl methacrylate-co-ethylenedimethacrylate) beads can be used as the replacement of silica gel, and are suitable as supports for the preparation of cellulose-based CSPs coated by simple adsorption and immobilized with a diisocyanate linker. However, the chiral recognition abilities of these CSPs shows no enhancement because the uses of cellulose-based selectors and preparation methods may completely cover the surface of polymer supports. Thus, the analytes have no access to the native surface of the support and non-specific interactions with the surface functionalities are not observed. [37].

4. The application of cellulose-based CSPs in enantioseparation of chiral pesticides

Chiral HPLC is a good method to separate enantiomers/stereoisomers of chiral pesticides because it facilitates the preparation of single enantiomers for study of enantiomeric bioactivity, toxicology and environmental fate. In recent years, cellulose-based CSPs prepared with different cellulose derivatives and methods resulted in their very broad application for chiral separation of pesticides such as organophosphates [38], organochlorine, triazole, synthetic pyrethroids, acylanilides, imidazolinones, phenoxypropanoic-acid herbicides and related compounds [39].Table 2 summarizes the resolution results of 79 chiral pesticides in current references.

As shown in Table 2, the stereoselective separations of most of chiral pesticides can be achieved on NP-HPLC and some on RP-HPLC using cellulose-based CSPs. The most efficient CSP with the highest chiral recognition ability is CDMPC, available under the commercial names of Chiralcel OD, Chiralcel OD-H, Chiralcel OD-R, Chiralcel OD-RH, Lux Cellulose-1 and Kromasil CelluCoatTM. The coated CDMPC on APS exhibited higher chiral discrimination for most of pesticides than the bonded type available under the commercial names of Chiralpak IB and Chiralpak IC. For example, the resolution factor (Rs) of systemic fungicide-metalaxyl on the coated CDMPC is 4.54 with hexane/IPA (80:20) as the mobile phase, which is significantly higher than that on the bonded CDMPC with an Rs of 0.632 using hexane/IPA (97/3) as the mobile phase.

The second most efficient CSP in terms of resolution is CTMB available under the commercial names of Chiralcel OJ, Chiralcel OJ-H, Chiralcel OJ-RH, Lux Cellulose-3. It exhibited higher chiral discrimination for some chiral pesticides than CDMPC. For example, the Rs of triazole fungicide-imazalil on Chiralcel OJ-H is 5.21, which is significantly higher than 1.51 obtained on Chiralcel OD-H using the same mobile phase of hexane/IPA (100/3) and the same flow rate of 0.8 mL/min on NP-HPLC. The combination of CDMPC and CTMB on NP-HPLC and RP-HPLC can separate most chiral pesticides listed in Table 2.

The separations on NP-HPLC were better than those on RP-HPLC for most chiral pesticides. The cellulose-based CSPs on NP-HPLC can generally give better resolution and yield a larger amount of a single enantiomer in one injection. However, its application is limited because some racemates are polar and difficult to dissolve in the weak polar solvents used as mobile phase on NP-HPLC. For this reason, the amount of racemates loaded on CSPs cannot be increased. The separation on RP-HPLC is sometimes less effective than on NP-HPLC, but it can use more methanol, acetonitrile or water in the mobile phase and can thus significantly improve the solubility of some racemates that will not readily dissolve in the hexane, heptane and isopropanol used in NP-HPLC. This is very helpful to prepare optically pure enantiomer of polar chiral compounds and obtain more enantiomer in a shorter time. Additionally, the use of HPLC in the reversed phase can easily be connected in tandem with mass spectrometry, which makes it possible to establish more sensitive and more efficient analytical methods for enantioselective studies of chiral pesticides [40-42].

No.	Pesticide	CSP or Chiral colum	Chromatographic condition[*1]	Separation effect[*2]	Elution order[*3]	Reference
1	amiprophos	Chiralcel OJ-H	hexane/IPA(100/5); 0.8mL/min; UV 254nm	Rs: 1.65		[43]
		Chiralcel OD-H	hexane/IPA(100/5); 0.8mL/min; UV 254nm	-		[43]
2	benalaxyl	CDMPC	hexane/IPA(97/3); 1.0 mL/min; UV 22nm	Rs>1.5	R-(-) /S-(+)	[44]
		ChiralpakIB; Chiralcel OJ-H	hexane(IPA or ethanol); 0.5 mL/min; UV 220 nm;			[45]
3	benzex	Chiralcel OJ	hexane/IPA(91/9); 0.5 mL/min			[46]

No.	Pesticide	CSP or Chiral colum	Chromatographic condition[*1]	Separation effect[*2]	Elution order[*3]	Reference
4	bifenthrin	Chiralcel OJ-H	hexane/ethanol(98/2); 1.0 mL/min; CD 230nm			[47]
5	bioallethrin	CDMPC	hexane/ethanol(99/1); 1.0 mL/min;	α: 1.27		[48]
		CMPC	hexane/ethanol(99/1); 1.0 mL/min;	α: 1.39		[49]
6	bitertanol	Chiralcel OD-H	hexane/IPA(100/3); 0.8mL/min; UV 254nm	Rs: 1.52		[50]
		Chiralcel OJ-H	hexane/IPA(100/10); 0.8mL/min; UV 254nm	Rs: 3.70		[50]
7	carfentrazone-ethyl	CDMPC	hexane/IPA(99.9/0.1); 1.0 mL/min; UV 230nm	Rs: 0.52		[51]
8	chlordane	Chiralcel OD	hexane; 1.0 mL/min		OR: TC(trans)+/- CC(cis)+/-	[52,53]
9	crotoxyphos	Chiralcel OJ	hexane/ethanol(90/ 10); 0.8mL/min; UV 230nm	Rs: 1.81	OR: -/+	[54,55]
10	crufomate	Chiralcel OD	heptane/ethanol(90/10); 1.0 mL/min;	Rs: 1.1	OR: +/-	[55]
		Chiralcel OJ	heptane/ethanol(99.4/ 0.6); 0.3mL/min; UV 203nm	Rs: 0.90	OR: -/+	[55]
11	cycloprothrin	Chiralcel OJ-H	hexane /IPA(70/30), 35°C, 1.0 mL/min, UV 254 nm			[56]
		Chiralcel OD-H	Hexane/IPA(90/10), 35°C, 1.0 mL/min, UV 254 nm			[56]
12	cypermethrin	CDMPC	hexane/IPA (90/10); 0.5mL/min; UV 230nm	seven peaks		[57]
13	alpha-cypermethrin	CDMPC	hexane/IPA (90/10); 0.5mL/min; UV 230nm	Rs: 1.53		[57]
14	theta-cypermethrin	CDMPC	hexane/IPA(99/1); 0.8mL/min; UV 230nm	Rs: >1.5	OR: -/+	[58,57]
15	beta-cypermethrin	CDMPC	hexane/IPA (99/1); 0.5mL/min; UV 230nm	four peaks		[57]
16	dialifos	Chiralcel OJ	heptane/ethanol(90/ 10); 0.9mL/min; UV 220nm	Rs: 3.12	OR: +/-	[55]
17	dichlorprop	Chiralcel OJ-H	hexane/IPA (90/10); 0.5mL/min; UV 228nm	Rs: 1.34	S/R	[59]
18	diclofop-methyl	CDMPC	hexane/IPA (95/5); 0.5mL/min; UV 270 nm	Rs: 11.8	S/R	[60-62]
		CDMPC	hexane/n-butyl alcohol (84/16); 0.5mL/min; UV 280 nm			[63]

No.	Pesticide	CSP or Chiral colum	Chromatographic condition[*1]	Separation effect[*2]	Elution order[*3]	Reference
		CDMPC	hexane/isobutanol (98/2); 1.0 mL/min; UV 230nm	Rs: 6.15	OR: -/+	[64,39,65]
		CDMPC coated on TiO2/SiO2	hexane/IPA(65/35), 1.0 mL/min	Rs: 1.50		[19]
		CDMPC	ACN/water (50/50); 0.8 mL/min; UV 230 nm	Rs: 1.53	OR: -/+	[66]
		CTMB	hexane/IPA (50/50); 0.5mL/min; UV 254 nm	Rs: 1.68	R/S	[67,68,63]
		CTB	hexane/n-butyl alcohol (84/16); 0.5mL/min; UV 280 nm			[63]
		CTPC	hexane/n-butyl alcohol (84/16); 0.5mL/min; UV 280 nm			[63]
		Chiralcel OJ-H	hexane/IPA/acetic acid (90/10/0.2); 0.5mL/min; CD 282nm	Rs: 5.49	R/S	[69,70]
19	diclofop acid	Chiralcel OJ-H	hexane/IPA/acetic acid (90/10/0.2); 0.5 mL/min; UV 230 nm			[70]
20	difenoconazole	Chiralcel OJ	hexance/ethanol(90/10); 0.6 mL/min; UV 230nm.	Rs: 3.79	OR: +/-/+/-	[71]
21	diniconazole	CDMPC; Chiralcel OD	hexane/n-butyl alcohol(98/2); 1.0 mL/min; UV 220nm	Rs: 1.53	OR: +/-	[72,61]
		Chiralcel OD	hexane/IPA(90/10); 0.6mL/min; UV 253nm	Rs: 1.17	OR: +/-	71
		Chiralcel OD-H	hexane/IPA(100/5); 1.0 mL/min; UV 225nm	α: 1.20	R(-)/S(+)	[73,50]
		Chiralcel OJ; Chiralcel OJ-H	hexane/IPA(100/3); 1.0 mL/min; UV 225nm	α: 1.14	R(-)/S(+)	[73,74]
		Lux Cellulose-1	ACN/water(70/30), MET/water(80/20); 1.0 mL/min; UV 220nm	Rs: 2.31, 2.62	OR: -/+	[75,66]
22	dioxabenzofos	Chiralcel OJ	hexane/IPA(95/5); 1.0 mL/min; UV 220nm	Rs: 1.56	OR: -/+	[76]
		Chiralcel OD	hexane/IPA(99.5/0.5); 1.0 mL/min; UV 220nm	Rs: 1.42	OR: -/+	[76]
23	epoxiconazole	Lux Cellulose-1; CDMPC	ACN/water(50/50), MET/water(80/20); 1.0 mL/min; UV 220nm	Rs: 2.04, 1.62	OR: −/+	[75,66]
24	ethofumesate	CDMPC	hexane/IPA (98/2); 1.0 mL/min; UV 230nm	Rs: 6.34		[77]
			hexane/IPA (93/7); 1.0 mL/min; UV 230nm	α: 1.58	OR; +/-	[78,79]

No.	Pesticide	CSP or Chiral colum	Chromatographic condition[*1]	Separation effect[*2]	Elution order[*3]	Reference
			hexane/isobutanol (95/5); 1.0 mL/min; UV 230nm	Rs: 7.05	OR: +/-	[64,80]
25	fenamiphos	Chiralcel OJ	heptane/ethanol(99.1/ 0.9); 0.5mL/min; UV 203nm	Rs: 1.08	OR: +/-	[55]
		CDMPC	ACN/water(70/30); 0.8mL/min; UV 230nm	α: 1.00		[81]
26	fenbuconazole	Lux Cellulose-1	ACN/water(90/10), MET/water(70/30); 1.0 mL/min; UV 220nm	Rs: 4.79, 3.96	OR: +/-	[75]
27	fenoxaprop-ethyl	CDMPC	hexane/ethanol (93/7); 0.5mL/min; UV 290nm	Rs: 1.83		[61]
			MET/water (80/20); 0.8mL/min; UV 265nm	Rs: 1.01	OR: +/-	[66]
			ACN/water (50/50); 0.8mL/min; UV 230nm	Rs: 1.53	OR: -/+	[66]
28	fensulfothion	Chiralcel OJ	heptane/ethanol(96/4); 0.8mL/min; UV 201nm	Rs: 1.21	OR: -/+	[55]
29	fenthiaprop	CDMP	ACN/water (50/50); 0.8mL/min; UV 230nm	Rs: 1.53	OR: -/+	[66]
30	fipronil	Chiralcel OD	isooctane/IPA(96/6); 6.0 mL/min;			[82]
		CDMPC	hexane/IPA(95/5); 1.0 mL/min; UV 230nm			[83]
31	flamprop-methyl	CDMPC	hexane/ IPA(97/3); 1.2mL/min; UV 230nm	Rs: 1.59	R/S	[84]
32	fluazifop-butyl	CDMPC	hexane/n-butyl alcohol (89/11); 0.5mL/min; UV 270nm	Rs: 2.55	S/R	[61]
33	fluazifop-p-butyl	CDMPC	hexane/ IPA (95/5); 0.5mL/min; UV 251nm	Rs: 3.80	S/R	[60]
		CHIRALPAK IC	hexane/IPA(90/10); 1.0 mL/min; UV 254nm			[85]
34	fluroxypyr-meptyl	CDMPC	hexane/ IPA(99/1); 0.5 mL/min; UV 230nm	Rs: 1.31		[86]
		CDMPC	MET/water(80/20); 0.5mL/min; UV 230nm	Rs: 1.07	OR: +/-	[66]
35	flutriafol	Chiralcel OD; Chiralcel OD-H; CDMPC	hexance/IPA(95/5); 0.6mL/min; UV 230nm	Rs: 1.37	OR: -/+	[71,50, 64]
		Lux Cellulose-1	ACN/water(70/30), MET/water(70/30); 1.0 mL/min; UV 220nm	Rs: 1.99, 1.39	OR: -/+	[75]
36	tau-fluvalinate	Chiralcel OJ	hexane/ethanol(90/10); 0.3mL/min; UV 210 nm	Rs: 1.59		[87]

No.	Pesticide	CSP or Chiral colum	Chromatographic condition[*1]	Separation effect[*2]	Elution order[*3]	Reference
		Chiralcel OG	hexane/IPA	Rs: <0.91		
		Chiralcel OD-R	MET/water; UV 210 nm	Rs: <0.91		
37	fonofos	Chiralcel OJ	heptane/ethanol(99.5/0.5);1.0 mL/min; UV 202nm	Rs: 2.1	OR: +/-	[55,54]
		Chiralcel OJ-H	hexane/IPA(100/10); 0.8mL/min; UV 254nm	Rs: 9.58		[43]
		Chiralcel OD-H	hexane/IPA(100/0.5); 0.8mL/min; UV 254nm	-		[43]
38	heptachlor epoxide	Chiralcel OD	hexane; 1.0 mL/min; UV 215nm			[53]
39	hexaconazole	CDMPC; Chiralcel OD	hexance/IPA(91/9); 0.5mL/min; UV 270.9nm	Rs: 4.79	OR: +/-	[61,66, 71,72]
		CDMPC	hexance/ tertiary butanol (95/5); 0.5mL/min; UV 270nm	Rs: 2.30		[88]
		Chiralcel OD-H	ACN/MET(98/2); 0.5mL/min; UV 254nm	Rs: 1.51		[89]
		Lux Cellulose-1	ACN/water(90/10), MET/water(80/20); 1.0 mL/min; UV 220nm	Rs: 2.25, 2.12	OR: +/-	[75]
40	imazalil	Chiralcel OD-H	hexane/IPA(100/3); 0.8mL/min; UV 220nm	Rs: 1.51		[50]
		Chiralcel OJ-H	hexane/IPA(100/3); 0.8mL/min; UV 220nm	Rs: 5.21		[50]
		Chiralcel OD	ACN/water(50/50); 0.8mL/min; UV 240nm	Rs: 0.91	OR: −/+	[66]
41	imazamox	Chiralcel OD-R	ACN/ PBS buffer(50mM)(20/80); 1.0 mL/min			[39]
		Chiralcel OJ	hexane(0.1%TFA)/IPA(60/40)	Rs: 0.89		[90]
42	imazapic	Chiralcel OJ	hexane/ alcohol/TFA (75/25/0.1); 1.0 mL/min; UV 254nm		OR: +/-	[90]
43	imazapyr	Chiralcel OJ	hexane/ IPA/acetic acid (84.6/15.4/0.1); 0.8mL/min; UV 275nm		OR: +/-	[91]
44	imazaquin	Chiralcel OJ-H	Hexane/IPA/Acetic acid(84.6/15.4/0.1); 0.8 mL/min; UV 275 nm		CD: +/-	[91,39]
		Chiralcel OD-R	ACN/ PBS buffer(50mM)(20/80); 1.0 mL/min	Rs: 2.44		[90]
45	imazethapyr	Chiralcel OJ	hexane/ethanol/ acetic acid (75/25/0.5); 1.0 mL/min; UV 250nm		OR: +/-	[92]

No.	Pesticide	CSP or Chiral colum	Chromatographic condition[*1]	Separation effect[*2]	Elution order[*3]	Reference
		Chiralcel OJ	hexane/ IPA/acetic acid (84.6/15.4/0.1); 0.8mL/min; UV 275nm			[90]
46	indoxacarb	Lux cellulose-1; Chiralcel OD	hexane/IPA(85/15) 0.8mL/min; UV 310 nm		OR: -/+	[93,94]
47	isocarbophos	CDMPC	hexane/IPA(98/2); UV 225nm	Rs: 2.42	OR: -/+	[64,51,95]
48	isofenphos	Chiralcel OG	heptane/IPA(98/2); 1.0 mL/min;	Rs: 1.1	OR: +/-	[55]
		Chiralcel OJ	heptane/ethanol(99.4/ 0.6); 0.3mL/min; UV 201nm	Rs: 1.11	OR: +/-	[55]
49	isofenphos-methyl	Chiralcel OJ-H	hexane/IPA(100/1); 0.8mL/min; UV 280nm	Rs: 1.59		[81]
		Chiralcel OD-H	hexane/IPA(100/1); 0.8mL/min; UV 280nm	Rs: 1.73		[43]
		CDMPC	ACN/water(70/30); 0.8mL/min; UV 230nm	α: 1		[81]
50	iso-malathion	Chiralcel OJ	hexane/IPA(97/3); 1.0 mL/min; UV 220nm			[66]
51	lactofen	CDMPC	hexane/IPA(99/1); 1.0 mL/min; UV 230nm	Rs: 1.87	OR: +/-	[64, 39]
		CDMPC	MET/water(75/25); 0.8mL/min; UV 265nm	Rs: 1.07	OR: -/+	[66]
		Chiralpak IC	hexane/ CH_2Cl_2/TFA (65/35/0.1)	Rs: 8.11		[96]
52	lambda-cyhalothrin	Chiralecl OD	Hexane/IPA(95/5), 0.5 mL/min; UV 236 nm		CD: -/+	[97]
		Chiralecl OJ	hexane; ethanol (95/5); 0.6 mL/min, UV 236 nm		CD: -/+	[97]
		Chiralecl OJ	Hexane/IPA(90/10); 0.4 mL/min, UV 236 nm		CD: -/+/+/-	[97]
53	malaoxon	Chiralcel OJ	hexane/IPA(96/4); 1.0 mL/min; UV 220nm	Rs: 4.06	R/S OR: +/-	[98]
54	malathion	Chiralcel OJ	heptane/ethanol(90/ 10); 0.9mL/min; UV 220nm	Rs: 4.11	OR: +/-	[55]
		Chiralcel OJ	hexane/IPA(97/3); 1.0 mL/min; UV 220nm	Rs: 3.35	OR: +/-	[98]
		CDMPC	hexane/IPA(99/1); 1.0 mL/min; UV 210nm	Rs: 1.44	OR: +/-	[65]
		CDMPC	ACN/water(70/30); 0.8mL/min; UV 230nm	α: 1.0		[81]
55	metalaxyl	CDMPC	hexane/IPA(80:20); 1.0 mL/min; UV 230nm	Rs: 4.54		[26,99]
		CDMPC coated on TiO_2/SiO_2	hexane/IPA(65/35); 1.0 mL/min	Rs: 2.97		[19]

No.	Pesticide	CSP or Chiral colum	Chromatographic condition*1	Separation effect*2	Elution order*3	Reference
		ChiracelOJ-H	hexane/IPA(90:10); 0.5 mL/min; CD 236 nm		S/R	[100]
		Bonded CDMPC	MET/water(50/50)	Rs: 0.506		[101]
		Bonded CDMPC	ACN/water(80/20)	Rs: 0.766		[101]
		Bonded CDMPC	hexane/IPA(97/3)	Rs: 0.632		[101]
		Bonded CDMPC	hexane/tertbutyl alcohol (95/5)	Rs: 0.918		[101]
56	metalaxyl acid	CDMPC	hexane/IPA/TFA(70/30/0.1%); 1.0 mL/min	Rs: 1.96	CD(-)/(+)	[102]
57	metalaxyl intermediate	CDMPC	hexane/IPA(99/1); 1.0 mL/min;	Rs: 1.85	CD (-)/(+) at 228 nm; CD(+)/(-) at 280nm	[102,26]
58	methami-dophos	CDMPC	hexane/IPA(90/10); 1.0 mL/min; UV 230nm	Rs: 1.54	OR: +/-	[65,103]
		Chiralcel OD	heptane/ethanol(90/ 10); 1.0 mL/min;	Rs: 1.7	OR: +/-	[55]
		Chiralcel OJ	heptane/ethanol(93.5/6.5); 0.8mL/min; UV 200nm	Rs: 1.56	OR: +/-	[55]
		CDMPC	ACN/water(70/30); 0.8mL/min; UV 230nm	α: 1.0		[81]
59	metolachlor	Chiralcel OD-H	hexane/diethyl ether (91/9); 0.8 mL/min; UV 230 nm			[104]
60	myclobutanil	CDMPC	hexane/IPA(73/26); 0.5mL/min; UV 221.5nm	Rs: 13.3		[61]
		CTPC	hexane/IPA(73/26); 0.5mL/min; UV 221.5nm	Rs: 1.54		[50]
		Lux Cellulose-1; CDMPC; Chiralcel OD	ACN/water(90/10), MET/water(90/10); 1.0 mL/min; UV 220nm	Rs: 5.10, 4.91	OR: +/-	[75, 66]
61	naproanilide	CDMPC	hexane/IPA(80/20); 1.0 or 0.5mL/min	Rs: 1.91		[105]
		Chiralcel OD-H; Chiralcel OJ-H	hexane; 1.0 mL/min; UV 254 nm		OR: +/−	[106]
		Bonded-CTB	hexane; 1.0 mL/min; UV 254 nm		OR: −/+	[106]
62	napropamide	Chiralpak OJ-H	hexane/IPA(80/20); 0.5mL/min; 40°C; UV 220nm			[107]
63	paclobutrazol	CDMPC	ACN/water(40/60); 0.8 mL/min; UV 230nm.	Rs: 1.93	OR: +/−	[108]
		Chiralcel OD; CDMPC	hexance/IPA(100/2); 0.8 mL/min; UV 225nm.	Rs: 1.83		[50, 61]
		OJ	hexance/IPA(100/10); 0.8 mL/min; UV 225nm.	Rs: 4.05		[50]

No.	Pesticide	CSP or Chiral colum	Chromatographic condition[*1]	Separation effect[*2]	Elution order[*3]	Reference
64	penconazole	Lux Cellulose-1 Chiralcel OD-H	ACN/water(50/50), MET/water(90/10); 1.0 mL/min; UV 220nm	Rs: 7.58, 2.29	OR: -/+	[75,89]
65	permethrin	Chiralcel OJ	hexane/ethanol(95/15); 0.3mL/min; UV 210 nm	Rs: 1.47		[87]
		Chiralcel OD-R	MET/water; UV 210 nm	Rs: <0.91		
66	phenthoate	Chiralcel OJ	hexane/IPA(90/10); .6mL/min; UV 230nm		OR: -/+	[109]
		CDMPC	ACN/water(70/30); 0.8mL/min; UV 230nm	α: 1.0		[81]
67	profenofos	CDMPC	hexane/IPA(99.5/0.5); UV 210nm	Rs: 1.35	OR: +/-	[64]
		Chiralcel OJ	heptane/ethanol(99.5/0.5); 1.0 mL/min; UV 202nm	Rs: 3.52	OR: +/-	[55]
		Chiralcel OJ	hexane; 0.8mL/min; UV 230nm	Rs: 1.12	OR: +/-	[54]
		CDMPC	ACN/water(70/30); 0.8mL/min; UV 230nm	α: 1.0		[81]
68	propiconazole	Chiralcel OD	hexance/IPA(90/10); 0.6 mL/min; UV 230nm.	Rs: 2.95/ 2.72/ 1.04	OR: +/-/+/-	[71]
69	prothiophos	Chiralcel OJ	heptane/ethanol(98/2); 15°C; 1.0 mL/min; UV 202nm	Rs: 1.6	OR: +/-	[55]
70	quizalofop-P-ethyl	CDMPC	hexane/NPA (91/9); 0.5mL/min; UV 332nm	Rs: 1.7	R/S	[61]
		Chiralcel OJ-H	hexane/MET/methylene dichloride(450/2/8); 1.0 mL/min; UV 290 nm		S/R	[110]
71	pyraclofos	Chiralcel OD	hexane/IPA(90/10); 1.0 mL/min; UV 254nm		OR: -/+	109
72	tebuconazole	CDMPC; Chiralcel OD	hexane/IPA(98/2); 1.0 mL/min; UV 220nm	Rs: 1.63	OR: -/+	[50,61, 64, 71]
		Chiralcel OJ-H; CTMB	hexane/IPA(100/10); 0.8mL/min; UV 225nm	Rs: 5.64		[50, 61]
		CTPC	hexane/IPA(91/9); 0.6mL/min; UV 269.8nm	Rs: 1.16		[61]
		Chiralcel OD-H	ACN/IPA(70/30); 0.5mL/min; UV 254nm	Rs: 0.67		[89]
73	tetraconazole	Lux Cellulose-1	ACN/water(90/10); 1.0 mL/min; UV 220nm	Rs: 1.39	OR: +/-	[75]
74	triadimefon	CDMPC; Chiralcel OD	hexane/IPA(99/1); 1.0 mL/min; UV 230nm	Rs: 1.47	OR: -/+	[64, 71]
		Chiralcel OD	hexane/IPA(100/5); 1.0 mL/min; UV 225nm	α: 1.20	R(-)/S(+)	[73]

No.	Pesticide	CSP or Chiral colum	Chromatographic condition[*1]	Separation effect[*2]	Elution order[*3]	Reference
		Chiralcel OJ	hexane/IPA(100/5); 1.0 mL/min; UV 225nm	α: 1.17	R(-)/S(+)	[73]
		Lux Cellulose-1 Chiralcel OD-H	ACN/water(70/30), MET/water(90/10); 1.0 mL/min; UV 220nm	Rs: 2.43, 2.73	OR: -/+	[75, 66, 89]
75	triadimenol	Chiralcel OD-H	hexane/IPA(100/3); 1.0 mL/min; UV 225nm	α: 1.81	1R,2S(+)/1S,2R(-)	[73]
		Chiralcel OD-H	hexane/IPA(100/2); 1.0 mL/min; UV 225nm	α: 1.03	1S,2S(-)/1R,2R(+)	[73]
		CDMPC	hexane/ethanol(99.2/0.8); 0.8mL/min; UV 278nm	Rs: 0.64/2.87/0.37		[61]
		Chiralcel OJ-H	hexane/IPA(100/3); 1.0 mL/min; UV 225nm	α: 1.16	1R,2R(+)/1S,2S(-)	[73]
		CTMB	hexane/n-butyl alcohol (89/11); 0.5mL/min; UV 278nm	Rs: 0.18/ 0.69/ 0.52		[61]
		CTPC	hexane/IPA(91/9); 0.5mL/min; UV 278nm	Rs: 1.53/ 0.88		[61]
		Lux Cellulose-1	MET/water(60/40); 0.5 mL/min; UV 220nm	Rs: 1.45/2.73/2.16	OR: (-)-A,/(+)-A/(-)-B/ (+)-B	[40]
76	trichlorfon	CDMPC	ACN/water(70/30); 0.8mL/min; UV 210nm	α: 1.0		[81]
77	trichloronate	Chiralcel OD	heptane; 1.0 mL/min	Rs: 1.1		[55]
		Chiralcel OJ	heptane; 1.0 mL/min; UV 205nm	Rs: 1.40		[55]
		Chiralcel OJ	hexane/heptane/ethanol (90/5/5); 1.0 mL/min	Rs: 4.03	OR: +/-	[111]
78	uniconazole	CDMPC	hexane/ n-butyl alcohol(89/11); 0.5mL/min; UV 268.6nm	Rs: 1.45		[61]
		CTPC	hexane/ ethanol (93/7); 0.5mL/min; UV 269.8nm	Rs: 2.16		[61]
79	vinclozolin	CDMPC	hexane/IPA(99/1); 1.0 mL/min; 2.0 UV 210nm	Rs: 1.46	OR: +/-	[65]

[*1] ACN, MET and IPA means acetronitrile, methanol and isopropanol respectively.

[*2] α and Rs means the separation factor and the resolution facotr respectively.

[*3] CD and OR means signals obtained from circular dichrism detector and optical rotation detector respectively.

Table 2. Summary of resolution results of chiral pesticides on cellulose-based CSPs

5. Conclusion

Cellulose derivatives have high chiral recognition abilities for racemates and have already become a very popular and useful source material for CSPs. Cellulose-based CSPs can be prepared by coating or bonding cellulose derivatives on decorative silica gel or other supports with various preparation methods. The coated CSPs exhibit higher discrimination abilities for chiral pesticides and are more popular than the bonded CSPs. However, the bonded CSPs can tolerate broader solvent ranges, including THF and $CHCl_3$, which cannot be used on coated CSPs as mobile phases because they have strong dissolution abilities that can dammage or destroy them. Coated CDMPC and CTMB had the broadest application in the stereoselective separations of chiral pesticides. For most pesticides, better separations were obtained on NP-HPLC than on RP-HPLC. However, RP-HPLC can improve the amount of racemates loaded on CSPs as it allows the use of more polar solvents to enhance the solubility of racemates in mobile phases. Additionally, it can be easily connected in tandem with MS, allowing for the development of more sensitive methods for analysis of enantiomers/stereoisomers. The cellulose-based CSPs on NP-HPLC and RP-HPLC provide very powerful tools to prepare individual enantiomers and study the activity, toxicity and environmental fates of chiral pesticides.

Author details

Jing Qiu, Shouhui Dai, Tingting Chai, Wenwen Yang, Shuming Yang and Hualin Zhao
Institute of Quality Standards & Testing Technology for Agro-Products, Chinese Academy of Agricultural Sciences, Beijing, China
Key Laboratory of Agri-Food Quality and Safety, Ministry of Agriculture, Beijing, China

Acknowledgement

The financial support from National Natural Science Foundation of China (Project number 21177156 and 20907073) for this work is hereby gratefully acknowledged.

6. References

[1] Okamoto Y, Ikai T. Chiral HPLC for efficient resolution of enantiomers. Chem Soc Rev 2008;37:2593-2608.

[2] Wang T, Chen YW. Application and comparison of derivatized cellulose and amylase chiral stationary phases for the separation of enantiomers of pharmaceutical compounds by high-performance liquid chromatography. J Chromatogr A 1999;855:411-421.

[3] Okamoto Y, Kaida Y. Resolution by high-performance liquid chromatography using polysaccharide carbamates and benzoates as chiral stationary phases. J Chromatogr A 1994;666: 403-419.

[4]   Hesse G, Hagel R. Eine vollständige Recemattennung durch eluitons-chromagographie an cellulose-tri-acetat. Chromatographia 1973;6(6):277-280.

[5]   Williams A. Opportunities for chiral agrochemicals. Pestic Sci 1996;46:3-9.

[6]   Racke KD, Skidmore M, Hamilton DJ, Unsworth JB, Miyamoto J, Cohen SZ. Pesticide fate in tropical soils. Pestic Sci 1999;55(2):219-220.

[7]   Franco P, Minguillón C, Oliveros L.Bonded cellulose-derived high- performance liquid chromatography chiral stationary phases III. Effect of the reticulation of the cellulose derivative on performance. J Chromatogr A 1997; 791:37-44.

[8]   Kasuya N, Kusaka Y, Habu N, Ohnishi A. Development of chiral stationary phases consisting of low-molecular-weight cellulose derivatives covalently bonded to silica gel. Cellulose 2002;9:263-269.

[9]   Daicel Chiral Technologies (China) Co., Ltd. The list of Chiral columns: http://www.daicelchiraltech.cn/hand/hand.asp (accessed 21 April 2012).

[10] Guangzhou FLM Scientific Instrument Co., Ltd. Chiral columns on HPLC: http://www.gzflm.com/product/ category _4_6_10.aspx (accessed 21 April 2012).

[11] Thunberg L, Hashemi J, Andersson S. Comparative study of coated and immobilized polysaccharide-based chiral stationary phases and their applicability in the resolution of enantiomers. J Chromatogr B 2008;875:72-80.

[12] Okamoto Y, Kawashima M, Hatada K. Chromatographic resolution : XI. Controlled chiral recognition of cellulose triphenylcarbamate derivatives supported on silica gel. J Chromatogr 1986;363:173-186.

[13] Chankvetadze B, Chankvetadze L, Sidamonidze S, Kasashima E, Yashima E, Okamoto Y. 3-fluoro, 3-chloro and 3-bromo-5-methylphenylcarbamates of cellulose and amylase as chiral stationary phases for high-performance liquid chromatographic enantioseparation. J Chromatogr A 1997;787:67-77.

[14] Tang SW, Liang XF, Wang F, Liu GH, Li YL, Pan FU. Synthesis and HPLC chiral recognition of regioselectively carbamoylated cellulose derivatives. Chirality 2012;24:167-173.

[15] Ikai T, Yamamoto C, Kamigaito M, Okamoto Y. Enantioseparation by HPLC using phenylcarbonate, benzoylformate, p-toluenesulfonylcarbamate, and benzoylcarbamates of cellulose and amylose as chiral stationary phases. Chirality 2005;17:299-304.

[16] Yamamoto C, Inagaki S, Okamoto Y. Enantioseparation using alkoxyphenylcarbamates of cellulose and amylose as chiral stationary phase for high-performance liquid chromatography. J Sep Sci 2006;29:915-923.

[17] Jin ZL, Hu FF, Wang Y, Liu GH, Wang F, Pan FY, Tang SW. Preparation and evaluation of amylose and cellulose tris(3-trifluoromethylphenyl- carbamates)-based chiral stationary phases. Chinese J Chromatogr 2011;29:1087-1092.

[18] Chang YX, Yuan LM, Zhao F. Effect of chiral additives in the preparation of cellulose-based chiral stationary phases in HPLC, and effect on enantiomer resolution. Chromatographia 2006;64:313-316.

[19] Ge J, Zhao L, Shi YP. Preparation and evaluation of a novel cellulose tris(n-3,5-dimethylphenylcarbamate) chiral stationary phase. Chinese J Chem 2008;26:139-142.

[20] Chen X, Yang L, Zou HF, Zhang Q, Ni JY. Preparation of the chemically bonded cellulose phenylcarbamates chiral stationary phases for the separation of enantiomers. Chinese J Anal Chem 2000;28:1074-1078.

[21] Katoh Y, Tsujimoto Y, Yamamoto C, Ikai T, Kamigaito M, Okamoto Y. Chiral recognition ability of cellulose derivatives bearing pyridyl and bipyridyl residues as chiral stationary phases for high-performance liquid chromatography. Polymer Journal 2011;43:84-90.

[22] Yang JH, Choi SH. Synthesis of a chiral stationary phase with poly[styrene-bcellulose 2,3-bis(3,5-dimethylphenylcarbamate)] by surface-initiated atom transfer radical polymerization and its chiral resolution efficiency. J Appl Polym Sci 2011,122:3016-3022.

[23] Oliveros L, Senso A, Minguillón C. Benzoates of cellulose bonded on silica gel: chiral discrimination ability as high-performance liquid chromatographic chiral stationary phases. Chirality 1997;9:145-149.

[24] Minguillón C, Franco P, Oliveros L, López P. Bonded cellulose-derived high-performance liquid chromatographic chiral stationary phases I. Influence of the degree of fixation on selectivity. J Chromatogr 1996;728:407-414.

[25] Zhang S, Ong TT, Ng SC, Chan HS. Chemical immobilization of azido cellulose phenylcarbamate onto silica gel via Staudinger reaction and its application as a chiral stationary phase for HPLC. Tetrahedron Letters 2007;48:5487-5490.

[26] Zhou ZQ, Qiu J, Jiang SR. HPLC separation of metalaxyl and metalaxyl intermediate enantiomers on cellulose-based sorbent. Anal Lett 2004; 37(1): 167-172.

[27] Yashima E, Sahahvattanapong P, Okamoto Y. Enantioseparation on cellulose tris(3,5-dimethylphenylcarbamate) as a chiral stationary phase: influences of pore size of silica gel, coating amount, coating solvent, and column temperature on chiral discrimination. Chirality 1996;8:446-451.

[28] Ikai T, Yamamoto C, Kamigaito M, Okamoto Y. Immobilized polysaccharide-based chiral stationary phases for HPLC. Polymer Journal 2006;38:91-108.

[29] Ikai T, Yamamoto C, Kamigaito M, Okamoto Y. Immobilized-type chiral packing materials for HPLC based on polysaccharide derivatives. J Chromatogr B 2008;875:2-11.

[30] Chen XM, Okamoto Y. Efficient immobilization of polysaccharide derivatives as chiral stationary phases via copolymerization with vinyl monomers. Macromol Res 2007;15:134-141.

[31] Tang SW, Okamoto Y. Immobilization of cellulose phenylcarbamate onto silica gel via intermolecular polycondensation of triethoxysilyl groups introduced with (3-glycidoxypropyl)triethoxysilane. J Sep Sci 2008;31:3133-3138.

[32] Qu HT, Li JQ, Wu GS, Shen J, Shen XD, Okamoto Y, Preparation and chiral recognition in HPLC of cellulose 3,5-dichlorophenylcarbamates mmobilized onto silica gel. J Sep Sci 2011;34:536-541.

[33] Okamoto Y, Aburatani R, Miura S, Hatada K. Chiral stationary phases for HPLC: cellulose tris(3,5-dimethylphenylcarbamate) and tris(3,5dichlorophenylcarbamate) chemically bonded to silica gel. J Liq Chromatogr 1987;10:1613-1628.

[34] Yashima E, Fukaya H, Okamoto Y. 3,5-Dimethylphenylcarbamates of cellulose and amylose regioselectively bonded to silica gel as chiral stationary phases for high-performance liquid chr. J Chromatogr A 1994; 677:11-19.

[35] Qiu W, Han XQ, Wei Y, Liu YH, Chang J, Liu WL. Preparation of covalenfly bonded cellulose-derived chiral stationary phase and enantioseparation. Chemical Engineer 2009;161:5-7.

[36] Qin F, Chen Xm, Liu YQ, Zou HF, Wang JD. Improved procedure for preparation of covalently bonded cellulose tris-phenylcarbamate chiral stationary phases. Chinese J Chem 2005;23:885-890.

[37] Ling F, Brahmachary E, Xu MC, Svec F, Jean MJ. Polymer-bound cellulose phenylcarbamate derivatives as chiral stationary phases for enantioselective HPLC. J Sep Sci 2003;26:1337-1346.

[38] Nillos MG, Gan J, Schlenk D. Chirality of organophosphorus pesticides: Analysis and toxicity. J Chromatogr B 2010;878:1277-1284.

[39] Ye J, Wu J, Liu WP. Enantioselective separation and analysis of chiral pesticides by high-performance liquid chromatography. Trends in Anal Chem 2009; 28(10):1148-1163.

[40] Liang HW, Qiu J, Li L, Li W, Zhou ZQ, Liu FM, Qiu LH. Stereoselective separation and determination of triadimefon and triadimenol in wheat, straw and soil by liquid chromatography-tandem mass spectrometry. J Sep Sci 2011; 34: 1-8.

[41] Qian MR, Wu LQ, Zhang JW, Li R, Wang XY, Chen ZM. Stereoselective determination of famoxadone enantiomers with HPLC-MS/MS and evaluation of their dissipation process in spinach. J Sep Sci 2011;34:1236-1243.

[42] Qiu J, Yang SM, Yu HX. A method to simultaneously detect enantiomers of chiral triazole pesticides. China: CN 102353740A; 2012.

[43] Wu T, Li CY, Li QL, Zhang BZ, Li JY. Study of enantiomeric separation of chiral pesticides by High-performance Liquid Chromatography. Journal of hebei normal university 2009;33:218-223.

[44] Qiu J, Wang QX, Zhu WT, Jia GF, Wang XW, Zhou ZW. Stereoselective determination of benalaxyl in plasma by Chiral High-Performance Liquid Chromatography with diode array detector and application to pharmacokinetic study in rabbits. Chirality 2007;19:51-55.

[45] Feng SL, Lin CM. Study on enantiomer separation of triazole chiral fungicides with supercritical fluid chromatography. Zhejiang University of Technology,2010:63-70.

[46] Liu WP, Zhang AP, Yuan HJ. Residues and enantiomeric fractions of organochlorine pesticides in agricultural soils from Zhejiang. Zhejiang University of Technology 2009:44-45.

[47] Liu HG, Zhao MR. Enantioselective eytotoxicity of the insecticide bifenthrin on a human amnion epithelial(FL) cell line. Toxicology 2008;253:89-96.

[48] Zhou ZQ, Liu J, Wang M, Jiang SR. Determination of optical purity of sr-bioallethrin enantiomers by chiral high performance liquid chromatography. Chinese J Chromatogr 2001;19(6):526-528.

[49] Zhou ZQ, Liu J, Wang M, Jiang SR. The Separation of Bioallethrin on Chiral Column. Chinese J Anal Chem 2002;30(4): 504.

[50] Wu T. Study on the separation and conversion of chiral pesticide enantiomers. University of science and technology of Hebei; 2009.

[51] Wang P, Jiang SR, Liu DH, Jia G, Wang Q, Wang P, Zhou ZQ. Effect of alcohols and temperature on the directchiral resolutions of fipronil, isocarbophos and carfentrazone-ethyl. Biomed Chromatogr 2005;19:454-458.

[52] Yuan HJ, Zhang AP. The analysis and detection of EF value of chiral chlorine pesticids in soil. The forth national conference on persistent organic pollutants 2009:39-40.

[53] William L, Champion JR, Lee J, Garrison AW. Liquid chromatographic separation of the enantiomers of trans-chlordane, cis-chlordane, heptachlor, heptachlor epoxide and α-hexachlorocyclohexane with application to small-scale preparative separation. J Chromatogr A 2004;1024:55-62.

[54] Nillos MG, Fuentes GR. Enantioselective acetylcholinesterase inhibition of the organophosphorous insecticides profenofos, fonofos and crotoxyphos. Environ Toxicol Chem 2007;26:1949-1954.

[55] Ellington JJ, Evans JJ, Rickett KB, Champion WL. High-performance liquid chromatographic separation of the enantiomers of organophosphorus pesticides on polysaccharide chiral stationary phases. J Chromatogr A 2001;928:145-154.

[56] Jiang B, Wang H, Fu QM, Li ZY. The chiral pyrethroid cycloprothrin: Stereoisomer synthesis and separation and stereoselective insecticidal activity. Chirality 2008; 20:96.

[57] Wang P, Zhou ZQ, Jiang SR, Yang L. Chiral resolution of cypermethrin on cellulosetris(3,5-dimethylphenylcarbamate) chiral stationary phase. Chromatographia 2004;59: 625-629.

[58] Wang QX, Qiu J, Zhu WT, Jia GF, Li JL, Bi CL, Zhou ZQ, Stereoselective degradation kinetics of theta-cypermethrin in rats. Environ Sci Technol 2006;40(3): 721-726.

[59] Ma Y. Study on the analysis of chiral pesticide enantiomers dichlorprop and enantioselective behavior in environment.Zhejiang University; 2005.

[60] Shen BC, Yuan JY, Wang HC. Enantioseparation of three aryloxyphenoxypropionic acids herbicides on CDMPC and vancomycin CSP Column. Journal of Instrumental Analysis 2008;27(12):1379-1382.

[61] Pan CX, Wu QZ, Shen BC, Zhang DT, Xu XZ. Enantioseparation of four aryloxyphenoxypropionic acid herbicides by high performance liquid chromatography on cellulose tris-(3.5-dimethylphenylcarbamate and(S,S)-Whelk-O 1. Chinese J Anal Chem 2006;34(2): 159-164.

[62] Li XJ, Di ZD, Ming YF, Zhao YF, Li YM, Chen LR. Enantiomeric resolution of diclofopmethyl by HPLC on Chiralcel OD column. Chemical Researches 2005;16(3):71-74.

[63] Pan CX. The separation of drug enantiomers on HPLC. Zhejiang University; 2005.

[64] Wang P. Study on the analysis of chiral pesticide enantiomers and enantioselective degradation behaviors in soil. China Agricultural University; 2006.

[65] Wang P, Jiang SR, Liu DH, Zhang HJ, Zhou ZQ. Enantiomeric resolution of chiral pesticides by high-performance liquid chromatography. J Agric Food Chem 2006;54:1577-1583.

[66] Tian Q. Study on the analysis and separation of chiral pesticide enantiomers on polysaccharide-type chiral stationary phases under reversed phase conditions. China Agricultural University; 2007.

[67] Chen P, Na PJ, Han XQ, Pan CP, Hou JG, Du XZ, Gao JZ. Enantiomeric resolution of diclofop-methyl by HPLC with chiral stationary phase. Journal of Instrumental Analysis 2003;22(3):39-41.

[68] Liu WL, Han XQ. Enantioseparation of dichlofop-methyl on chemically bonded chiral stationary phase. Chemical reagents 2009;31(10): 825-827.

[69] Cai XY. Effects of MCD and humic acid on aquatic toxicity and bioavailability of the chiral herbicide diclofop methyl. Zhejiang University; 2006.

[70] Lin KD, Cai XY, Chen SW, Liu WP. Simultaneous determination of enantiomers of rac-diclofop methyl and rac-diclofop acid in water by high performance liquid chromatography coupled with fluorescence detection. Chinese J Anal Chem 2006;34(5):613-616.

[71] Zhou Y, Li L, Lin K, Zhu XP, Liu WP. Enantiomer separation of triazole fungicides by high-performance liquid chromatography. Chirality 2009;21:421-427.

[72] Wang P, Jiang SR, Liu DH, Wang P, Zhou ZQ. Direct enantiomeric resolutions of chiral triazole pesticides by high-performance liquid chromatography. J Biochem Biophys Methods 2005;62:219-230.

[73] Li CY, Zhang YC, Li QL. Chiral separation and enantiomerrization of triazole pesticides. Chinese J Anal Chem2010;38: 237-240.

[74] Wu YS, Lee HK, Li SFY. High-performance chiral separation of fourteen triazole fungicides by sulfated b-cyclodextrin-mediated capillary electrophoresis. J Chromatogr A 2001;912 :171-179.

[75] Qiu J, Dai SH, Zheng CM, YangSM, Chai TT, Bie M. Enantiomeric separation of triazole fungicides with 3-μm and 5-μm particle chiral columns by Reverse-Phase High-Performance Liquid Chromatography. Chirality, 2011;23: 479-486.

[76] Zhou SS. Stability and stereoselective biological activities of chiral organophosphorus pesticides.China, Zhejiang: Zhejiang University; 2009.

[77] Wang P, Jiang SR, Zhou ZQ. Study on the chiral separation of the enantiomers ethofumestate. Chemical World 2004;11:581-582.

[78] Wang P. Studies on enantiomeric activity and stereoselective behavior of chiral pesticide ethofumesate in organism and environment. China Agricultural University; 2005.

[79] Wang P, Jiang SR, Qiu J, Wang QX, Wang P, Zhou ZQ. Stereoselective degradation of ethofumesate in turfgrass and soil. Pesticide Biochemistry and Physiology 2005;82:197-204.

[80] Zhang XJ, Shen BC, Chen JJ, Xu BJ, Xu XZ. The sthdy of separation and thermodynamics of adsorpation of the enantiomers of ethofumasate on a modified cellulose. Anal Lett 2006;39:1451-1461.

[81] Tian Q, Lv CG, Wang P, Ren LP, Qiu J, Li L, Zhou ZQ. Enantiomeric separation of chiral pesticides by high performance liquid chromatography on cellulose tris-3,5-dimethyl carbamate stationary phase under reversed phase conditions. J Sep Sci 2007;30:310-321.

[82] Bo ZB. The separation of efficacy fipronil enantiomers. World Pesticides 2004;26:14-15.

[83] Lu DH, Liu DH, Gu X, Diao JL, Zhou ZQ. Stereoselective metabolism of fipronil in water hyacinth (Eichhornia crassipes). Pestic Biochem Phys 2010;97:289-293.

[84] Hou SC. The preparation of Liquid chromatography chiral stationary phase and its performance study. China Agricultural University; 2003.

[85] Qu LF, Hu JL, Zhan B. A study on analytical methods of the chiral isomers of fluazifop-p-butyl. Hangzhou Chemical Industry 2010;40(1):40-42.

[86] Qiu J, Li L, Zhou ZQ, Jiang SR, Zhao HX. Separation of fluroxypyr-meptyl enantiomers by High Performance Liquid Chromatography. Chinese J Pestic Sci 2004; 6(2):84-86

[87] Yang GS, Vázquez PP, Frenich AG, Martínez JL and Aboul-Enein HY. Separation and simultaneous determination of enantiomers of tau-fluvalinate and permethrin in drinking water. Chromatographia 2004;60:523-526.

[88] Han XQ, Wen XG, Guan YH. Influences of mobile phase composition and temperature on chiral separation of some triazole pesticides. Chinese Journal of Applied Chemistry 2004;21:140-143.

[89] Han XQ, Wei Y. The enantioseparation of some triazole pesticides on chiralcel OD-H column. Chemical Reagents 2010;32:74-76.

[90] Lao WJ, Gan JY. High-performance liquid chromatographic separation of imidazolinone herbicide enantiomers and their methyl derivatives on polysaccharide-coated chiral stationary phases J Chromatogr A 2006;1117:184.

[91] Lin KD, Xu C, Zhou SS, Liu WP, Gan J. Enantiomeric separation of imidazolinone herbicides using chiral high-performance liquid chromatography. Chirality 2007;19:171-178.

[92] Zhou QY. Study on the enantioselective inhibition of Phytotoxicity of Imazethapyr to corn roots and its mechanism. Zhejiang University; 2010.

[93] Sun DL, Qiu J, Wu YJ, Liang HW, Liu CL, Li L. Enantioselective degradation of indoxacarb in cabbage and soil under field conditions. Chirality 2012, DOI: 10.1002/chir.22047.

[94] Xu Q, Liu KT, Tang FR, Wang JB. Determination of indoxacard 15% SC by HPLC. Modern Agrochemicals 2010,9:31-35.

[95] Lin K, Liu WP, Li L, Gan J. Single and joint acute toxicity of isocarbophos enantiomers to daphnia magna. J Agric Food Chem 2008;56:4273-4277.

[96] Zhang T, Nguyen D, Franco P, Isobe Y, Michishita T, et al. Cellulose tris-(3,5-dichlorophenylcarbamate) immobilised on silica: a novel chiral stationary phase for resolution of enantiomers. J Pharmaceut Biomed 2008;46:882-891.

[97] Xu C, Wang JJ, Liu WP, Sheng GD, Tu YJ, Ma Y. Separation and aquatic toxicity of enantiomers of the pyrethroid insecticide lambda-cyhalothrin. Environ Toxicol Chem 2008;27:174.

[98] Zhang AP. Effect of cyclodextrins on environmental behavior of selected chiral pesticides. Zhejiang University; 2006.

[99] Ming YF, Zhang HL, Li YM. The separation of the enantiomers of metalaxyl. The first academic exchange conference on chromatography of the Midwest China 2006;8:145-148.

[100]   Chen SW, Lin KD, Liu WP. Determination of metalaxyl enantiomeric purity by nonchiral high performance liquid chromatography with circular dichroism detector. Chinese J Anal Chem 2006;4(34):525P-528.

[101]   Han XQ, Li J, Yun CL, Sun YY, Wang HS. Chiral Separation of Compounds on Covalently Bonded Cellulose Chiral Stationary Phase by NO1Tflal and Reversed Phase HPLC. Journal of Instrumental Analysis 2007;26:55-58.

[102]   Qiu J, Wang QX, Zhou ZQ, Yang SM. Enantiomeric separation and circular dichroism detection of metalaxyl acid metabolite by Chiral High Performance Liquid Chromatography. Asian J Chem 2009;21(8): 6095-6101.

[103]   Lin K, Zhou SS, Xu C, Liu WP. Enantiomeric resolution and biotoxicity of methamidophos. J Agric Food Chem 2006;54:8134–8138.

[104]   Polcaro CM, Berti A, Mannina L, Marra C, Sinibaldi M, Viel S. Chiral HPLC resolution of neutral pesticides. J Liq Chromatogr Relat Technol 2004;27:49-61.

[105]   Cai XJ, Li Z, Xu XZ. Study on Enantioseparation of several herbicides on chiral stationary phases with HPLC. Journal of Instrumental Analysis 2007;26(6):891-894.

[106]   Lin CM, Liu JY, Zhang DT. Enantioseparation of naproanilide on cellulose derivatives chiral stationary phases. Agrochemicals 2007;46(8):526-528.

[107]   Chen SW, Cai XY, Liu WP. Characterization of napropamide enantiomers by CD and determination of the enantiomeric ratios in water. Spectrosc Spect Anal 2009;26(9):1649-1652.

[108]   Burden RS, Carter GA, Clark T, Cooke DT, Croker SJ, Deas AHB, Hedden P, James CS, Lenton JR. Comparative activity of the enantiomers of triadimenol and paclobutrazol as inhibitors of fungal growth and plant sterol and gibberellin biosynthesis. Pestic Sci 1987;21:253-267.

[109]   Yang HY. Study on environmental behavior of insecticide pyraclofos. Zhejiang University; 2008.

[110]   Pan H, Duan R, Liao Y, Xu BM. Chromatographic separation of chiral pesticides and their intermediates. Chemistry & Bioengineering 2007;24(8):76-78.

[111]   Liu WP, Lin K, Gan JY. Separation and aquatic toxicity of enantiomers of the organophosphorus insecticide trichloronate. Chirality 2006;18:713-716.

Chapter 13

Antiquarian Books as Source of Environment Historical Data

Jürgen Schram, Rasmus Horst, Mario Schneider, Marion Tegelkamp, Hagen Thieme and Michael Witte

Additional information is available at the end of the chapter

1. Introduction

In recent times historical environment situations became inevitable also for modern environmental analytic. They alone allow to evaluate an anthropogenic impact into the environment compartments. All prognoses to future in environmental sciences at least base on interpolations deduced from historical situations in combination of present situations. Mostly glaciology [1], [2] is a source of information about historical environment. Unfavourable for such a proceeding is the fact, that glacier mostly are situated at inhabited places [3]. So they reveal only global or supra-regional data. Data from cultural centre's nowadays only are taken by imission measurements, e.g. bones [4], monitoring the exposure of an individual to its environment.

We approach the task to receive information about the historical environment situation in settlement areas by examining historical well dated and localisable artefacts of the *homo faber*.

To make a product acceptable to such a strategy, it must fulfil some requirements, listed in the following:

- Property of an analytical passive sampler
- Common and widespread product
- Reproducible way of production (guild)
- Defined state at the begin of contamination
- Damage / manipulation must be evident
- Exact dating of the product
- Regional allocation possible

In the work introduced here, historical paper like antiquarian books are used as such a source of environment historical data specially for the environmental compartment of air and water. Antiquarian books mostly are fulfilling all listed requirements.

The selected objects of study further must have a complete block with original binding/cover. They must be produced before 1840 ("handmade paper").

The paper of historical books can be allocated by their watermark, which is characteristic for the paper mill. It can be dated by the printing date of the book – lean production was necessary, because the amount of paper for one book project was so big, that long time storing was nearly impossible

The absence of water spots and/or conservatorial treatment makes sure that there is no unwanted contamination of the book (Figure 1).

The work is divided into two parts, one dealing with airborne contaminants [5], one with water-borne contaminants [6]. In both cases antiquarian books are used as source for environmental historical data.

2. Antiquarian books and environmental airborne contaminants

The degradation of books and papers in the last decade has been focussed at from science and the public. Historical storage conditions were taken into account as factors, but they are experimental not to be simulated under real conditions, because it is impossible to simulate realistically long time storing.

The first part of this publication herein precisely starts at this point. An attempt was made to detect the historical immigration of airborne pollutants specifically in books, so as to provide the basis for a long-term correlation between pollutants in the paper, the corresponding concentration of atmospheric trace components and the damage of the paper.

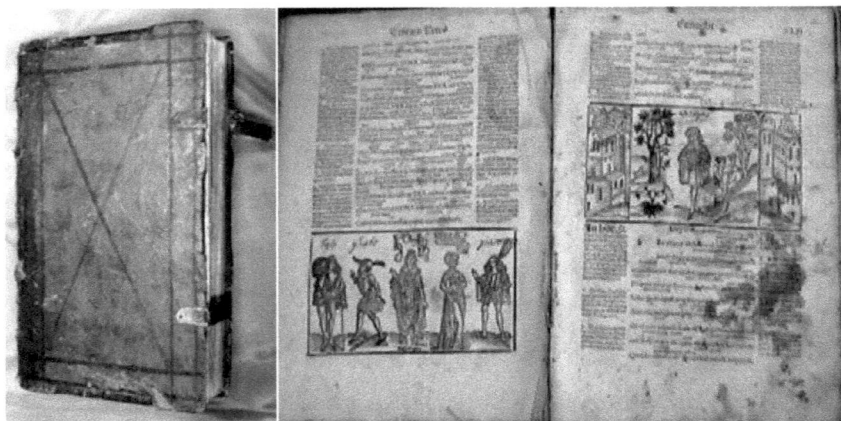

Figure 1. Ancient books, Terentius, 1499, not usable as source of historical environmental data because of obvious contamination

2.1. Volatile pollutants in paper and decay of paper

The degradation of paper and books in the last decade strongly has been brought into interest in science and the public. The importance of environmental factors during storage in the libraries was highlighted and partially documented with current environmental analytical data. Historical storage conditions were discussed as important factors, however, both were only discussed in models – but no proof of the intrance of historical pollutants by atmosphere was given.

As part of the work described herein was set precisely at this point. An attempt was made to record the historical input of airborne pollutants specifically in bound antiquarian books [5].

In the focus of this study there are no exemplary studies on the adsorbability of harmful substances and at least their degradation mechanisms in the paper [7], [8]. Rather, as part of a comprehensive project, which is still running, the pollutants really trapped by the historical paper of the books were determined. Later the pollutants experimentally found to be present in library environment (via indoor measurements) in combination with kinetics of the uptake of pollutants in the book, should allow to estimate the situation of historical indoor environment and historical sources of contaminants.

The starting point for this study was the hypothesis that the paper of antiquarian books can serve as a passive sampler for air pollutants.

Components such as the cellulose of the paper play a role as a trap for a great variety of contaminants. This property of the cellulose is often used in the environmental analysis of gas samples for sample enrichment of various polar and medium polar contaminants such as Atmospheres [9], [10], [11]. This process will take place also in paper writing material.

This passive sampler system should contain information about the pollution that a book was exposed to, although of course not only the current pollutant inputs, but also historical contaminants / pollution must be considered.

Historical damage, e.g. of gas gangrene of leather materials due to the gas lighting of 19.century have to be taken into account during these investigations.

It is assumed that the significantly not air-permeable cover of books has a shielding effect for the passage of harmful gases.

So as a criterion for detecting a contaminant sampled via this mechanism is a concentration difference between the centre of an individual page and its edges, which are more in contact with the air surrounding the page.

One part of the overall project is to try to find a way that allows a correlation between pollutants in the paper, the damage to the paper and for modern inputs of pollutants in the concentration of atmospheric trace components of the current storage locations.

2.2. Contaminations in paper

The decay of writing materials such as paper provides a non-negligible danger to the existence of entire libraries, and thus our cultural knowledge at all.

The damage is often catalysed by impurities in the writing surface either directly or causally induced [12]. Just trace elements are involved in a variety of ways in decay processes of cellulose. Such impurities can already produce in very small trace concentrations harmful effects.

They can be attributed to different sources. In principle, three mechanisms of entry of pollutants into books are to be distinguished:

Entry to the manufacturing process

Already in the production of written support materials, traces have been introduced into the material of pollutants. These can be direct (black-colored parchment) or as a result of their decomposition processes (ink corrosion) have a detrimental effect [13].

Entry through restoration measures

Conservation processes often bring chemicals into the paper. This entry can be wished (magnesium-Wei T'o the entry process, pesticides) or technically unavoidable (chloramine T residues after bleaching processes) done [14].

Input of pollutants from the environment during storage

Writing materials generally have a large internal surface that is covered with chemically reactive groups. They are therefore able to adsorb and thus to enrich pollutants from phases in contact with the material. The absorbed substances can cause damage (damaged paper in the field of cutting edge books) [15], [8].

2.3. Mechanisms of pollutant entry into paper

Of these damage processes in this part especially the latter damage mechanisms are investigated. The air surrounding the books is to be considered as a contaminating factor. For books which were stored for centuries, especially the historical situation is to be seen as the mostly important.

The possible components in the gaseous state, or bound to dust particles or aerosol droplets from the surrounding air migrate into the cellulose of the paper. These pollutants can be divided into two main groups:

- Molecular, stable in aqueous systems, usually non-polar compounds (pesticides, dyes, solvents)
- Polar with water ion forming compounds (SO_x, NO_x, acids, amines)

The first group (molekular unpolare organic trace substances) according to today's knowledge play only a relatively minor role in as initial substances of decay mechanisms. On the other hand polar inorganic and organic substances or ions in aqueous systems are very important. They seem to be the main cause of various damage mechanisms. Examples therefore include cellulose degradation processes by acidic components or oxidative degradation of cellulose, catalysed by traces of heavy metal [12].

In the project described here, therefore, an attempt was made to determine in the paper of selected books the concentration of some ionic impurities with a high damage potential.

As particularly significant thereby the material region nearby to the cutting edges is to be seen. There the ambient air comes into most intensive contact with the cellulose of the paper. These cutting edge regions are often already damaged, which visually clearly can be recognized.

A difference in the corresponding ion content between the centre and the cutting edges of a page reveals a adsorption of these compounds from the surrounding atmosphere in the border region of the books – so acting as a passive sampler.

Such a difference in concentration can´t be attributed to the production of paper (Figure 2). The multiple folding of the individual printed sheets of paper before the binding of a book coincides that the edge region of such a sheed Is no longer the cutting edge of the book page.

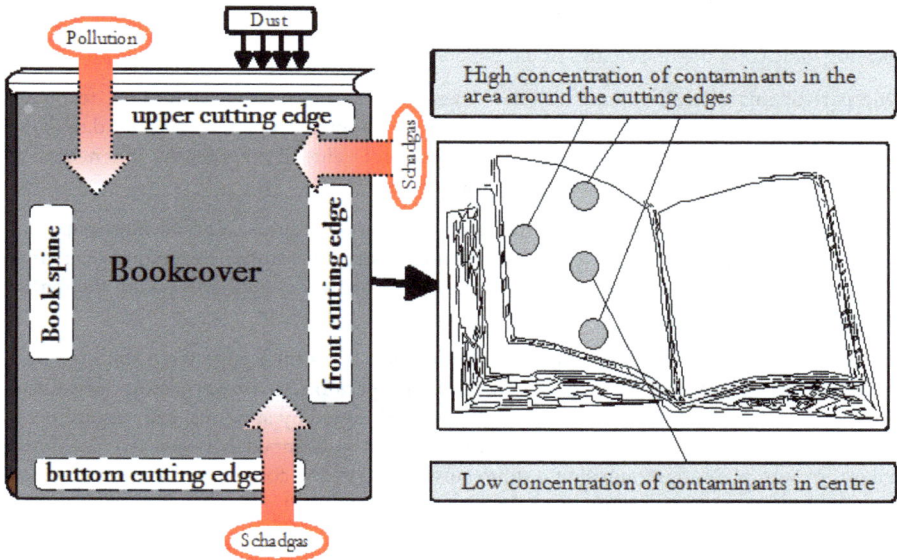

Figure 2. Mechanism of incorporation of contaminants into a book

As the source of this adsorption process some pollutant gases are obvious. Mostly they hydrolyse to the corresponding ionic components which can be detected in the paper:

• Acid gases

NO_x → HNO_2, HNO_3

SO_x → H_2SO_3, H_2SO_4

HCl   →   HCl (aqu.)

CO_2 →   "H_2CO_3"

CH_3COOH → CH_3COOH

R-COOH   →   R-COOH

- Basic gases

NH_3 →   "NH_4OH"

2.4. Analysis of pollutants after the entry into the paper

To solve the corresponding chemical analysis problem initially encountered an appropriate problem-matched sampling system was developed that meets the requirements for the analysis of unique cultural objects.

Since using reflective spectroscopy methods directly applied on antiquarian paper does not provide the necessary information (detection limits) specially with respect to the acid-forming ionic components, eluates of the papers had to be examined.

One part of this work therefore was the development of techniques that allow to obtain aqueous eluates of well-defined locations on written antiquarian papers, without damaging the delicate system irreversible. Because of the cultural value of these books, it is necessary to work absolutely non-destructive with those objects.

Therefore some investigations had to be done with model papers. Normal filter paper, as used in laboratory was used to develop the non-destructive elution technique. Hereafter historical sheets of paper were utilised to develop the analytical methods, taking the matrix interference into account.

Finally a method was developed wherein the selected spots on a page were stamped by a cyclic rubber stamp with a saturated solution of paraffin to prevent scum rings and chromatographic effects. The encircled area was conditioned by spraying with a solution of 30 % w/w ethanol in water for 10 min. Afterwards it is clamped between two meandric polypropylene plates, guaranteeing a uniform washing of the conditioned area. It is rinsed thoroughly with a elution solution of deionised water (2 x 1.0 mL) by a peristaltic- pump with a flow rate of 0.27 mL/min – so eluting the adsorbed ions. The flushed area was 8,5 cm².

Up to four clamp arms were provided with appropriate meanderic suction plates to eluate one page at three spots simultaneous. So - after appropriate pre-swelling of the paper - it is thus possible to eluate the antiquarian paper without permanently visible irreversible changes of the condition of the paper (Figure 3). An additional burden on the papers by an additional chemical entry will not occur with this method - on the contrary, the studied site is cleaned by washing with distilled water of pollutants.

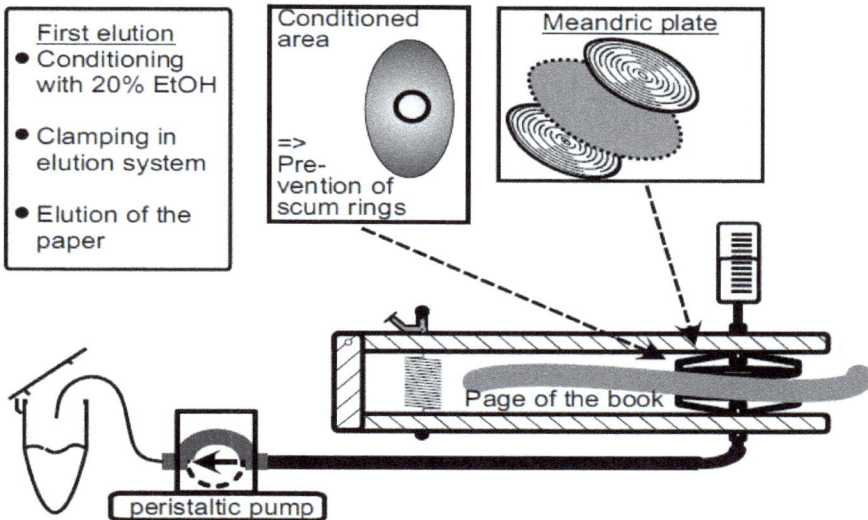

Figure 3. Elution system to determine the first elution of historical paper

The resulting solutions were analysed by capillary electrophoresis and ion exchange chromatography concerning its contend on harmful ions [16], [12].

These systems require sufficient small sample volumes (0.8 mL) and thus provide a careful and nondestructive method for the analysis in cultural and historical samples because each sampling at least represents a unavoidable interference in the cultural object [7], [17]. The detection limits of the system (0.5 mg/L SO_4^{2-}, 0.2 mg/L NO_3^-) were sufficiently small to prove the relevant concentrations.

2.5. Objects of study concerning airborne pollutants

We studied books of the 17th-18th century, that had been made from rag paper and not industrially produced papers.

Only those made from fermented and then shredded rags rag papers (up to ca.1840) allow the described procedure to determine the entry of acidifying substances via the gas phase. - in contrast to modern manufactured paper, for there are many sulphur containing chemicals used in the production process (e.g. sulphite and sulphate pulp).

In total 11 books of the production period from 1607 to 1753 were examined. All books except one had an intact cover, which came from the manufacturing time. In particular the following books were examined from different sources:

- Lucas, F.; Concordantiae biblorum, Lyon 1612
- Spanheimius, F.; operum miscellaneorum, Leiden 1703

- Luther, M. /, Klemmen, C. Die heilige Schrift Neuen Testaments, Tübingen 1729, Part I
- Luther, M. /, Klemmen, C. Die heilige Schrift Neuen Testaments, Tübingen 1729, Part II
- De Lambertini, P.C., De servorum beatificatione et beatorum canonizatione; Patavia 1743
- Wilischen, CF; Biblia parallelogram - harmonico exegetica, Leipzig 1753

In a second phase of 5 books from a library [17] were investigated, which were kept for about 300 years ago in the same room:

- Faber, M., Opus concionum Tripartitum; Verdussen, Antwerp 1663
    - Volume 1
    - Volume 2
    - Volume 3
- Adriani, H.; Catholycke seremoonen op all Epistelen end Gospels van de soundings end heylighe daghen van den gheheelen Jare; Verdrussen, Antwerp 1620
- Viguerius, J.; Institutiones thiological it sacris literis Conciliis doctoribus ecclesiasticis, Thomas Aquinas praecique D., Walter, Cologne, 1607

In each book different pages at four points were eluted (Figure 2) and analysed towards the content of Cl-, SO_4^{2-} and NO_3^-. One sampling point was the centre of the page, one in the middle of the top cutting line, one in the middle of lower cutting line and one in the middle of front cutting line. In some cases 6 points (Figure 4) were examined.

2.6. Results and discussion of the adsorption of airborne pollutants in historical books

All the books examined with intact cover showed a clear accumulation of pollutants NO_3^- and SO_4^{2-} at the cutting edges of the book. The centre of each side was in average significantly lower contaminated with pollutants than the average range (by a factor 2 to 10).

The highest concentrations were apparent in the region of the head section, followed by the front and finally the foot trim. Diagram 1 shows exemplary values. For the borders of a paper in the book this difference can not be explained by the paper production but only by the entry of airborne pollutants into the cellulose of the edges of books. It is obvious here that the ions detected by hydrolysis of gaseous pollutants - have been formed - primarily through contact with the atmosphere. They are in good agreement with the above theories to the entry mechanism from the gas phase.

Apart from the adsorption of pollutants into the cellulose of the book, there are further possible processes, that can cause differences of concentrations of contaminants at diverse areas of a book. So the printing ink could be a source as well as the degradation of the paper. Furthermore the book cover could be a source of an input of chemicals into the cellulose. Comparing the distribution of contaminants from the first to the last pages in combination with the distribution on a single page allows to differentiate between these processes. Figure 4 illustrates this coherence.

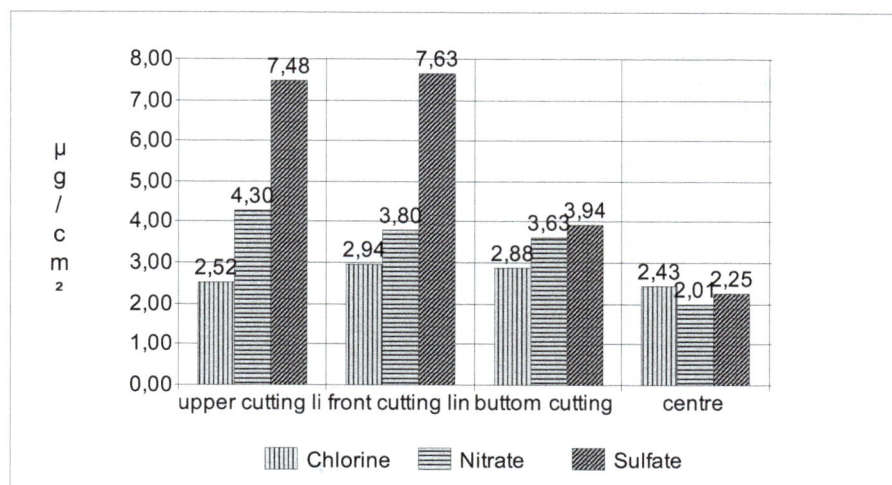

Diagram 1. Pollutant concentrations in De Lambertini, P.C.; Patavia 1743

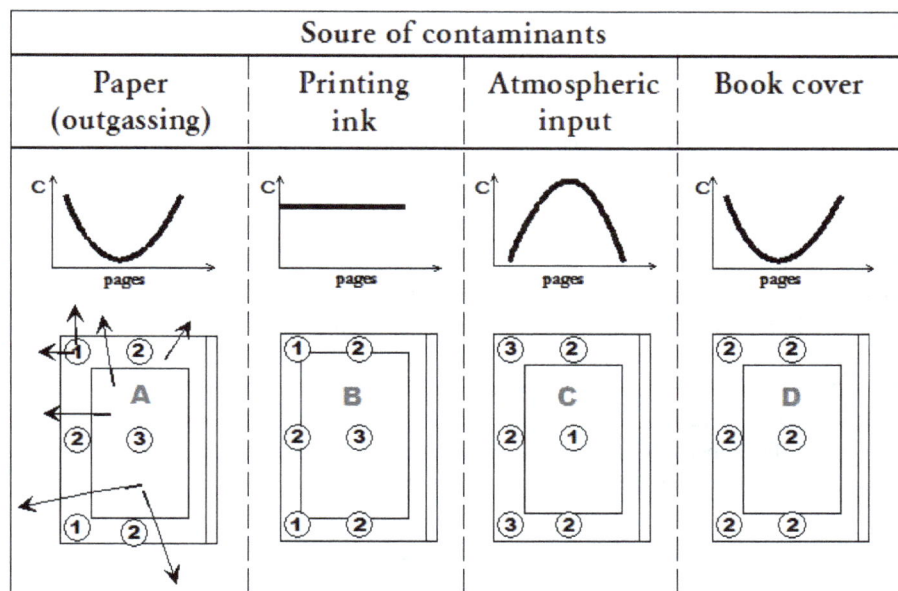

Figure 4. Causes for a variation of concentrations of contaminants within a book. Concentration differentiation at a single page and over the whole book allows to deduce the source of the contaminant. The numbers 1, 2, 3 symbolize the concentrations. The numbers are proportional to the concentrations.

Looking at the distribution (Diagram 2) within the different pages of a book unsually an increase in the concentrations of contaminants is observed in the direction of the middle layers. This result can be understood by the shielding effect of book cover, which is more pronounced in the middle layers.

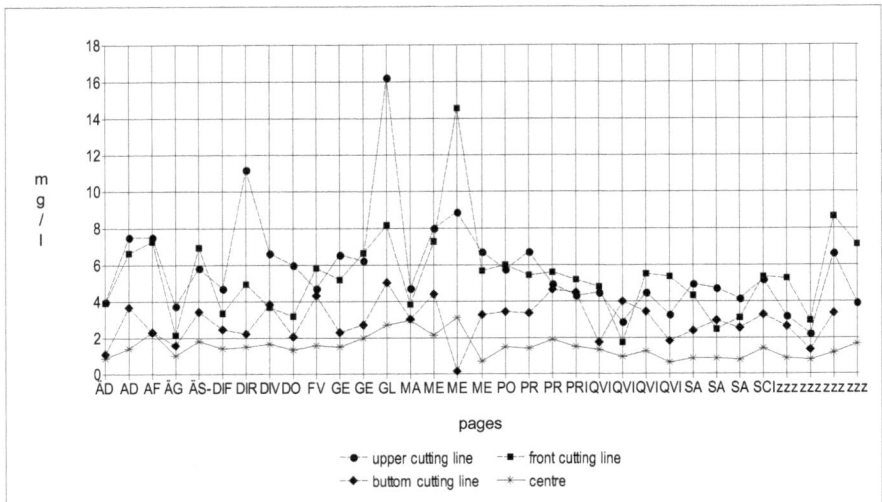

Diagram 2. Page depending distribution of nitrate concentration in the eluate [mg/L] in Lucas, F., Lyon 1612

Page depending distribution of nitrate concentration in the eluate [mg/L] in Lucas, F., Lyon 1612. The letters on horizontal axis shows the pagination of the book.

So for the first time, the entry of NO_x, SO_x, and HCl from the atmosphere into the finished book could be proved [5]. The concentration of the hydrolytic produced sulphate, nitrate and chloride ions were much lower in the edges of the pages of bound books, than in the centre (by a factor of 2-10) of the same pages. Thus it can be assumed that the cellulose of the book here acts as a trap for contaminants and so enriches over longer periods of time these pollutants. The upper cut edge of the books showed in all cases the highest pollutant concentrations. Studies of uptake of organic acids mostly showed a similar behaviour[15].

3. Cellulose of antiquarian books as source for historical water data

3.1. Paper production and water contamination

The production of paper in ancient times was a well defined technical process [18], which was "quality assured" by the strict regulations of the guilds. It was started with rags, which were sorted and cut. After fouling 1 month in water the resulting textile fragments were crushed with a pounding-machine to fibres. Cleaning of the fibre with river water happened in the HOLLÄNDER, a mills/shredder which allowed to produce a pulp of fibres. Finally the paper was produced with a sieve, containing the watermark, in thin layers. Water was pressed out and the paper was dried between layers of felt.

Paper production from pulp in a vat (e.g.16.Century) means at least the contact of cellulose with huge amounts of water of river (production water). It was the solvent of the pulp. With the energy of water mills the rag milling machines were powered. Therefore paper mills always were situated directly at a river. Nearly all pictures, showing the production process of paper, in the background show a river.

3.2. Use of paper as source for historical water data

The new approach to get environmental data uses the characteristics of production process described above. Cellulose has selective ion exchange properties – so it is able to trap (heavy) metal cations from water during production [19], [20]. The fibres adsorb contaminants (traces of metals) from the (river) water, forming a dynamic adsorption equilibrium.

This ion exchange process may be caused by traces of e.g. carbocylates, amines and amino-acids fixed to the fibres of the paper. It could be proofed, that there is a great variety of selectivity of different modern papers towards heavy metals, while alkalines and earthalkalines are much less attracted. Each metal can be characterized by an individual *Langmuir / Donnan*-plot on every special paper. These facts allows us to take into focus individual heavy metal -ions and their interaction with the paper.

This is stopped and conserved when the pulp is separated from the production water. With drying the heavy metal content is conserved within the paper and so stores in historical books.

To get information about the heavy metal content of historical water, it is first necessary to determine the adsorbed heavy metal content in the paper at some blank area (First elute). This process should be done without any destruction in a flow system. In the elutes, the heavy metal content is determined by Inverse Stripping Voltammetry (Pb and Cd) and Graphite Furnace Atomic Absorption Spectrometry (GFAAS) (Zn and Cu).

In a second step the eluted paper area is equilibrated in a non-destructive flow system under defined conditions with different concentration of heavy metals (Cu, Pb, Zn, Cd) to plot the adsorption isotherm. For the initially determined first elute, it is now possible to calculate the concentration of heavy metals in the production water of the pulp – and so of the corresponding river. A similar approach to nowadays contaminants was done with biofilm, allowing to determine illegal discharger in waster water technology [21], [22], [23].

Our approach is shown in (Figure 5).

Method:
1.) Determination of the adsorbed heavy metal content in the paper
2.) Characterisation of the adsorption equlibrium in that very paper

3.) Calculation of the historical contamination of the production water with heavy metals

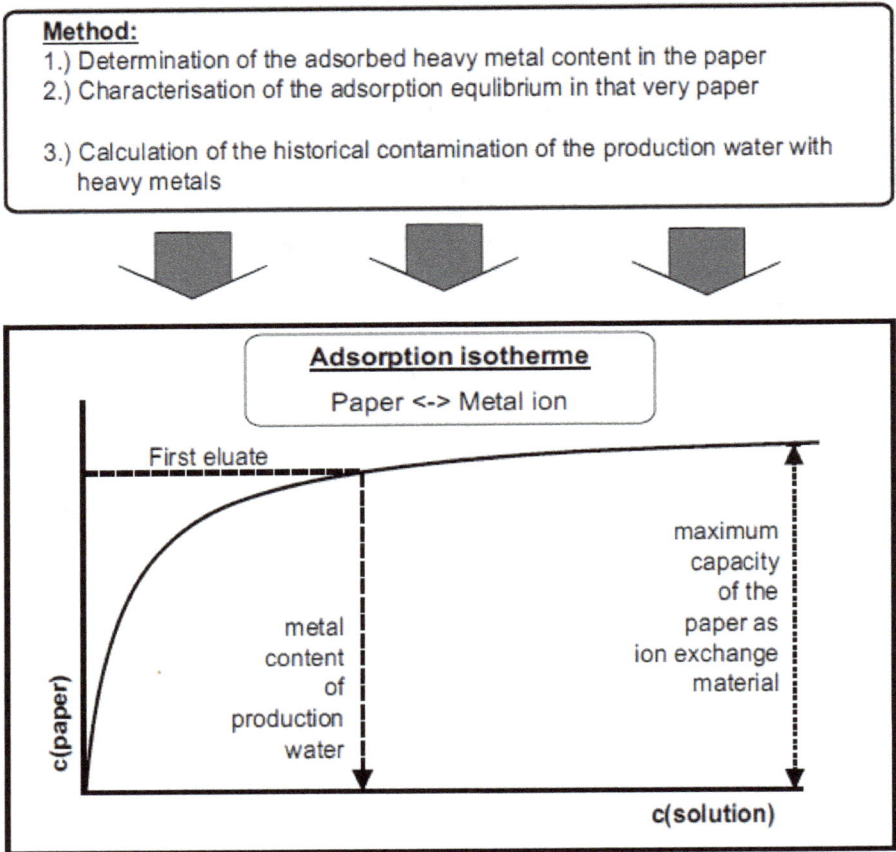

Figure 5. Principal approach to get environmental data from historical paper

3.3. Experimental approach to determine historical water pollution

Antiquarian books reveal a new source of environmental historical data. Because of their cultural value, it is necessary to work absolutely non-destructive with those objects.

Therefore some investigations had to be done with model papers. Normal filter paper, as used in laboratory was used to develop the non-destructive elution technique. hereafter historical sheets of paper were utilised to develop the analytical methods, taking the matrix interference into account.

Historical books without any damage and manipulation, fulfilling requirements above described, were selected in close cooperation with some libraries (Table 1).

Book 1	Ioannis Maldonati Societatis Iesu - Theologi commentarii in quattuor Evangelistas Stefanus Mercatoris; Pontmousson, 1596
Book 2	Paciuchelli Lectiones morales in Ionam Prophetam- Band II Hieronymus Verdussen; Amsterdam/Antwerpen, 1680
Book 3	J. Meyer Sive historiae Rerum Belgicarum Sigismund Feyerabend; Frankfurt a.M., 1580
Book 4	Nicephori Callisti Xantopuli Scriptoris vere catolici, ecclesiasticae historiae libri decem et octo ..... Sigismund Feyerabend; Frankfurt a. M., 1588
Book 5	Jan Jakub Scheuchzer Geestelyke Natuurkunde Petrus Schenk; Amsterdam, 1735
Book 6	Estius (Hessels van Est). / Petrus Lombardus In quatuor libros sententarium commentaria, Vol 1 Jean de Nully; Paris, 1697
Book 7	Estius (Hessels van Est). / Petrus Lombardus In quatuor libros sententarium commentaria, Vol 2 Jean de Nully; Paris, 1697
Book 8	Estius (Hessels van Est). / Petrus Lombardus Commentariorum in epistolas apostolicas N. Boucher, Rouen, 1709

Table 1. Books selected for the determination of process waters.

On one of the pages of the historical books 3 spots are selected. As already above mentioned they were stamped by a cyclic rubber stamp (d = 40 mm) with a saturated solution of paraffin (RT, mp. 80°C) in n-hexane to prevent scum rings and chromatographic effects. The encircled area is conditioned by spraying with a solution of 30 % w/w ethanol in water for 10 min. afterwards it is clamped in an elution system (figure 3).

The paper is clamped between two meandric polypropylene plates, guaranteeing a uniform washing of the conditioned area. It is rinsed thoroughly with a aqueous solution of 0,1 mol/L $Mg(NO_3)_2$ (pH 3.7 adjusted with HCl) by a peristaltic pump with a flow rate of 0,27 mL/min.

There are taken several fractions (each 10 mL) of the elute (n = 4-8). They are analysed separately to indicate the completeness of the extraction (Figure 6).

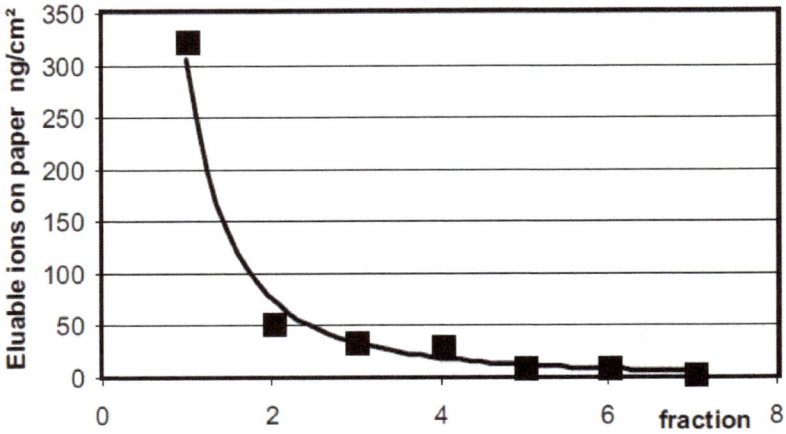

Figure 6. First elution. Concentration of the separated fractions

Afterwards in the flow-system the eluted paper spot is equilibrated for 100 min by cycling different concentration of heavy metals (mixture of Cu^{2+}, Pb^{2+}, Zn^{2+}, Cd^{2+}, each 10, 50, 100, 150, 250 µg/L) to determine the adsorption isotherm. The adsorption of metals on paper is done under defined conditions with deionized water (pH = 5.7), 100 mL containing 0.0422 g $Ca(NO_3)_2$ x 4 H_2O (Figure 7). The solution cycling in the flow system was at least 25.0 mL, which means a great excess of ions in relation to the ion exchange capacity of the paper. All these experiments were done at RT.

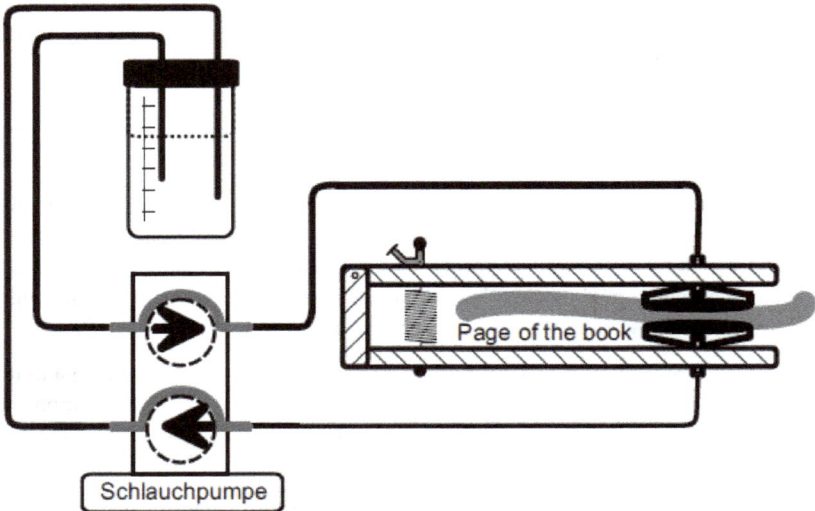

Figure 7. Equilibration system to determine the adsorption isotherms of historical paper

Several samples of 100 μL are taken during this process and analysed, revealing the kinetic of the extraction (Figure 8).

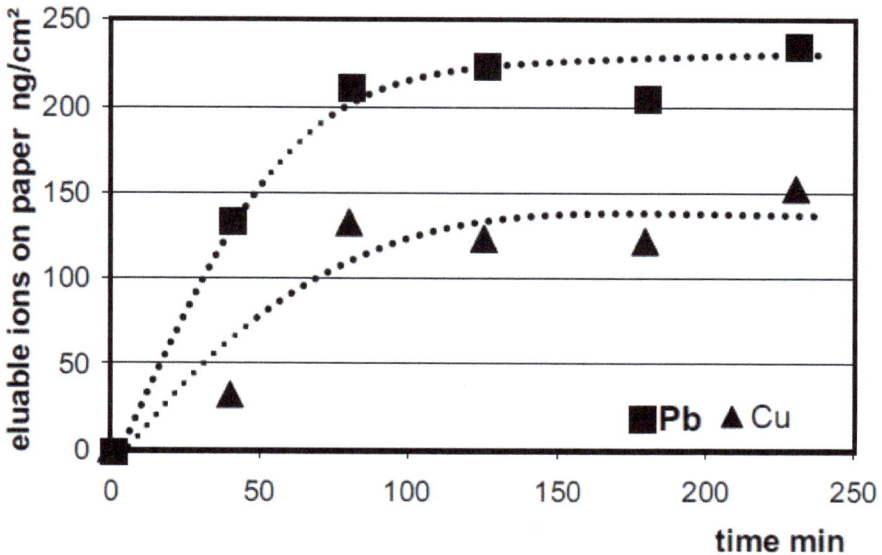

Figure 8. Kinetic of the adsorption equilibrium on filter paper

The determination of the elutes were done in case of Cu^{2+} and Zn^{2+} by ETV-AAS and in case of Pb^{2+} and Cd^{2+} with Inverse Stripping Voltammetry.

Atomic absorption was done using the Shimadzu atomic absorption spectrometer AA-6800, equipped with a high speed self reversal background compensation method (HSR). The injection volume was 20 μL. Only self reversal technique [24] allows to compensate background in the elutes of historical paper, which is invisible to deuterium technique. The instrumental operating conditions were standard reference methods for ETAAS (Cu^{2+}: 342,8 nm, Zn^{2+}: 213,9 nm)

Differential pulse voltammetric determinations were done on the system 757 VA of Metrohm at a hanging mercury electrode. In NH_4NO_3 electrolyte (10mmol/l) the deposition was done for 90 s at – 1,15 V, the sweep from – 1,15 V to 0,05 V with a sweep rate of 0,0595 V/s and a pulse amplitude of 0,05 V.

All calibrations were done by standard addition (Two additions).

3.4. Results and discussion

Some graphic examples for the resulting adsorption isotherm are shown in Figure 9. In the corresponding concentration rage, the adsorption isotherm is nearly linear. Using of the values of the "first elute", the process water concentration can be calculated.

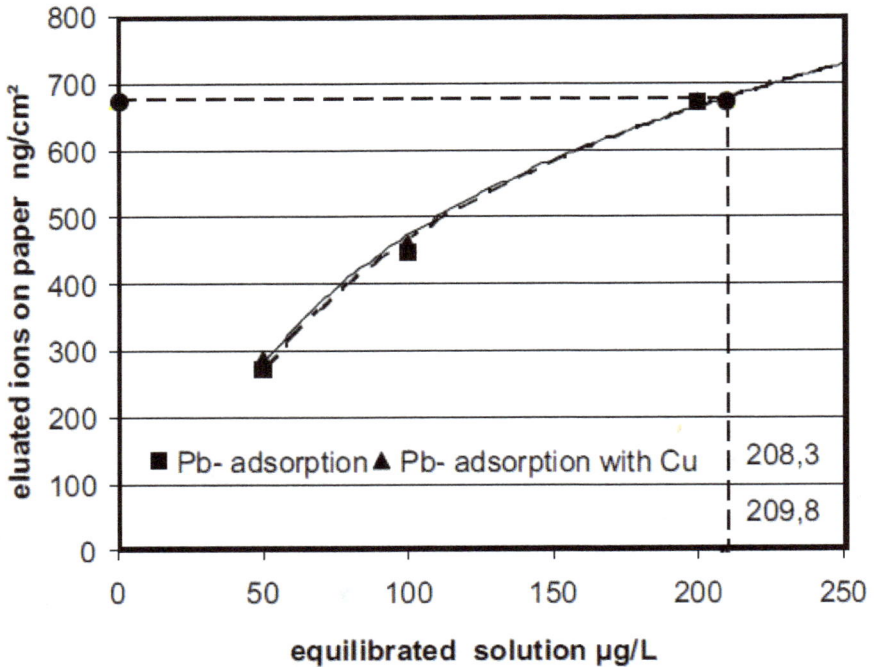

Figure 9. Graphical calculation of the composition of pulp process water of historical paper. The neglecting influence of Cu^{2+}- concentration (200 µg/L) on the adsorption of Pb^{2+} on 17.century paper is shown as well.

3.5. Influence of matrix ions on the adsorption of heavy metals

The basis of the determination of heavy metal contamination of historical (river) waters is the equilibrium between these metal ions and the fibre of cellulose of the paper. It works as a weak acid ion exchanger.

Ions like Ca^{2+} or Mg^{2+}, present in normal surface water, compete against the heavy metal ions like Cu^{2+} or Pb^{2+} and so shift the balance of adsorption for those compounds towards the solution.

To be able to estimate the content of heavy metal ions in the paper production water, it first should be necessary to investigate the influence of these matrix compounds in the water on the adsorption isotherm

We could show, that the content of Ca^{2+} has no influence on the amount of Pb^{2+} and Cu^{2+} ions adsorbed on historical paper (Figure 10). There is no cross-selectivity for adsorption of heavy metals in presence of Ca-matrix > 50 mg/L Ca^{2+}. In consequence it is not necessary to know exactly the chemical composition concerning Ca^{2+} and Mg^{2+} of the water used at that time.

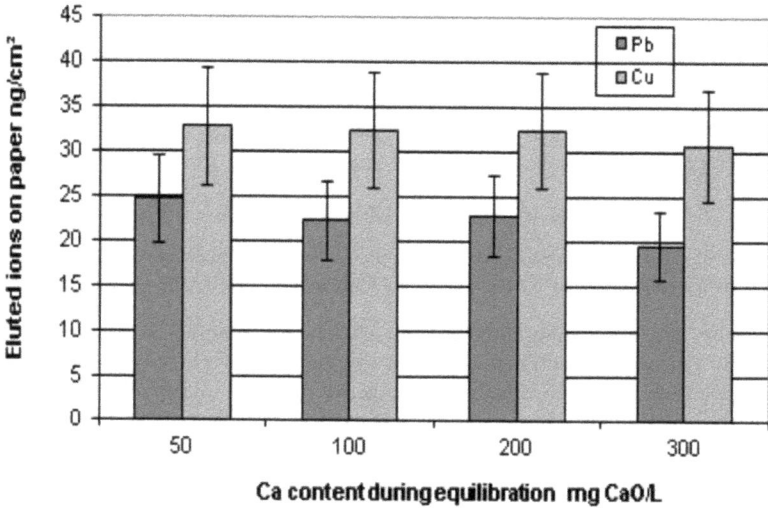

Figure 10. Influence of the Cu^{2+} - concentration (measured as mg(CaO)/Lon the adsorption of Pb^{2+} and Cu^{2+} traces (each 10 μg/L) on 17.century paper

To determine the adsorption isotherm the paper is equilibrated with well defined concentrations of a multi element solution. This is only useful if the adsorption each heavy metal ion is independent of the other heavy metal ions, present during the equilibration 13 shows, that there is no dependence of the adsorption isotherm for Pb^{2+} ions of the presence of Cu^{2+} ions.

3.6. Validation of the procedure

New chemical methods normally are validated by the analysis of certified standard reference material or – if not available - by a reference to a well established analytical method [25], [26].

Both is impossible with the method presented here. Neither exists exists a "reference-paper" exists allowing the analysis of a "historical reference production water", nor a reference method analysing these water yet is established.

In consequence there are only the plausibility of the results as well as the plausibility of the process which can provide security for results or at least a low residual error probability.

For every point of the adsorption isotherm, a mass balance was calculated, showing that there was no irreversible adsorption in the paper/system or any contamination by the system. This mass balance furthermore ensures the plausibility of the concept.

Second the result of the calculated production water concentration should be the same within one page and in one book. Both could be proved, in several cases (Table 2).

Page	Spot	Pb µg/L	Cu µg/L	Zn µg/L
553	1	107	10,3	54
553	2	83	10,3	69
553	3	117	12,6	52
561	Mean(1-3)	70	15	48

Table 2. Calculated M $^{2+}$ concentration of historical production water on different spots of one page and of different pages of "Geestelyke Natuurkunde" of J.J.Scheuchzer, Amsterdam at Petrus Schenk 1735

These results are astonishing for the first eluates for example on some pages were significantly different concentrated. But these varying values were compensated by the fact, that all spot showed varied adsorption isotherm (Figure 11), smoothing the difference significantly. The isotherm is taken / used here as measured – no attempts were made to fit them to the BET-theories.

The standard deviation of the metal concentrations of the "first eluates" is much greater than the standard deviation of the concentrations calculated for the process water. This fact reveals local differences of the ion exchange characteristics of the paper, caused by different thickness or varied chemical micro structure and so – not at least - verifies indirectly the results of the corresponding production water. Different ion exchange systems in contact with the same water reveal - according our theory - the same water concentration.

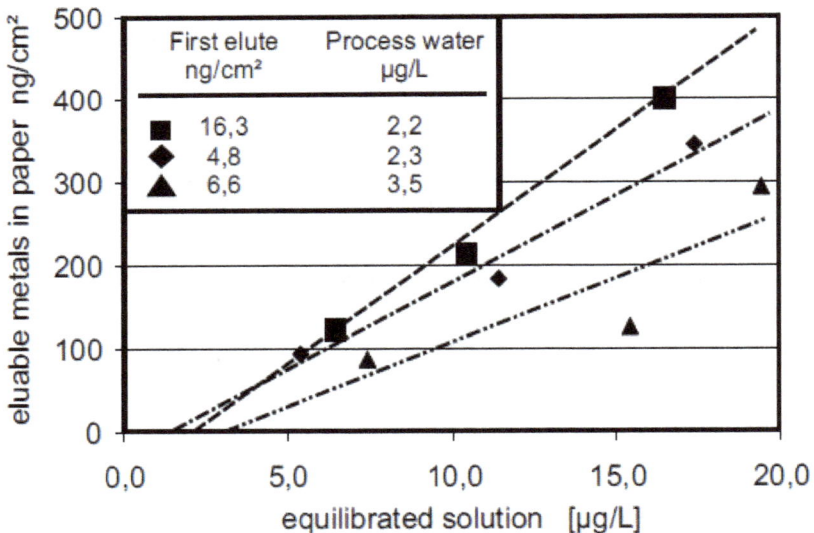

Figure 11. Inhomogeneity of different eluted spot of a page is compensated by the varied adsorption isotherms.

At least the plausibility of the results can be assured by the books themselves. We looked into two books, being volume 1 and volume 2 of the same work "In quatuor libros sententarium commentaria" from Estius (Hessels van Est). Both volumes were printed in Rouen (Rothomagni) in the year 1709 by N. Boucher. Both papers had the same watermark, which yet couldn't be identified. It is very probably that both volumes were printed from the same lot of paper, having at least the same production water. There was sufficient compliance between the results concerning the heavy metal content of the production water (Table 3:).

Book	Dated	Pb		Cu		Zn		Cd	
		µg/L	+/-	µg/L	+/-	µg/L	+/-	µg/L	+/-
Estius Vol 1	1697	23,1	3,0	23,3	7,0	25,6	6,3	4,2	2,0
Estius Vol 2	1697	17,4	3,3	35,7	6,0	28,7	8,2	5,3	2,2

Table 3. Comparison of the heavy metal content of a production water

The limits of detection (LOD) were calculated from the adsorption isotherm for each ion involving the LOD's of matrix containing elutes with the ETVAAS respectively Inverse Stipping Voltammetry using the 3σ - criteria. The LOD allow to determine historical waters even in the range of modern instrumental analytic (Table 4:).

Ion (2+)	Pb	Cu	Zn	Cd
	µg/L	µg/L	µg/L	µg/L
LOD (3σ)	0,93	0,92	0,35	1
LOQ (5σ)	1,6	1,5	0,4	1,7

Table 4. Limit of Detection (LOD) and Limit of quantification (LOQ) for the determination of historical process water with the help of antiquarian paper.

3.7. Determination of historical production water.

Historical books from the 16. to the 18. century, fulfilling the above described requirements, were selected and eluted in the above described manner (Figure 2). All books had at least folio format (width > 25 cm, height > 35 cm).

On selected pages, mostly from the middle of the books and without any printing the "first elutes" were leached out parallel at three spots of one page. In some books several pages were analysed.

In all selected papers it was possible to determine the adsorption isotherm for one (Pb^{2+}) or several elements (Cu^{2+}, Pb^{2+}, Zn^{2+}, Cd^{2+}).

So heavy metal contents of production waters were calculated from the paper of books of the 16.-18. century (Table 5:).

Book	Page	Date	Pb		Zn		Cu		Cd
		.of print	µg/L		µg/L		µg/L		µg/L
1	952	1596	9	+/- 3	21	+/-5	3	+/- 0,7	< LOD
2	463	1690	18	+/- 7	67	+/-13	1	+/-1,4	< LOD
3	93	1580	15	+/- 2	4	+/- 5	< LOD +/- 2		< LOD
4	802	1588	23	+/- 6	28	+/- 4	4	+/- 1,5	< LOD
	862		34	+/- 4	21	+/- 11	3	+/-1,5	< LOD
5	553	1735	102	+/-18	61	+/-12	11	+/-1,3	< LOD
	561		70	+/-7	48	+/-32	15	+/-1,4	< LOD
6		1697	23,1 +/- 3		25,6 +/- 6,3		23,3 +/- 7,0		4,2 +/- 2,0
7		1697	17,4 +/- 3,3		28,7 +/- 8,2		35,7 +/- 6,0		5,3 +/- 2,2
8		1709	51 +/- 17		182 +/- 28		41 +/- 18		34 +/- 22

Table 5. Concentrations of heavy metals found in historical process waters.

The results contained from process waters from the Rhine, Main and Moselle region around 1580-1700 are comparable to modern uncontaminated surface water (Pb^{2+}~ 1-10 µg/L), sometimes even higher than today [27]. Remarkable high values were determined in the process waters of the Netherlands around 1680-1735. Here concentrations up to 50 µg/L Zn^{2+} and 100 µg/L Pb^{2+} are to be found (Table 6). Specially in the region of Amsterdam and Antwerp, there are no fast flowing rivers as they are found at many other paper producing sites. It is known, that Dutch paper mills used windmills for energy supply instead of water mills. The water in those economical booming regions were only slowly flowing waters - in consequence of the geographical situation of the Netherlands. In consequence the waters of highly populated regions may be heavily loaded with hazardous pollutants. Lead – for example - at this time was commonly used, eg. for the production of tools, bullets, colours, metal type and water technology. Astonishing high values of heavy metals were found in a book, printed in Rouen in the year 1709. Specially the concentration of cadmium calculated from this paper seems to be very high – although this metal at this time was unknown. But Zn^{2+} in minerals[28] often is accompanied by Cd^{2+}. If it is no modern contamination, the water came in contact with a (mineral) compound liberating a lot of toxic metals.

After the whole analytical process, there is only a small imprint rested in the book. The stamped paraffin easily can be removed by heating with a flat iron between blotting paper. Apart from a water cleaned spot, there are no relicts (Figure 12) in the paper. So the procedure is conform to the strict requirements demanded for cultural heritage.

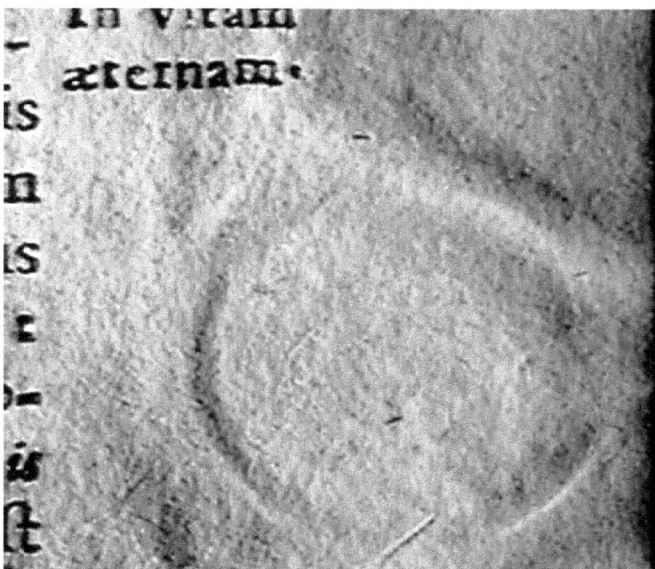

Figure 12. Picture of the paper after finishing the elution procedure

4. Conclusion

Historical contamination of environmental compartments can be seen as an experiment done by mankind. It allows – in its interpretation – to deduce answers for present environmental questions. So for example in presence the consequences of a change of climate is estimated by the data gained at the shelf ice layer in Arctic and Antarctica regions [29].

In the present work first steps to receive information about the historical environment situation in settlement area were done. We succeeded to determine the historical environmental compartment water by examining historical well dated and localisable commodities of the *homo faber* – old books and its paper.

It could be proved, that the atmosphere surrounding a book for centuries leaves traces in the cellulose of these books. In the center of the books, most anion forming acids are in lower concentrations than cutting edges, being in close contact with the atmosphere. The in historical times often emitted SO_x are in much higher concentrations to be found, than the actual dominant NO_x, which in former times were very rare, for they are only produced wir modern high temperature burning processes.

So the acid catalysed decay of cellulose of antiquarian books becomes not only a consequence of modern production techniques, but as well of historical environmental conditions. If we succeed to understand the kinetics of the "passive sampler" book, we will even be able to describe these historical environments.

At present, the distribution of organic acids with the same technique and organic components with the principle of thermal extraction as well as its adsorption kinetics are under investigation in antiquarian books. In another long-term framework of this project is to explain the basis of the data thus obtained material is a correlation between damage type, degree of damage and concentration of individual pollutants present in the respective storage areas.

The information thus obtained will then provide the basis to develop in later projects concepts that allow to keep the content to be at particularly harmful gas components in library spaces as small as possible. Thought is the use of specific sorbent into the air intake side of the usually existing climate systems. From the experience gained here is a concept for the safe storage of records can be developed.

In addition, new contributions made to the environmental history, it might even succeed in using the information obtained to infer historical air pollution.

Ancient papers from well dated books are eluted without destruction of their paper and the resulting solution is analysed by ETAAS and Invers Stripping Voltammetry to determine the historical impact of metals. Afterwards in a flow-system the eluted paper spot is equilibrated with different concentration of heavy metals (Cu^{2+}, Pb^{2+}, Zn^{2+}, Cd^{2+}) to plot the adsorption isotherm of that very spot.

It could be shown that the gained values concerning the content of some heavy metal ions in the pulp production water are plausible and seem to be realistic.

Yet there seems to be no contradiction to define the process water to be identical with the river water.

Such a correlation would be very essential for environmental history, Therefore it is necessary to discuss how realistic this approach could be. Nearly all contemporary pictures of the 16. -18. century show a beck or river directly "behind the windows of the building". So the water for the pulp production in most cases will be directly taken from the river / beck into the production container. In this filling process an additional contamination is not very probable. Contamination is more supposable by desorption from the container material. All bequeathed pulp mills use wooden casks or granite vats for the hand paper production. Both material should not have any significant influence on the heavy metal content of the production water. So to our opinion it is allowed to draw the conclusion that the production water is – concerning its contamination with heavy metal ions – identical with the water of the corresponding river.

So a new field of environmental data is accessible, allowing to evaluate historical environmental pollution. In moment one antiquarian book edition, stored in different libraries is in investigation to assure the principle.

5. Summary

Environmental historical data alone allow to evaluate the anthropogenic impact into environment. It is often forgotten, that all scientific predication of future development of environment is extrapolated from historical data. To receive information about the historical environment situation in inhabited regions, we approach to this task examining historical well dated and locatable products of the homo faber.

The work introduced here uses books as a source of environment historical data specially for the environmental compartment of water and air.

The paper of historical books is well dated by their printing and allocated by their watermark [30]. Often its owner signs reveal the history of the rooms were it was situated.

In the Instrumental Analysis cellulose is often used as a trap of traces of analyts. So in gaseous phase it enriches middle polar up to polar compounds as NO_x, SO_x, HCl, organic acids or esters. In aqueous phases they additionally have to a certain extend the properties of ion exchanging polymers. Both processes can be used to make a new approach to aquire environmental historical data. Historical papers mostly are produced from cellulose – e.g. of a pulp produced by old tatters. For this process is well documented, it is possible to use them as passive samplers for airborne and aqueous contaminants – so opening an new method to get environmental historical data of the compartments air and water.

Airborne contaminants enter a book only by its cutting edge and not by the air-impermeable cover. In consequence a gradient between the centre of the pages and its border, representing an transportation process of the contaminants like in a analytical passive sampling equipment.

The cellulose of historical paper is a trap for traces of heavy metals and other compounds contaminating their production water in historical times. Great amounts of water were brought into contact with the paper pulp in historical paper mill process. The cellulose of the pulp acts as ion exchange material for heavy metals, forming a dynamic equilibrium.

Ancient papers from well dated and located books are eluted without destruction of their paper and the resulting solution is analysed towards its contaminants, caused either by the storing of the books in library (airborne contaminants) or by adsorption from its production water (water-borne contaminants) with modern methods of instrumental analysis.

In case of airborne contaminants like NO_x, SO_x, HCl, organic acids it is possible to prove that acids in the paper are not only introduced by production, but to a great extend as well trapped from environment. We further on show, that some organic acid are set free by the books into the environment.

A well defined pulp production process, starting with used clothes, so allows to estimate the historical heavy metal concentration (Cu^{2+}, Pb^{2+}, Zn^{2+}, Cd^{2+}) in the production water.

In a flow-system the paper is first eluted and then the eluted paper spot is equilibrated with different concentration of heavy metals to plot the adsorption isotherm of that very spot.

Both data together allow a calculation of the heavy metal content of the historical river. For different waters of Germany and the Netherlands of the 16.-18. century the heavy metal load could be estimated. The resulting concentrations mostly were similar to the level of modern surface waters, but in the case of the dutch waters of the 17. century, they were e.g. for Pb^{2+} significantly higher than modern values.

In summary this new field of research opens a new source of environmental historical data, which may at least help to understand local environmental processes and the long time stability of paper.

Author details

Jürgen Schram[*], Rasmus Horst, Mario Schneider, Marion Tegelkamp, Hagen Thieme and Michael Witte

Instrumental & Environmental Analysis, Dep. Chemistry, Niederrhein University of Applied Sciences, Germany

Acknowledgement

This work was funded by the Research fund of the University of Applied Sciences Niederrhein in Krefeld / Germany. The authors are grateful to the staff of Shimadzu Europe, Duisburg, Germany, specially to Uwe Oppermann who promoted the work with technical support concerning atomic absorption spectroscopy. Thanks to the library of St.Josef, Krefeld, Germany (C. Zettner) and the Brigitten Bibliothek, Kaldenkirchen, Germany (K.J. Dors †) for the possibility to measure books out of their book inventory.

6. References

[1] P. J. Barrett, M. Sarti, M. Wise and W. Sherwood, Terra Antartica, 2000, 7(1/2), 209,

[2] Ch. R. Fielding, Ch. Thomson, MRA, Terra Antartica, 1999, 6(1/2), 173 pp ,

---

* Corresponding Author

[3] B. Wennrich, M. Wagner, P. Melles and P. Morgenstern, Int J Earth Sci (Geol Rundsch), 2005. 94: 275,

[4] S. Baron, M. Lavoie, A. Ploquin, J. Carignan, M. Pulido, and J.-L. De Beaulieu; Environ. Sci. Technol., 2005, 39, (14), 5131,

[5] J. Schram, T. Schmidt, M. Tegelkamp, Nachrichten aus Chemie, Technik und Laboratorium; NCTLDI, 1999, 47 (12) 1430,

[6] J. Schram, M. Schneider, R. Horst and H. Thieme, J. Environ. Monit., 2009, 11, 1101–1106
,

[7] J. Havermans, Environmental influences on the deterioration of paper, Rotterdam 1995,

[8] J. Havermans, Restaurator 1995, 16, 209,

[9] Guo,-YH; Zhao,-XL; Liu,-CM; Fenxi-Huaxue. Mar 1996; 24(3): 341-343,

[10] [10] Pekol,-TM; Cox,-JA; Environ-Sci-Technol. Jan 1995; 29(1): 1-6,

[11] [11] Horvath,-Z; Lasztity,-A; Zih-Perenyi,-K; Levai,-A.; Microchem-J. Nov 1996; 54(4): 391-401,

[12] Richardson,-DE; Jewell,-IJ; Appita. Mar 1985; 38(2): 113-118,

[13] Petherbridge, G.; Conservation of Library and archive Materials and the graphic Arts; London 1986,

[14] Smith, R.D.; The paper conservator, 12 (1988), 31,

[15] J.Schram; Th. Schmidt; "Aniquarische Bücher als Senke luftgetragener Schadstoffe"; Jahrestagung "Archäologie und Denkmalpflege" der GDCH; Vortrag/Tagungsband Würzburg 1998, S.52,

[16] Cox,-D; Jandik,-P; Jones,-W.; Pulp-Pap-Can. Sep 1987; 88(9): 90-93,

[17] Catalogue of the Library of the former Brigittenklosters St.Clemens in Kaldenkirchen/ Germany,

[18] Ch. J. Biermann, Handbook of Pulping and Papermaking. San Diego: Academic Press. 1993,

[19] T. Schönfeld und E. Broda, Microchimica Acta, 1951, 36,1, 537,

[20] Z. Horvath, A. Lasztity, K. Zih-Perenyi, A. Levai, Microchem-J. Nov 1996; 54(4), 39,

[21] J. Kintrup, G. Wünsch, Wasserwirtschaft, Abwasser, Abfall, 2001, 48(8), 1068,

[22] D. Laschka, m. Trumpp, Korrespondenz Abwasser, 1991, 38, 495,

[23] N. v. Richthofen, Einsatz von Multielementanalysen für Sielhäute und Klärschämme zur Indirekteinleiteridentifizierung, Universität Hannover Inst.f. Siedlungswasserwirtsch. u. Abfalltechn., 2005 ,

[24] S. B. Smith, Jr and G. M. Hieftje, Applied Spectroscopy, 1983, 37 (5), 419,

[25] Funk, Werner / Dammann, Vera / Donnevert, Gerhild: Qualitätssicherung in der Analytischen Chemie, Wiley-VCH Verlag, Weinheim 2005(2),

[26] Miller, James M. / Crowther, Jonathan B. (Hrsg.): Analytical Chemistry in a GMP Environment, John Wiley & Sons, Inc., New York 2000,

[27] H. Özmen, F. Külahcı, A. Çukurovalı and M. Dogru, Chemosphere, 2004, 55,3, 401,

[28] B.J. Alloway, T. Reimer, R. Eis, Schwermetalle in Böden, Springer, 1999,

[29] W. F, Vincent, J.A.E. Gibson, M.O. Jeffries, Polar Record, 2001, 37 (201): 133,

[30] Wasserzeichensammlung Piccard, Piccard online, Hauptstaatsarchiv Stuttgart, http://www.piccard-online.de,

Permissions

The contributors of this book come from diverse backgrounds, making this book a truly international effort. This book will bring forth new frontiers with its revolutionizing research information and detailed analysis of the nascent developments around the world.

We would like to thank Theo van de Ven and Louis Godbout, for lending their expertise to make the book truly unique. They have played a crucial role in the development of this book. Without their invaluable contribution this book wouldn't have been possible. They have made vital efforts to compile up to date information on the varied aspects of this subject to make this book a valuable addition to the collection of many professionals and students.

This book was conceptualized with the vision of imparting up-to-date information and advanced data in this field. To ensure the same, a matchless editorial board was set up. Every individual on the board went through rigorous rounds of assessment to prove their worth. After which they invested a large part of their time researching and compiling the most relevant data for our readers. Conferences and sessions were held from time to time between the editorial board and the contributing authors to present the data in the most comprehensible form. The editorial team has worked tirelessly to provide valuable and valid information to help people across the globe.

Every chapter published in this book has been scrutinized by our experts. Their significance has been extensively debated. The topics covered herein carry significant findings which will fuel the growth of the discipline. They may even be implemented as practical applications or may be referred to as a beginning point for another development. Chapters in this book were first published by InTech; hereby published with permission under the Creative Commons Attribution License or equivalent

The editorial board has been involved in producing this book since its inception. They have spent rigorous hours researching and exploring the diverse topics which have resulted in the successful publishing of this book. They have passed on their knowledge of decades through this book. To expedite this challenging task, the publisher supported the team at every step. A small team of assistant editors was also appointed to further simplify the editing procedure and attain best results for the readers.

Our editorial team has been hand-picked from every corner of the world. Their multi-ethnicity adds dynamic inputs to the discussions which result in innovative outcomes. These outcomes are then further discussed with the researchers and contributors who give their valuable feedback and opinion regarding the same. The feedback is then collaborated with the researches and they are edited in a comprehensive manner to aid the understanding of the subject.

Apart from the editorial board, the designing team has also invested a significant amount of their time in understanding the subject and creating the most relevant covers. They scrutinized every image to scout for the most suitable representation of the subject and create an appropriate cover for the book.

The publishing team has been involved in this book since its early stages. They were actively engaged in every process, be it collecting the data, connecting with the contributors or procuring relevant information. The team has been an ardent support to the editorial, designing and production team. Their endless efforts to recruit the best for this project, has resulted in the accomplishment of this book. They are a veteran in the field of academics and their pool of knowledge is as vast as their experience in printing. Their expertise and guidance has proved useful at every step. Their uncompromising quality standards have made this book an exceptional effort. Their encouragement from time to time has been an inspiration for everyone.

The publisher and the editorial board hope that this book will prove to be a valuable piece of knowledge for researchers, students, practitioners and scholars across the globe.

List of Contributors

Andrew J. Spiers, Ayorinde O. Folorunso and Kamil Zawadzki
The SIMBIOS Centre, University of Abertay Dundee, Dundee, UK

Yusuf Y. Deeni
School of Contemporary Sciences, University of Abertay Dundee, Dundee, UK

Anna Koza
Novo Nordisk Foundation Center for Biosustainability, Hørsholm, Denmark

Olena Moshynets
Laboratory of Microbial Ecology, Institute of Molecular Biology and Genetics of the National Academy of Sciences of Ukraine, Kiev, Ukraine

Ana Baptista, Isabel Ferreira and João Borges
CENIMAT/I3N and Materials Science Department,
Faculty of Science and Technology of New University of Lisbon (FCT/UNL), Portugal

Anne Jokilammi, Risto Penttinen and Erika Ekholm
Department of Medical Biochemistry and Genetics, University of Turku, Turku, Finland

Miretta Tommila
Department of Medical Biochemistry and Genetics, University of Turku, Turku, Finland
Department of Anaesthesiology, Intensive Care, Emergency and Pain Medicine, Turku University
Hospital, Turku, Finland

John Rojas
Department of Pharmacy, School of Pharmaceutical Chemistry, The University of Antioquia, Medellín, Colombia

Javad Shokri and Khosro Adibkia
Faculty of Pharmacy, Tabriz University of Medical Sciences, Tabriz, Iran

Shilin Liu
College of Chemical and Material Engineering, Jiangnan University, Wuxi, Jiangsu, China

Xiaogang Luo
Key Laboratory of Green Chemical Process of Ministry of Education, Hubei Key Laboratory of Novel
Chemical Reactor and Green Chemical Technology, Wuhan Institute of Technology, Wuhan, China

Jinping Zhou
Department of Chemistry, Wuhan University, Wuhan, China

Jun Xi and Wenjian Du
Drexel University, Department of Chemistry, Philadelphia, USA

Linghao Zhong
Penn State University, Mont Alto, USA

Flávia Dias Marques-Marinho and Cristina Duarte Vianna-Soares
Department of Pharmaceutical Products, Federal University of Minas Gerais, Belo Horizonte, MG, Brazil

Shenqi Wang
College of Life Science and Technology & Advanced Biomaterials and Tissue Engineering Center, Huazhong University of Science and Technology, 1037 Luoyu Road, Wuhan, P.R. China

Yaoting Yu
The Key Laboratory of Bioactive Materials, Ministry of Education, Nankai University, Tianjin, P.R. China

Zdenka Persin, Tina Maver and Karin Stana Kleinschek
Laboratory for Characterisation and Processing of Polymers, Faculty of Mechanical Engineering, University of Maribor, Maribor, Slovenia

Zdenka Persin, Alenka Vesel, Tina Maver, Uros Maver and Karin Stana Kleinschek
Centre of Excellence for Polymer Materials and Technologies, Ljubljana, Slovenia

Alenka Vesel
Jozef Stefan International Postgraduate School, Ljubljana, Slovenia

Miran Mozetic
Jozef Stefan Institute, Ljubljana, Slovenia

Justyna Fraczyk, Beata Kolesinska, Inga Relich and Zbigniew J. Kaminski
Institute of Organic Chemistry, Lodz University of Technology, Zeromskiego, Lodz, Poland

www.ingramcontent.com/pod-product-compliance
Lightning Source LLC
Chambersburg PA
CBHW072253210326
41458CB00073B/1152

Jing Qiu, Shouhui Dai, Tingting Chai, Wenwen Yang, Shuming Yang and Hualin Zhao
Institute of Quality Standards & Testing Technology for Agro-Products, Chinese Academy of Agricultural Sciences, Beijing, China
Key Laboratory of Agri-Food Quality and Safety, Ministry of Agriculture, Beijing, China

Jürgen Schram, Rasmus Horst, Mario Schneider, Marion Tegelkamp, Hagen Thieme and Michael Witte
Instrumental & Environmental Analysis, Dep. Chemistry, Niederrhein University of Applied
Sciences, Germany